Design and Analysis of Ecological Experiments

Design and Analysis of Ecological Experiments

SECOND EDITION

Edited by
Samuel M. Scheiner *and*
Jessica Gurevitch

OXFORD
UNIVERSITY PRESS
2001

OXFORD
UNIVERSITY PRESS

Oxford   New York
Athens   Auckland   Bangkok   Bogotá   Buenos Aires   Calcutta
Cape Town   Chennai   Dar es Salaam   Delhi   Florence   Hong Kong   Istanbul
Karachi   Kuala Lumpur   Madrid   Melbourne   Mexico City   Mumbai
Nairobi   Paris   São Paulo   Shanghai   Singapore   Taipei   Tokyo   Toronto   Warsaw

and associated companies in
Berlin   Ibadan

Copyright © 2001 by Oxford University Press, Inc.

Published by Oxford University Press, Inc.
198 Madison Avenue, New York, New York 10016

Library of Congress Cataloging-in-Publication Data
Design and analysis of ecological experiments / edited by Samuel M. Scheiner and
Jessica Gurevitch.—2nd ed.
     p.   cm.
   Includes bibliographical references.
   ISBN 0-19-513187-8; 0-19-513188-6 (pbk.)
   1. Ecology—Statistical methods.   2. Experimental design.   I. Scheiner, Samuel M.,
1956–   II. Gurevitch, Jessica.
QH541.15.S72 D47 2000
577'.07'27—dc21   00-035647

9 8 7 6 5 4 3
Printed in the United States of America
on recycled acid-free paper

For Edith, Judy, *and* Kayla
and the memory of
my father, Sayre

S.M.S.

Dedicated with much appreciation to
my father, Louis Gurevitch,
and in loving memory of
my mother, Esther Gurevitch

J.G.

We note with deep sadness
the passing of one
of our contributors,
Tom Frost.

Preface to the Second Edition

This second edition of our book on advanced statistical techniques for ecologists comes out seven years after the first. We are very pleased with the reception of the first edition, which has been widely read and used in many graduate seminars on statistics in ecological research. The enthusiasm expressed for the book was part of the motivation for this new edition, which updates and expands on the original. We like to think that this book has played a role in introducing many previously unfamiliar approaches to ecologists and in setting higher standards for the use and reporting of more conventional analyses.

This new edition is designed to help continue and propel these trends. Nearly all of the chapters included from the first edition are revised, several substantially, to bring them up to date. We added four new chapters that reflect recent advances in statistical practice, particularly within ecology. The changes result from a combination of new statistical theory, new computer software and hardware capabilities, and evolving practices and improved statistical standards among ecologists.

We are also taking advantage of another important advance in information technology, the World Wide Web. Oxford University Press has established a companion Website [http://www.oup-usa.org/sc/0195131878/] that contains the source code and data sets for many of the chapters. By moving the source code to a Website, we were able to add a chapter to the book. Readers can now check their results against the sample data sets on-line.

It is exciting to be publishing a book in 2001. Ecological research and the use of statistics in ecology continue to evolve rapidly. We hope that this new edition will prove helpful to established researchers and to those in training, who will bring new standards and new insights into the service of the science in the twenty-first century.

Samuel M. Scheiner
Jessica Gurevitch

Preface to the First Edition

The genesis of this book was the result of a conversation while eating a bento (a Japanese box lunch of rice, sushi, pickles, and other delicacies) on a sidewalk in Yokohama in 1990. The conversation had turned to statistics and both of us were commenting on statistical issues and techniques that were either underused or misused by ecologists. For some time, Jessica had been contemplating a book on statistical techniques in experimental design and analysis, written for ecologists. The goal of such a book was both to encourage the correct use of some of the more well-known approaches, and to make some potentially very useful but less well known techniques available to ecologists. We both felt strongly that such a book was timely, and would be useful to ecologists working on applied as well as basic problems. Had Sam not intervened, this idea would undoubtedly have met the fate of many other fine ideas that have never made it into the light of day.

It was apparent to both of us that we did not have the skills to write such a book ourselves, nor the time. However, we were able to compile a list of topics and potential contributors whom we knew were knowledgeable about those topics. (An initial outline for the book was composed while sitting for hours, stalled in traffic going to and from Mount Fuji. At the next INTECOL meeting keep your eyes out for "I survived the trip to Mt. Fuji" t-shirts.) We batted (actually e-mailed) these ideas back and forth for nearly a year. The success of a symposium organized by Phil Dixon at the 1991 annual meeting of the Ecological Society of America on the design of ecological experiments encouraged us to continue our endeavors, and we managed to secure commitments from many of our contributors during that week. The enthusiasm for this undertaking expressed by colleagues we spoke with buoyed us, as did the interest and encouragement of Greg Payne from Chapman and Hall. Therefore, despite warnings about the travails of editing a book, we forged ahead. So—beware of the dangers of conversation over raw fish.

Samuel M. Scheiner
Jessica Gurevitch

Acknowledgments

We wish to thank a number of people who contributed to the production of this volume. We are grateful to those colleagues who painstakingly reviewed chapters: Marti Anderson, Norma Fowler, Charles Goodnight, Ed Green, Jeff Hatfield, Ray Hilborn, Ed Heske, Charles Janson, Martin Lechowicz, Bryan Manly, Jim McGraw, Tom Meagher, Scott Menard, Tom Mitchell-Olds, Patrick Phillips, Alan Tessier, Joel Trexler, Art Weis, and Neil Willits. We offer special thanks to Greg Payne, book editor of the first edition, and Kirk Jensen, book editor of the second edition, for their efforts. A fellowship from the Arnold Arboretum of Harvard University, responsible for sending J.G. to Japan and thus for the inception of this book, is gratefully acknowledged. S.M.S. was supported for his trip to Japan by funds from the Department of Biological Sciences, the College of Liberal Arts & Sciences and the Graduate School of Northern Illinois University, and the City of Yokohama; the Department of Biological Sciences provided support during the editorial process. S.M.S. thanks Judy Scheiner for all of her patience and forbearance during the production of this book. J.G. extends many thanks to Todd Postol for being in her corner all of these years.

Contents

Contributors

Steven J. Beaupre
Department of Biological Sciences
University of Arkansas
Fayetteville, Arkansas 72701

James S. Clark
Departments of Botany and Earth
    & Ocean Sciences
Duke University
Durham, North Carolina 27708

Noel Cressie
Department of Statistics
Iowa State University
Ames, Iowa 50011

Philip M. Dixon
Department of Statistics
Iowa State University
Ames, Iowa 50011

Arthur E. Dunham
Department of Biology
University of Pennsylvania
Philadelphia, Pennsylvania 19104

Aaron M. Ellison
Department of Biology
Mount Holyoke College
South Hadley, Massachusetts
01075

Ted Floyd
Great Basin Bird Observatory
One East First Street, Suite 500
Reno, Nevada 89501

Marie-Josée Fortin
School of Resource and Environmental
    Management
Simon Fraser University
Burnaby, British Columbia
    V5A 1S6
Canada

Gordon A. Fox
Department of Biology
University of South Florida
Tampa, Florida 33620

Thomas M. Frost†
Center for Limnology
University of Wisconsin
Madison, Wisconsin 53706

Deborah E. Goldberg
Department of Biology
University of Michigan
Ann Arbor, Michigan 48109

Jessica Gurevitch
Department of Ecology and Evolution
State University of New York–
    Stony Brook
Stony Brook, New York 11794

Larry V. Hedges
Department of Education
University of Chicago
Chicago, Illinois 60637

Dennis M. Heisey
Madison Academic Computing Center
University of Wisconsin
Madison, Wisconsin 53706

Steven A. Juliano
Department of Biological Sciences
Illinois State University
Normal, Illinois 61761

Michael Lavine
Institute of Statistics and Decision Sciences
Duke University
Durham, NC 27708

Randall J. Mitchell
Department of Biology
University of Akron
Akron, Ohio 44325

Erik V. Nordheim
Department of Statistics
University of Wisconsin
Madison, Wisconsin 53706

Peter S. Petraitis
Department of Biology
University of Pennsylvania
Philadelphia, Pennsylvania 19104

Catherine Potvin
Department of Biology
McGill University
Montreal, Québec H3A 1B1

Paul W. Rasmussen
Bureau of Research
Wisconsin Department of Natural
    Resources
Madison, Wisconsin 53706

Samuel M. Scheiner
Division of Environmental Sciences
National Science Foundation
4201 Wilson Boulevard
Arlington, Virginia 22230

Robert J. Steidl
School of Renewable Natural Resources
University of Arizona
Tucson, Arizona 85721

Len Thomas
Mathematical Institute
University of St Andrews
St Andrews, Scotland KY16 9SS
United Kingdom

Jay M. Ver Hoef
Alaska Department of Fish and Game
1300 College Road
Fairbanks, Alaska 99701

Carl N. von Ende
Department of Biological Sciences
Northern Illinois University
DeKalb, Illinois 60115

†deceased 2000

Design and Analysis of Ecological Experiments

1

Theories, Hypotheses, and Statistics

SAMUEL M. SCHEINER

Picture men in an underground cave, with a long entrance reaching up towards the light along the whole width of the cave. . . . Such men would see nothing of themselves or of each other except the shadows thrown by the fire on the wall of the cave. . . . The only truth that such men would conceive would be the shadows.

Plato, The Republic, Book VII

I hold that philosophy of science is more of a guide to the historian of science than to the scientist.

Lakatos (1974)

1.1 The Purposes of This Book

Ecology is more and more an experimental science. Ecologists are increasing their use of experiments to test theories regarding organizing principles in nature (Hairston 1989; Resetarits and Bernardo 1998). However, ecological experiments, whether carried out in laboratories, greenhouses, or nature, present many statistical difficulties. Basic statistical assumptions are often seriously violated. Highly unbalanced designs are often encountered as a result of the loss of organisms or other biological realities. Obstacles such as the large scale of ecological processes or cost limitations hinder treatment replication. Often it is difficult to appropriately identify replicate units. Correct answers may require complex designs or elaborate and unusual statistical techniques. To address these problems, we have fashioned this book as a toolbox containing the equipment necessary to access advanced statistical techniques, along with some cautionary notes about their application. These chapters are meant to serve as introductions to these topics, not as definitive summaries. Interested readers are provided with an entrée to the literature on each topic and encouraged to read further.

Most ecologists leave graduate school with only rudimentary statistical training; they are either self-taught or have taken a basic one-semester or one-year statistics course. Thus they lack familiarity with many advanced but useful methods for addressing ecological issues. Some methods can be found only in statistics journals that are inaccessible to the average ecologist. Other methods are presented in very general or theoretical terms that are difficult to translate into the actual computational steps necessary to analyze real data. Developments in applying these specialized approaches to ecological questions cannot be found in conventional statistical texts, nor are they explained sufficiently in the brief "Methods" sections of ecological articles. Such barriers put these methods out of the hands of the majority of ecologists, leaving them unavailable to all but a few.

One result of this gap between need and training has been a spate of notes and articles decrying the misuse of statistics in ecology. The criticisms include the improper identification of the nature of replicates and pseudoreplication (Hurlbert 1984; Gurevitch and Chester 1986; Potvin et al. 1990b), the improper use of multiple comparison tests (Day and Quinn 1989), the use of ANOVA when more powerful techniques are available (Gaines and Rice 1990), the misuse of stepwise multiple regression (James and McCulloch 1990), the improper reporting of statistical parameters and tests (Fowler 1990), and the overuse of significance tests (Yoccoz 1991).

This book is designed as a how-to guide for the working ecologist and for graduate students preparing for research and teaching careers in the field. The ecological topics were chosen because of their importance in current research and include competition (chapters 5, 11, 16, and 17), plant–animal interactions (chapter 12), predation (chapters 6, 9, and 10), and life-history analyses (chapters 6, 7, 8, and 13). The statistical techniques were chosen because of their particular usefulness or appropriateness for ecological problems, as well as their current widespread use or likely use in the future. Some may be familiar to ecologists but are often misused (e.g., ANOVA, chapters 4 and 5), other approaches are rarely used when they should be (e.g., power analysis, chapter 2; MANOVA, chapter 6; repeated-measures analysis, chapter 7; nonlinear curve fitting, chapter 10), and others are either newly developed or unfamiliar to most ecologists and yet are important in this field (e.g., nonparametric ANCOVA, chapter 7; spatial analysis, chapters 15 and 16; Bayesian analysis, chapter 17; meta-analysis, chapter 18). Some of the statistical approaches presented here are well established within the field of ecology, whereas others are new or controversial. We have attempted to mix the tried and true with the innovative. Each statistical technique is demonstrated by applying it to a specific ecological problem; however, that does not mean that its use is in any way limited to that problem. Clearly, many of the approaches will be widely applicable. With this book, we hope to encourage investigators to expand their repertoire of statistical techniques.

This volume deals primarily with the use of statistics in an experimental context, reflecting the increasing use of experimental studies in ecology, especially in natural settings. We emphasize manipulative experiments, although many of the techniques are also useful in the analysis of observational experiments (chapters 6, 11, 12, and 18) and nonexperimental observations (chapters 2, 3, 9, 10, 11, 14, 15, and 16). In doing so, we intend no slight to other forms of ecological experiments and research programs. Many other books are available that deal with the analysis of nonexperimental observations (e.g., Ludwig and Reynolds 1988; Digby and Kempton 1987; Manly 1990b, 1997). The list of ecological problems covered herein is not meant to be prescriptive, but illustrative of the wide range of problems amenable to experimental analysis. For example, chapter 9 deals with large-scale phenomena not usually covered by manipulative experiments. Our hope in compiling this book is to demonstrate that even relatively advanced statistical techniques are within the grasp of ecologists, thereby improving the ways that ecological experiments are designed and analyzed.

1.2 Theories, Hypotheses, and Statistics

1.2.1 An Error-Statistical Scientific Process

The vast majority of scientists are, consciously or unconsciously, metaphysical and epistomological realists. That is, we believe that there is an objective universe that exists independent of us and our observations. Science is, then, a process of building theories or models of what that universe consists of, how it is put together, and how it functions. Those theories consist of a set of assumptions about the universe (Pickett et al. 1994, box 3.2), pieces of the grand truth about the universe that we strive to discover, with our accumulated theories being the closest that we can ever come to that truth. In building and testing the theories, we believe that we are making successively closer approximations to the truth, although we can never be certain that we have ever actually arrived at the truth. We progress by carefully sorting through and eliminating possible sources of error in our logic and evidence. Science is an experimental process of learning from error.

Statistics plays a central role in this scientific process. The role of statistics is explicated by the philosopher Deborah Mayo (1996) in what she terms an error-statistical approach. Her treatment of this topic consumes a 493-page book, and I can do no more than sketch her thesis. If you have any interest in the philosophical underpinnings of our enterprise, I strongly recommend this book. It comes closer to capturing actual scientific practice than any other account of the philosophy of science that I have read. Most current treatments of the philosophy of science by ecologists (e.g., Peters 1991; Pickett et al. 1994) still espouse the falsificationism of Popper (1959) and the social constructivism of Kuhn (1962), views that are out of step with the realism of actual scientific practice.

The heart of the error-statistical approach is the recognition that scientists move forward in their search for truth by learning from error using error-probability statistics. This approach derives, in part, from what is termed Neyman–Pearson statistics, although it differs in significant ways. The approach can be summarized as a hierarchy of models:

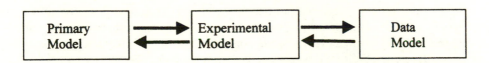

This hierarchy is deliberately shown horizontally, rather than vertically, because you can move in either direction, from the primary model to the data model or back. You should not be tempted to equate these directions with deductive and inductive processes.

This hierarchy is a framework for delineating the complex relationship between raw data and scientific hypotheses. Primary models concern the process of breaking down the principal question into a set of tests of hypotheses and estimates of theoretical quantities. Experimental models relate those questions to the particular experiment at hand. Data models focus on the generation and modeling of raw data.

Experimental models and their link with data models are the subject of this book. We explore a range of types of experiments. One type is the manipulative experiment, the more usual sense of an experiment, including laboratory studies of behavior (chapter 10), manipulations of growth conditions in environmental chambers or greenhouses (chapter 4), creation of seminatural environments, and imposition of experimental treatments on natural environments (chapters 5, 9, 15, 16, and 18). A second type is the observational experiment. This latter type of experiment is especially important in ecology because the scale of phenomena addressed or ethical considerations often preclude manipulative experiments. For example, theory may predict that prairies will be more species-rich than forests. One could test this hypothesis by measuring (observing) species numbers in each of those habitats. This procedure is a valid test of the theory, as long as the observations used for the test differ from the original, nonexperimental observations used to produce the theory. Intermediate between manipulative and observational experiments are natural experiments in which one predicts the outcome of some naturally occurring perturbation (Diamond 1986; time series analysis, chapter 9). [Some (e.g., Hairston 1989) would lump natural experiments in with observational experiments.]

Statistics plays a key role by linking the parts of the model hierarchy. Mayo identifies three tasks for error statistics: (1) Statistics provides techniques for data generation and assumption testing. For example, power analysis (chapter 2) and examination of residual distributions (chapter 3) fall in this category. (2) Statistics provides error probabilities (most of this book). (3) Statistics provides ways for checking on possible errors, such as mistakes about chance correlations, parameter estimates, or experimental assumptions. From a statistical point of view, a severe test is one in which both the Type I and Type II error probabilities are small (section 1.2.2; chapter 2). When our data pass these tests, we have learned something. We have answered one question with high reliability and added one more piece to the puzzle.

Central to this discussion are three different uses of the concept of "assumption." First are explanatory assumptions, those that we make about the universe (e.g., the assumption that conspecifics will compete more than distantly related species). These assumptions make up the theory that an experiment is attempting to test. Second are simplifying assumptions, those made for analytic convenience (e.g., the assumptions that mating is completely random). These assumptions either were tested by previous experiments or are known not to seriously bias results if violated. Third are statistical assumptions, those that underlie the statistical procedure used (e.g., the assumption in parametric statistics that the tested variable has a normal distribution). The statistical assumptions are the basis for the stated probability (the P-value) that the deviation of the observed parameter value from the predicted value is due only to random sampling effects. Meeting these assumptions is critical if the statistical procedure is to provide a correct answer. Unfortunately, these statistical assumptions are often not understood or are ignored by the experimentalist, resulting in incorrect conclusions (e.g., chapters 7, 9, and 10). Throughout this book, critical assumptions underlying statistical procedures are presented, along with warnings about the consequences of a failure to meet those assumptions. Knowing the consequences of violating such assumptions can be very useful because sometimes a reasonably robust answer can be obtained despite violations of some assumptions.

This one-step-at-a-time process, with statistical inference playing a key role at each step, characterizes scientific practice. However, statistical inference does not equate with scientific inference. Nor is scientific inference simply a brand of formal logic. The error-statistical process is much more complex because it encompasses the use of scientists' professional judgment as to which possible sources of error are plausible and likely. It is also important to recognize that an error-statistic framework is not a prescription for scientific practice. Unlike most philosophers of science, Mayo does not claim to present a set of global rules. Rather, she emphasizes that actual scientific practice is much more piecemeal and proceeds one question at a time.

1.2.2 Type I and Type II Errors

Two types of errors can arise over the decision to reject a hypothesis being tested. A Type I error is the declaration of the hypothesis to be false when it is actually true; a Type II error is the failure to falsify the hypothesis when it is actually false. Statistical procedures are designed to indicate the likelihood or probability of those errors. Most familiar is the probability of Type I errors, the α and P-values reported in "Results" sections in scientific articles. Somewhere along the way, a value of $P < 0.05$ became a magic number: reject the hypothesis if the probability of observing a parameter of a certain value by chance alone is less than 5%, and do not reject (or accept) if the probability is greater. Although some sort of objective criterion is necessary so that we simply do not alter our expectations to meet the results (throwing darts against the wall and then drawing a target around them), three additional matters must be considered.

First, we must consider the power of the test: the β-value or probability of a Type II error (chapter 2). (Technically, the power equals $1 - \beta$.) Often in ecology, because of the scale of the phenomena under study, large numbers of replicates of experimental treatments are often not used and sometimes not possible. In such situations, it may be acceptable to use a higher α-level, such as $\alpha = 0.1$, as a cutoff for a decision of statistical significance because the power of the test is low. See Toft and Shea (1983), Young and Young (1991), and Shrader-Frechette and McCoy (1992) for discussions of this issue especially with regard to the relative importance of each type of error in decision making. A very different approach to this problem is offered by the Bayesian school of statistics (Bayes

1763; chapter 17). Its proponents believe that the conventional procedure for decision making on the basis of error rates is wrong, and they use a different approach for statistical analysis and the evaluation of results [see Edwards et al. (1983) and Berger (1985) for more general discussions of Bayesian statistics].

Second, it is important to keep in mind the ultimate goal of any scientific investigation: deciding which of the hypotheses about the universe are correct. Ultimately these are yes or no decisions. What the statistical test should indicate is "Yes, the hypothesis is almost certainly true," "No, the hypothesis is almost certainly false," or "Maybe, another experiment must be performed." Thus, the P-values are only guidelines, and results that lie in the region $0.011 < P < 0.099$ (to pick two arbitrary values) should be judged with caution.

Third, it is important to distinguish between theory building and theory testing when we choose the criteria for making a decision. When looking for patterns and building theories, we want to be less conservative and use a higher α-value so that we do not overlook promising directions of research. On the other hand, when testing theories, we should take a more conservative stance; this conforms to Mayo's need to put a hypothesis to a severe test. (A more conservative procedure, in statistical parlance, is a procedure that is less likely to reject the null hypothesis.) The cost of making Type I or Type II errors may also be a consideration: if a harmful effect on an endangered species is tested with low statistical power, it is more cautious to be statistically liberal than statistically conservative (see chapter 2).

Sometimes it is difficult to determine the actual Type I error rate. This rate is supposed to be how often, if the experiment were repeated many times, the hypothesis would be rejected incorrectly when it was true. A problem arises if there is more than one estimation of the same parameter in the course of a study. For example, suppose we wish to determine whether two plant populations differ. We might choose some number of individuals in each population, measure 10 different traits of each individual, and compare the estimated parameters. Or, suppose we measure three different species to test the hypothesis in question. Under what conditions is it necessary to adjust the α-value? Of concern here is whether a single hypothesis is being independently tested more than once. If that is the case, then some sort of correction may be warranted, such as the Bonferroni procedure (Simes 1986; Rice 1990). Another solution, especially if the responses are not independent, is to use multivariate statistics, reducing the problem to a single statistical test (chapters 6 and 8).

There has been extensive debate over this issue with regard to the comparison of multiple means (Jones 1984). One side is concerned that the declaration that two samples are different can depend on what other samples are included in the analysis. Advocates of this position say that only individual pairs of means should be compared and no multiple sample tests such as ANOVA or Tukey's test should be employed (Jones and Matloff 1986; Saville 1990). The other side is concerned that such a procedure will cause an inflation of the Type I error rate (Day and Quinn 1989). Both sides have valid positions, and each investigator must decide (1) what is the question being addressed and (2) how conservative to be. For example, consider an experiment that examines the effects of three pesticides on corn earworm populations in six agricultural fields. If the question being addressed is, "How do corn earworms respond to pesticides?" then the samples are meant to represent a random sample of a larger set, all populations of the species. In that instance, a multiple comparison procedure, such as ANOVA, would be appropriate. If instead the question being addressed is, "Which of these three pesticides is best for controlling corn earworm?" then each sample is of individual interest, and individual comparisons are warranted. Regardless of the position we take, in published articles it is crucial to state clearly the question being addressed, the statistical procedure being used to answer it, the magnitudes of the Type I and Type II error rates, and the criteria for choosing the procedure and rates.

Keep in mind that statistical significance is not equivalent to biological significance. A large experiment can detect very small effects, which, in a world where stochastic factors are present, will just get swamped by environmental variation. For example, an experiment testing pairwise competitive interactions could be devised to detect a 0.001% advantage by one species. Such an effect would be meaningless in a situation in which the actual encounter rate is low and large disturbances perturb the community every few years. On the other hand, we must be cautious about failing to recognize the importance of weak processes that can become consequential over very large scales of space and time. For example, macroparasites (mites, ticks, and so on) in the aggregate can have large effects on the population dynamics of their host over long time periods, yet those effects could be difficult to detect because of low parasitism rates on any given individual. One technique for detecting such weak processes is meta-analysis (chapter 18), a method of combining the results from different experiments.

1.2.3 Pseudoreplication versus Nonindependence

An important opening salvo in the war against bad statistics in ecology was the article by Hurlbert (1984) in which he coined the term *pseudoreplication*, the statistical treatment of experimental units as independent when they are not. Since then the problem of pseudoreplication in ecology has decreased, although it has not entirely disappeared (Heffner et al. 1996). However, we now face the opposite problem, seeing pseudoreplication where it does not exist. I have frequently heard the statement, "Oh, that is pseudoreplication" when, in actuality, these are simply nonindependent observations that can be accounted for in the statistical model.

What is the distinction? Consider an experiment to examine the effects of site conditions on the size distribution of a gall-making insect. Galls form on the stems of bushes, with each gall containing multiple chambers each with a larva. Galls are collected and larva measured in a hierarchical fashion: 3 larva are measured from each gall, 5 galls are collected from each bush, 20 bushes are sampled from each site, and 4 sites are sampled. What is the minimal level of true replication? The answer is, "It depends on the question." For larva within galls, the growth of one affects the growth of the other. This is a case of pseudoreplication because the residual errors of the observations are not independent. However, for the galls growing on different branches, although the observations are not independent, this is not a case of pseudoreplication because the residual errors are independent. Instead, the two observations have a common external factor (bush) that can be accounted for statistically (that is, included in the model) in this case by nesting "bush" within "site" in the ANOVA (chapter 4). Other statistical techniques, such as spatial analysis (chapters 15 and 16), can also be employed. Hurlbert's original essay on pseudoreplication did not distinguish between these two types of nonindependence. As a result, many ecologists have taken too extreme a view of the problem.

I also encourage a more moderate stance on two other issues raised by Hurlbert (1984), extrapolation and demonic intrusion. Extrapolation deals with the problem of the lack of replication over ecological units. For example, chapter 9 deals with whole-lake manipulations. Because of cost and feasibility limitations, the experiment is done in only a single lake. One school of thought would claim that the researchers have no warrant to extrapolate their results to other lakes because they have no information on lake-to-lake variation in the experimental response. I contend that such an extrapolation is warranted if the researchers can provide enough additional information to convince us that the treatment response seen in this lake is similar to what would be seen in other lakes. This is an example of what Mayo calls the use of error statistics to test data and experimental assumptions.

Demonic intrusion is a related issue. Even if the researchers provide information showing that the manipulated lake is typical of other lakes in the region based on a long list of measured variables, how do they know that some other, unmeasured variable does not differ? More typically, this issue is raised with respect to artificial environments, for example, the use of a single pair of controlled-environment greenhouses with all replicates of one temperature treatment in one greenhouse and all replicates of another temperature in the other greenhouse (chapter 4). In this case, practical considerations prevent replication. I contend that at some point we have to let common sense rule. We can measure as many environmental variables in each greenhouse as possible to convince ourselves that they are otherwise identical. We can carry out subsidiary experiments (e.g., grow plants in both greenhouses at the same temperature) to assess among-greenhouse variability. If that variation is small relative to the treatment effects, we can safely ignore it. Yes, we can all come up with highly unlikely situations that might mislead us, and we might even have experienced some of these situations. But practical considerations often intervene in our attempts to design the ideal experiment. Scientific progress demands that we allow extra-statistical information to play a role in how we interpret our experimental results. The use of this additional information is why Mayo states that scientific inference is not identical to statistical inference.

1.2.4   Using and Reporting Statistics

Ultimately, statistics are just one tool among many that we, as scientists, use to gain knowledge about the Universe. To be used well, these tools must be used thoughtfully—not applied in an automatic fashion. Our message is that there is often no single right way to use statistics. On the other hand, there are many wrong ways. Several chapters present alternative procedures for analyzing similar problems (e.g., chapters 4, 7, 15, and 16). Which procedure is best depends on several factors, including how well the data meet the differing statistical assumptions of the various procedures. Most importantly, choosing the statistical procedure depends on the question being asked. Procedures differ in the types of question that they are best equipped to address. For example, von Ende (chapter 8) discusses several different types of repeated-measures analysis that are alternatively useful for asking such questions as whether treatments differ in their overall means or whether they differ in the shapes of their response curves. This book should provide a sense of the wide variety of statistical procedures available for addressing many common ecological issues. It should encourage a greater use of less common techniques and a decrease in the misuse of common ones.

In the design of experiments, it is critical to consider the statistical analyses that will be used to evaluate the results. Statistical techniques can go only so far in salvaging a poorly designed experiment. These considerations include not only the number and types of treatments and replicates (e.g., repeated-measures analysis, chapter 8), but also how experimental units are physically handled. For example, Potvin (chapter 4) shows how the physical arrangement of experimental blocks relative to the scale of environmental heterogeneity can affect the ability to detect a significant treatment effect. Consultation with a statistician at this stage is highly recommended.

Equally important to using statistics correctly is thorough reporting. Probably the most common statistical sin is not describing procedures or results adequately in publications (Fowler 1990; Gurevitch et al. 1992). It is important to explicitly report what was done. A scientific publication should permit a reader to draw an independent conclusion about the hypothesis being tested. Although the ecological, or other biological, assumptions behind an experiment are usually explicitly laid out, statistical assumptions are often not addressed, or worse, the exact statistical procedure is not specified. Thus, when a conclusion is reached that the hypothesis has been falsified, the reader does not know whether it is due to errors in the ecological assumptions or in the statistical assumptions. This need for accurate reporting is even more important when nonstandard statistical procedures are used. No one could hope to publish an experiment that manipulated light and nutrient conditions without specifically describing each treatment. Yet, the only statement about statistical procedures is commonly "ANOVA was used to test for treatment effects." Ellison (chapter 3) discusses some of the simple types of information that should be reported (and often are not). For example, often only P-values are reported but information on sample sizes and error variances is not. If the analysis is done with a computer, identify the software package used. These packages use different algorithms that can reach different conclusions even with the same data. Or, at some future date, a bug may be found in a software package invalidating certain analyses. I hope that an increase in the overall level of statistical sophistication of writers, reviewers, and editors will bring an increase in the quality of how statistical procedures are reported.

1.3 Using this Text

The material presented here presumes a basic knowledge of statistics comparable to that acquired in a one-semester introductory course. Any basic statistics book (e.g., Siegel 1956; Snedecor and Cochran 1989; Sokal and Rohlf 1995; Zar 1996) can be used as a supplement to the procedures presented here and as a source of probability tables. As this is largely a how-to book, the underlying statistical theory is not detailed, although important assumptions and other considerations are presented. Instead, references to the primary statistics literature are presented throughout the text. Readers are encouraged to delve into this literature, although it is sometimes opaque, especially when it concerns a technique that is central to their research program. The effort taken to really understand the assumptions and mathematical properties of various statistical procedures will more than pay for itself in the precision of the conclusions based on them.

Each chapter is designed as a road map to address a particular ecological issue. Unlike most conventional statistics books, each statistical technique presented here is motivated by an ecological question to establish the context of the technique. The chapters are written by ecologists who have grappled with the issues they discuss. Because statistical techniques are interrelated, each chapter provides cross-references to other relevant chapters, but each can be used independently. Also, a given statistical technique may be pertinent to a number of ecological problems or types of data. Each chapter presents a step-by-step outline for the application of each technique, allowing the reader to begin using the statistical procedures with no or minimal additional reference material. Some techniques, however, are difficult and sophisticated, even for someone with more experience. Readers are encouraged to use the techniques presented here in consultation with a statistician, especially at the critical stage of experimental design. Because most statisticians are not familiar with ecological problems and the optimal way to design and analyze ecological experiments, they would not necessarily arrive at these methods themselves, nor make them understandable to ecologists. Thus, this book can be used as a bridge between ecologists and their statistics colleagues.

We also recognize that this is the age of the computer. Many problems can be solved only by computer, either because of the enormous mass of data or because of the extensive number of mathematical steps in the calculations. Each chapter, where appropriate, has an appendix containing the computer code necessary for applying a specific statistical procedure. For this edition, we have removed much of this material from the printed text and placed it at a Website [http://www.oup-usa.org/sc/0195131878/] along with all of the data sets used as examples in this book. We chose the SAS (SAS Institute Inc. 1989a,b, 1990, 1992, 1996) statistical package because of its widespread availability on both mainframe and personal computers and its extensive use by ecologists. In addition, the syntax of SAS is easy to understand and thus to translate to other statistical packages. For example, SYSTAT (SPSS, Inc., Chicago, Illinois), a common personal computer package, has a syntax very similar to that of SAS. In some instances, procedures are not available in SAS; in those cases, another statistical package is suggested. In no instances are system commands given. Readers unfamiliar with mainframe or personal computers are advised to check manuals and their local computer consultants, mavens, and hackers. Good sources for information on statistical packages are the Technological Tools feature in the Bulletin of the Ecological Society of America (e.g., Ellison 1992), software reviews in the *Quarterly Review of Biology*, and computer magazines.

Acknowledgments   I thank Jessica Gurevitch, André Hudson, Steve Juliano, Marty Lechowicz, and Mark VanderMuelen for suggestions that greatly improved the previous version of this chapter.

2

Power Analysis and Experimental Design

ROBERT J. STEIDL

LEN THOMAS

2.1 Introduction

Ecologists conduct research to gain information about ecological patterns and processes (chapter 1). An underlying, fundamental goal for all research, therefore, is to generate the maximum amount of information from a given input of effort, which can be measured as time, money, and other similarly limited resources. Consequently, after we develop a clear set of questions, objectives, or hypotheses, the most critical aspect of ecological research is design.

Designing a new study involves making a series of interrelated choices, each of which will influence the amount of information gained and, ultimately, the likelihood that study objectives will be met. For a manipulative experiment, choices must be made about the number of treatments to apply and the way in which treatments are assigned to experimental units. Similarly, for an observational study, samples must be selected in some way from the larger population of interest. In both types of research, a critical decision is the number of replicates (experimental units receiving the same treatment) or samples to choose. When considering these issues, it is helpful to have a tool to compare different potential designs. Statistical power analysis is one such tool.

In this chapter, we focus on the use of power analysis in research design, called prospective (or a priori) power analysis. We review some basic theory and discuss the practical details of doing prospective power analyses. We also consider the usefulness of calculating power after data have been collected and analyzed, called retrospective (or a posteriori or post hoc) power analysis. Power analysis is most appropriate when data are to be analyzed using formal hypothesis-testing procedures. However, parameter estimation is often a more appropriate and informative approach by which to make inferences, so we discuss related techniques when estimation is the main goal of the study. Our discussion stays within the frequentist statistical paradigm; issues within the likelihood and Bayesian frameworks are considered elsewhere (chapter 17; Berger 1985; Royall 1997; Burnham and Anderson 1998; Barnett 1999).

Power analysis is increasing in popularity, as evidenced by the spate of introductory articles recently published in the biological literature (e.g., Hayes 1987; Peterman 1990; Muller and Benignus 1992; Taylor and Gerrodette 1993; Searcy-Bernal 1994; Thomas and Juanes 1996; Steidl et al. 1997). These all provide somewhat different perspectives and, in some cases, different background material than we present here. Unfortunately, power is given only cursory treatment in many biometry textbooks (e.g., Sokal and Rohlf 1995; Steel et al. 1996), although this has been changing to some extent (e.g., Rao 1998; Zar 1996). In addition, some specialized texts (Kraemer and Thiemann 1987; Cohen 1988; Lipsey 1990) and an excellent introductory monograph (Nemac 1991) focus on implementing power analysis using SAS. We provide other selected references throughout this chapter.

2.2 Statistical Issues

2.2.1 Statistical Hypothesis Testing

The theory of power analysis falls within the larger framework of statistical hypothesis testing (the so-called Neyman–Pearson approach; Neyman and Pearson 1928; Barnett 1999). In this framework, a research question is phrased as a pair of complementary statistical hypotheses, the null (H_0) and alternative (H_a) hypotheses. The finding that would be of interest to the researcher is stated typically as the alternative hypothesis, and a negative finding is stated as the null hypothesis. For example, suppose we were interested in assessing whether the average amount of plant biomass harvested per plot differs between control and treatment plots subjected to some manipulation. Typically, the null and alternative hypotheses of interest (in this case as two-tailed hypotheses) would be phrased as

H_0: $\mu_T = \mu_C$, which represents the case of equal population means

H_a: $\mu_T \neq \mu_C$, which represents the case of unequal population means

Imagine that we have collected data from 20 plots, 10 treatment and 10 control. We can use these data to calculate a test statistic that provides a measure of evidence against the null hypothesis of equal means. If we make a few assumptions about the distribution of this statistic, we can calculate the probability of finding a test statistic at least as extreme as the one observed, if the null hypothesis is true. This probability is often called the P-value or significance level of the test. Lower P-values suggest that the test statistic we calculated would be an unlikely result if the null hypothesis were indeed true, whereas higher P-values suggest that the test statistic would not be an unlikely result if the null hypothesis were true.

In this example, imagine that the amount of plant biomass averaged 113 kg/ha on treatment plots and 103 kg/ha on control plots (i.e., estimates of true population means μ_T and μ_C are $\bar{y}_T = 113$ and $\bar{y}_C = 103$, respectively). From these data, we also determined $s = 15$, which is an estimate of the pooled population standard deviation, σ. With this information and presuming the data approximate the necessary assumptions, we can use a two-sample t-test to generate a test statistic for the null hypothesis of $\mu_T = \mu_C$:

$$t = \frac{\bar{y}_T - \bar{y}_C}{\dfrac{s}{\sqrt{n}}} = \frac{113 - 103}{\dfrac{15}{\sqrt{20}}} = 2.98$$

This test statistic, based on a sample size of 20 (and therefore 18 degrees of freedom for this test) is associated with a two-tailed $P = 0.008$, indicating that the probability of obtaining a test statistic at least as extreme as the one we observed (2.98) if the null hypothesis of equal means is true is about 8 in 1,000—a reasonably unlikely occurrence.

If we apply the hypothesis-testing framework rigorously (which we do not advocate, but which is necessary for this discussion), we would use the value of the test statistic as the basis for a dichotomous decision about whether to reject the null hypothesis in favor of the alternative hypothesis. If the test statistic exceeds an arbitrary threshold value, called a *critical value*, then we conclude that the null hypothesis is false because evidence provided by the data suggests that attaining a test statistic as extreme as the one we observed is unlikely to occur by chance. The critical value is the value of the test statistic that yields $P = \alpha$, where α is the Type I error rate established by the researcher before the experiment is performed (see subsequent discussion). In this example, if we had chosen a Type I error rate of $\alpha = 0.05$, then $t_{crit} = 2.10$. Because the observed t-value is greater than the critical value, we reject the null hypothesis, which is a "statistically significant" result at the given α-level.

There is always a chance, however, that no matter how unlikely the test statistic (and therefore, how low the P-value), the null hypothesis may still be true. Therefore, each time a decision is made to reject or not reject a null hypothesis, there are two types of errors that can be made (table 2.1). First, a null hypothesis

Table 2.1 Possible outcomes of statistical hypothesis tests[a]

	Decision and result	
Reality	Do not reject null hypothesis	Reject null hypothesis
Null hypothesis is true	Correct $(1 - \alpha)$	Type I error (α)
Null hypothesis is false	Type II error (β)	Correct $(1 - \beta)$

[a]Probabilities associated with each decision are given in parentheses.

that is actually true might be rejected incorrectly (a Type I error; a false positive). As in the previous example, the rate at which a Type I error will be accepted is the α-level and is established by the researcher. Second, a null hypothesis that is actually false might not be rejected (a Type II error; a false negative). The probability of a Type II error is denoted as β. Statistical power is equal to $1 - \beta$ and is defined as the probability of correctly rejecting the null hypothesis, given that the alternative hypothesis is true (figure 2.1).

The statistical power of a test is determined by four factors in the following ways: power increases as sample size, α-level, and effect size (difference between the null and alternative hypothesis) increase; power decreases as variance increases. Some measures of effect size incorporate variance, leaving only three components. Effect size is the component of power least familiar to many researchers; we discuss this in detail in the next section.

2.2.2 Measures of Effect Size

In the context of power analysis, effect size is defined broadly as the difference between the null hypothesis and a specific alternative hypothesis. The null hypothesis is often one of no effect, and in these cases effect size is the same as the alternative hypothesis. For example, in the plant biomass experiment, the null hypothesis is no difference in mean biomass between treatment and control plots. One specific alternative hypothesis states that a 20 kg/ha difference between treatment and control plots exists. Effect size, in this case, is $(20 - 0) = 20$ kg/ha. However, other measures of effect size could have been used.

Choosing a meaningful effect size (or range of effect sizes) to use in experimental planning is usually the most challenging aspect of power analysis. In gen-

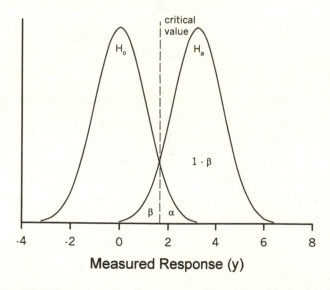

Figure 2.1 Graphical illustration of α, β, power $(1 - \beta)$, and the critical value for a statistical test of a null (H_0) versus alternative (H_a) hypothesis.

eral, we want to use effect sizes that are biologically important in the context of the study. In the plant biomass experiment, a difference of 1 kg/ha is not likely to be biologically important, but a difference of 20 kg/ha may be, depending on the goals of the study.

The effect size concept can also be used to quantify the results of experiments (chapter 18). In these cases, effect size is then defined broadly as the treatment response (for manipulative experiments) or the degree to which the phenomenon of interest is present in the population of interest (for observational studies). Effect size used in power analysis is not a population parameter; rather it is a hypothetical value that is determined by the null and alternative hypotheses as specified by the researcher. This point is critical and causes a great deal of confusion when power is examined retrospectively (section 2.3.2).

When discussing results of power analyses, the particular measure of effect size used must be specified explicitly because there usually are several measures of effect size available for a given statistical test (Richardson 1996). We introduce several measures of effect size and comment on their use.

Simple effects.   When the question of interest can be reduced to one about a single parameter, such as questions about the difference between two population means (or any other parameter) or the difference between a single population mean and a fixed value, then establishing a meaningful measure of effect size is straightforward. Most apparent are measures of *absolute effect size* (or raw effect size), which are stated as departures from the null hypothesis and have the same units as the parameters of interest. For example, in a two-sample setting comparing two population means, the null hypothesis typically would be stated as H_0: $\mu_1 - \mu_2 = 0$. A useful measure for specifying absolute effect size, therefore, is the difference between population means (or equivalently, the difference between the null and alternative hypotheses): $|\mu_1 - \mu_2|$. We used this measure of effect size when establishing the effect size of 20 kg/ha in the previous example. Similarly, in simple linear regression, one measure of absolute effect size is the difference between the slope of the regression line and a slope of zero (or any other fixed, meaningful value, such as the annual rate of change in a monitored population that would trigger management action). In logistic regression, a measure of absolute effect size is the deviation from an odds ratio of 1 (chapter 11). Because absolute effect sizes are related directly to measurements made by researchers, they are the easiest to specify and interpret.

In research studies with a temporal or spatial control, measures of *relative effect size* are useful because they represent the change in the response variable due to a treatment relative to the control $(\mu_T - \mu_C)/\mu_C$. Relative effect sizes are usually expressed as percentages, for example, the percentage increase in population size due to treatment. In the plant biomass example, we could specify that we are interested in a 20% increase in yield. This would correspond to a yield of 120 kg/ha if the true average harvest in the control plot were 100 kg/ha ((120 − 100)/100 = 20%). Finally, *standardized effect sizes* are measures of absolute effect size scaled by variance (or a related estimate of variation) and therefore combine these two components of hypothesis testing. In the two-sample setting, a standardized measure of effect size is $|\mu_1 - \mu_2|/\sigma$, where σ is the pooled within-population standard deviation. In the plant biomass example, if the population standard deviation were 15 and the true yield from control plots were 100 kg/ha, then an absolute effect size of 20 kg/ha would correspond to a standardized effect size of $|120 - 100|/15 = 1.33$. Standardized measures are unitless and therefore comparable across studies. They can be useful in research planning as a way to specify effect size when no previous data exist (for example, when there is no information about σ). However, they may be more difficult to interpret in terms of biological importance, so we prefer specifying absolute or relative measures where possible and considering the variance component of power analysis separately.

Complex effects. Establishing a meaningful effect size when an experiment includes multiple factors or multiple levels of a single factor is considerably more challenging. For example, if the plant biomass experiment were extended to include additional treatment levels, one possible null hypothesis for this new experiment would be H_0: $\mu_1 = \mu_2 = \cdots = \mu_k$, where μ_1 is the control yield and μ_2 through μ_k are the yield of the $(k-1)$ treatment levels (a one-factor fixed-effect ANOVA). In this context, a useful absolute effect size can be based on the variance of the population means:

$$\sigma_\mu^2 = \frac{1}{k} \sum_{i=1}^{k} (\mu_i - \overline{\mu})^2$$

Subsequently, $(\sigma_\mu^2)^{1/2}$ or σ_μ provides an absolute measure of effect size in the same units as the original data, much like the two-sample measure $|\mu_1 - \mu_2|$ discussed previously. Unlike the two-sample case, however, biological interpretation of σ_μ with more than two groups can be challenging.

Four approaches have been used to establish effect sizes for these more complex situations. The first approach is to specify all of the cell means (the μ_i's). In an experiment with three treatments and a control, for example, we might specify that we are interested in examining power given a control yield of 100 kg/ha and treatment yields of 120, 130, and 140 kg/ha. This approach requires researchers to make an explicit statement in terms of the experimental effects they consider to be biologically important. Although this exercise is challenging, these statements are easily interpretable. The second approach is to use measures of effect size such as σ_μ, but to seek to understand their meaning by experimenting with different values of the μ_i's. For example, yields of 100, 120, 130, and 140 kg/ha, correspond to a σ_μ of 7.4. After some experimentation with different yields, we may conclude that $\sigma_\mu \geq 7$ represents a biologically important effect. The third approach is to simplify the problem to one of comparing only two parameters. For example, in a one-factor ANOVA, we could define a measure of absolute effect size as $(\mu_{max} - \mu_{min})$, which places upper and lower bounds on power, each of which can be calculated (Cohen 1988; Nemac 1991). The fourth approach is to assess power at prespecified levels of standardized effect sizes (e.g., σ_μ/σ for the previous ANOVA example or $|\mu_1 - \mu_2|/\sigma$ for a two-sample t-test) that have been suggested for a range of tests (Cohen 1988). In the absence of other guidance, power can be calculated at three levels as implied by the adjectives small, medium, and large (Cohen 1988). These conventions are used widely in psychology and other disciplines, where a medium standardized effect size may correspond to median effect sizes used in psychological research (Sedlmeier and Gigerenzer 1989). There is no guarantee, however, that these standardized effect sizes have any meaning in ecological research, so we recommend this approach only as a last resort.

2.3 Types of Power Analyses

2.3.1 Prospective Power Analyses

Prospective power analyses are performed when planning a study. They are exploratory in nature and provide the opportunity to investigate—individually or in some combination—how changes in study design and the components of power (sample size, α, effect size, and within-population variance) influence the ability to achieve study goals. Most commonly, prospective power analyses are used to determine (1) the number of replicates or samples (n) necessary to ensure a specified level of power for tests of the null hypotheses, given specified effect sizes, α, and variance, (2) the power of tests of the null hypothesis likely to result when the maximum number of replicates possible is constrained by cost or logistics, given the effect sizes, α, and variance, and (3) the minimum effect size that can be detected, given a target level of power, α, variance, and sample size.

   Example 1. Sample sizes necessary to achieve a specified level of power, where population variance is known. Imagine that we are planning a new plant biomass experiment. Assume from previous work that field plots yielded an average of 103 kg/ha under control conditions and that the population standard deviation, σ, was 16. In this new experiment, we decide to consider the treatment effective if it increases or decreases plant biomass on plots by an average of 20% (i.e., we will use a two-tailed test). The relative effect size of interest, therefore, is 20%, and the absolute effect size is 20% of control plots or 103 kg/ha $\times 0.20 =$ 20.6 kg/ha. After some consideration of the relative consequences of Type I and Type II errors in this experiment (section 2.5.4), we establish $\alpha = \beta = 0.1$, so the target power is $1 - \beta = 0.9$. Because the population standard deviation is known, we use a Z-test for analysis. We then calculate that 22 samples are required (11 controls and 11 treatments) to meet a power of 0.9 for 20% effect size and $\sigma = 16$ (see http://www.oup-usa.org/sc/0195131878/).

   In addition to the challenges involved in choosing biologically meaningful effect sizes (section 2.3.2), this example illustrates similar challenges establishing the relative importance of Type I and Type II errors (α and β, respectively, table 2.1) in power analyses, which we explore in section 2.5.4.

   The previous example is somewhat unrealistic because we assumed that the population variance was known in advance, a rare scenario. In real-world prospective analyses, we need methods to estimate variance. Undoubtedly, the preferred method for obtaining a priori estimates of variance for power analysis is to conduct a pilot study. Pilot studies offer a number of other advantages, including the opportunity to test field methods and to train observers. We recommend using not just the variance estimated from the pilot study, but also the upper and lower confidence limits of the variance estimate to assess the sensitivity of the results and to provide approximate "best case" and "worst case" scenarios (see example 2). A second method for obtaining variances is to use those from similar studies performed previously on the same or similar systems, which can be gathered from colleagues or from the literature. Again, it is useful to repeat the power analyses using the highest and lowest variances available (preferably based on confidence intervals) or values somewhat higher and lower than any single estimate gathered. If variance estimates are obtained from systems other than the study system, the intervals used should be correspondingly wider. Finally, if no previously collected data are available, then the only choice is to perform power analyses using a range of plausible values of variance and hope that these encompass the true value.

Example 2. Sample sizes necessary to achieve a specified level of power, where a previous estimate of population variance is used. In almost all cases, the population variance used for prospective analyses will not be known. In example 1, assume the estimated standard deviation is still 16 but was based on a previous study where the sample size was 20. Because we are not assuming that the variance is known, we use a t-test for analysis. This means that the sample size required given a population standard deviation of 16 will be slightly higher than in example 1—in this case 24 rather than 22 (see http://www.oup-usa.org/sc/0195131878/).

To assess the sensitivity of this result, we can recalculate the required sample size using, for example, 90% confidence limits, 12.63 and 22.15, on the estimate of the standard deviation (see the appendix). These lead to required sample sizes of 14 and 44, respectively. If the population variance in the new experiment is the same as in the previous study, then the probability of obtaining a variance larger than the upper 90% confidence limit is $(1 - 0.9)/2 = 0.05$. Therefore, if we are conservative and use the larger sample size, we have a 95% chance of obtaining the level of power desired. Substituting confidence limits on variance into power calculations in this way leads to exact confidence limits on power for any t-test or fixed effect F-test (Dudewicz 1972; Len Thomas, 1999, unpublished ms).

If instead we are constrained to a maximum sample size of 24, we can use confidence limits on the variance estimate to calculate confidence limits on the expected power given a fixed sample size. Using this approach and $n = 24$, 90% confidence limits on power are 0.72 and 0.98. If a power of 0.72 is not acceptable, then we must use a higher α-level or perhaps reevaluate and increase treatment intensity (and therefore the likely effect size) used in the study.

Most prospective power analyses are more complex than these examples because many components of the research design can be altered, each of which can influence the resulting power of the effort (singly or in combination with other components). In addition, study goals often are not defined so sharply, and power analysis begins as an investigation of what is possible. At this stage, therefore, we recommend producing tables or graphs displaying the interactions among plausible levels of the design components and their effects on power (e.g., figure 2.2). Further, multiple goals or hypotheses are usually considered in most research

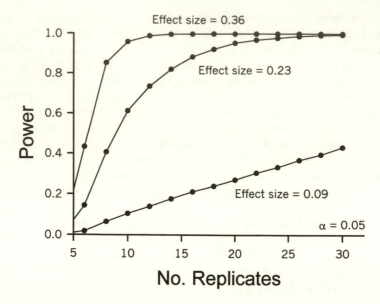

Figure 2.2 The influence of number of replicates on statistical power to detect small (0.09), medium (0.23), and large (0.36) effect sizes (differences in the probability of predation) between six large and six small trout using a Wilcoxon signed-ranks test. Power was estimated using a Monte Carlo simulation.

designs. This entails multiple, related power analyses, and the consideration of the relative importance of these different goals. Finally, study design is usually more complex than comparing samples from two populations. In these cases, considering alternative possible designs is important, as power often can be increased within a fixed budget and sample size by imaginative design decisions (e.g., Steidl et al. 1997). These include increasing the likely effect size by increasing the range or intensity of treatment levels, reducing experimental error by blocking (chapter 4) or measuring covariables, and selecting an efficient method of assigning treatments to experimental units (usually the number of replicates should be highest in treatment combinations where the variance is expected to be highest). Using a statistical model for data analysis that is consistent with the design can also have a strong influence on power (Hatfield et al. 1996; Steidl et al. 1997). These and other techniques for increasing efficiency are discussed in texts on experimental and sampling design (e.g., Thompson 1992; Kuehl 1994).

2.3.2 Retrospective Power Analyses

Retrospective power analyses are performed after a study has been completed and the data analyzed. At this point, the outcome of the statistical test is known: either the null hypothesis was rejected or it was not. If it was not rejected, we may be concerned with committing a Type II error if the statistical power of the test was low. At this point, all the information necessary to calculate power is available, the design, sample size, and α-level used are known, and the effect size and variance observed in the sample provide estimates of the effect size and variance in the population. Whether any of this information is appropriate for estimating retrospective power is controversial (Thomas 1997). Some researchers believe that the whole concept of retrospective power is invalid within the hypothesis-testing framework (Goodman and Berlin 1994; Zumbo and Hubley 1998; Gerard et al. 1998; Hoenig and Heisey 2001). Others feel that it is informative if the effect size observed is not used in the calculations (Rotenberry and Weins 1985; Cohen 1988; Peterman 1990; Muller and Benignus 1992; Searcy-Bernal 1994; Steidl et al. 1997; Thomas 1997). Although we believe that retrospective power analyses have a place, we favor alternative approaches such as the use of confidence intervals, for reasons discussed here and in section 2.5.2.

Retrospective power is a concern most often when a statistical test has failed to provide sufficient evidence to reject the null hypothesis. In this case, we wish to distinguish between the two reasons for failing to reject the null hypothesis: (1) the true effect size was not biologically important and therefore the null hypothesis was true or nearly true, and (2) the true effect size was biologically important but we failed to reject the false null hypothesis (i.e., we committed a Type II error). To make this distinction, we can calculate the power to detect a minimum biologically important effect, given the sample size, α-level used, and variance estimated in the study. If power at this effect size is large, then true effect sizes of the magnitude of the minimum biologically important effect would likely lead to statistically significant results. Given that the test was not significant, we can infer that the true effect size is likely not this large. Using similar logic, if power at this effect size is small, we can infer that the true effect size could be large or small, so the results are inconclusive.

Example 3. Retrospective power analysis. In an attempt to explain regional differences in reproductive success of ospreys (*Pandion haliaetus*), thickness of eggshells (an indicator of organochlorine contamination) was compared between a colony with low reproduction and one with normal reproduction (Steidl et al. 1991). A two-tailed, two-sample t-test comparing mean thickness between colonies yielded $t_{49} = 1.32$, $P = 0.19$, which is not statistically significant at any reasonable α-level (low reproduction: $\bar{y} = 0.459$ mm, $n = 10$; normal reproduction: $\bar{y} = 0.481$ mm, $n = 41$; pooled $s = 0.0480$).

In this case, failing to detect a biologically important difference in eggshell thickness could lead to incorrect conservation decisions. One way to address the possibility that a biologically important difference existed but was not detected by the statistical test is through retrospective power analysis. This raises the question of what comprises a biologically important difference. In this case, assume that previous research has suggested that a 5% reduction is eggshell thickness would likely impair reproduction. This would translate into an absolute difference of 0.024 mm (0.481×0.05), which gives an estimated power of 0.29 for a two-tailed t-test using the observed value of s and $\alpha = 0.05$, with 95% confidence limits of 0.20 to 0.39 (see http://www.oup-usa.org/sc/0195131878/). Of course, power to detect a larger 10% difference (0.048 mm) in eggshell thickness is higher at 0.80, with 95% confidence limits of 0.61 and 0.92.

Another relevant approach is to estimate the minimum detectable effect size for a given level of power, which is the minimum effect size that would have yielded $P \leq \alpha$. In this example (5% reduction) and with $\alpha = 0.05$, $s = 0.048$, and power $= 0.8$, the minimum detectable eggshell thickness is 0.048 mm (95% confidence limits of 0.040 and 0.060). Similarly, you could also estimate the sample size that would have been necessary to detect the observed effect size. In this example, the sample size necessary to detect the observed effect size (0.481 − 0.459 = 0.022 mm) would have been 128 (approximate 95% confidence limits of 90 and 197) (see http://www.oup-usa.org/sc/0195131878/).

Although the use of retrospective power analysis when the null hypothesis is not rejected has been recommended broadly, a number of problems are commonly ignored. First, we assume implicitly that the estimate of power at a given effect size (or effect size for a given power) can be translated into a statement of confidence about the true effect size. For example, "given the null was not rejected and that retrospective power for effect size x is $1 - \alpha$, then we have at least $(1 - \alpha)100\%$ confidence that the interval $(-x, x)$ contains the true effect size." However, such a statement has never been justified formally (Hoenig and Heisey 2001). Second, performing retrospective power calculations only when the null hypothesis is not rejected compromises these analysis. Third, confidence intervals about the estimates of power or the detectable effect size are conservative (i.e., too wide), although there are methods for correcting them (Muller and Pasour 1997). Fourth, because retrospective power calculations do not use information about the observed effect size, they are inefficient compared to the inferences that can be drawn using standard confidence intervals about the estimated effect size (section 2.5.2). Because of these problems, we believe that estimating power retrospectively is rarely useful, and instead we recommend the use of confidence intervals about estimated effect size.

One situation in which retrospective power analysis is never helpful is when power is estimated with the effect size observed in the study (sometimes called the observed power). The calculated value of power is then regarded as an estimate of the "true" power of the test, i.e., the power given the underlying population effect size. Such calculations are uninformative and potentially misleading (Steidl et al. 1997; Thomas 1997; Gerard et al. 1998). First, they do not take into account the biological significance of the effect size used. Second, the observed power estimates are simply a reexpression of the P-value: low P-values lead to high power and vice versa. Third, even as estimates of "true" power, they are biased and imprecise.

2.4 Statistical Solutions: Calculating Power

2.4.1 Power Analysis Using Standard Tables or Software

Power can be estimated for common statistical tests using tables or figures in statistics texts (e.g., Rao 1998; Zar 1996) or specialized monographs (Kraemer and Thiemann 1987; Cohen 1988; Lipsey 1990). This approach can provide an .

easy way to obtain quick, approximate results but is not ideal for an in-depth study of power for two reasons. First, estimates of power and related parameters (such as minimum detectable effect size) often are inaccurate, either because they must be interpolated from tabulated values or read from a graph, or in some cases because the tabulated values are themselves based on approximations (Bradley et al. 1996). Second, an in-depth study requires calculating the desired statistics at many levels of the other parameters and graphing the results, which is laborious if done by hand.

Alternatively, a growing number of computer programs perform power analysis (http://www.oup-usa.org/sc/0195131878/). These range from "freeware" programs to large, relatively sophisticated commercial packages. Further, some general-purpose statistical and spreadsheet software packages have built-in power analysis capabilities or add-on modules or macros (e.g., the SAS module Unify-Pow; O'Brien 1998).Thomas and Krebs (1997) performed a detailed review of 29 programs, comparing their breadth, ease of learning, and ease of use. Although their specific recommendations about packages will become increasingly outdated as new software is released, the criteria they used and their general comments remain relevant. They considered an ideal program to be one that (1) covers the test situations most commonly encountered by researchers; (2) is flexible enough to deal with new or unusual situations; (3) produces accurate results; (4) calculates power, sample size, and detectable effect size; (5) allows easy exploration of multiple values of input parameters; (6) accepts a wide variety of measures of effect size as input, both raw and standardized; (7) allows estimation of sampling variance from pilot data and from the sampling variability statistics commonly reported in the literature; (8) gives easy-to-interpret output; (9) produces presentation-quality tables and graphs for inclusion in reports; (10) allows easy transfer of results to other applications; and (11) is well documented. They recommended that beginner to intermediate users consider the specialized commercial power analysis programs nQuery Advisor, PASS, or Stat Power, whereas those on a budget try some of the freeware packages such as GPower and PowerPlant (see http://www.oup-usa.org/sc/0195131878/ for an up-to-date list of available software).

2.4.2 Programming Power Analysis Using General-purpose Statistical Software

Most statistical tests performed by ecologists are based on the Z-, t-, F-, or χ^2-distributions. Power analysis for these tests can be programmed in any general-purpose statistical package that contains the appropriate distribution functions (http://www.oup-usa.org/sc/0195131878/). The advantage of this approach is that power analyses can be tailored exactly to the experimental or sampling design being considered. This is particularly useful for relatively complex designs that are not supported by most dedicated power-analysis software. This approach may be most convenient for those who already own a suitable statistics package.

Programming your own power analyses for the t-, F-, and χ^2-tests requires an understanding of noncentral distributions and noncentrality parameters. The parametric distributions commonly used for testing are known as *central* distributions, which are special cases of more general distributions called *noncentral* distributions. Whereas central distributions describe the distribution of a statistic under the assumption that the null hypothesis is true, noncentral distributions describe the distribution under any specified alternative hypothesis. Compared to central distributions, noncentral distributions contain one additional parameter, called a *noncentrality parameter*, which corresponds to the relevant measure of effect size. For example, the noncentrality parameter, δ for the noncentral t-distribution, assuming a two-sample t-test, is

$$\delta = \frac{|\mu_1 - \mu_2|}{\sigma} \sqrt{\frac{n_1 n_2}{n_1 + n_2}}$$

The exact way in which software packages define noncentrality parameters can vary. As this formula illustrates, the noncentrality parameter can be considered as a measure of standardized effect size (in this case $|\mu_1 - \mu_2|/\sigma$), with an additional term that depends on the way in which sample units are allocated to treatments (O'Brien and Muller 1993). When the noncentrality parameter is zero ($\delta = 0$), noncentral distributions equal their corresponding central distributions. In general, the more false the null hypothesis, the larger the noncentrality parameter (Steiger and Fouladi 1997). Programming power analyses involves translating the measure of effect size used into a noncentrality parameter, then using this value in the appropriate noncentral distribution function. Only central distributions are required for power analysis using Z-tests or random effects F-tests (Sheffé 1959, p. 227). SAS code for the examples in this chapter and other relevant SAS probability functions are provided at http://www.oup-usa.org/sc/0195131878/.

2.4.3   Power Analysis Using Simulation

Sooner or later, we encounter a statistical test for which the previous two approaches are not appropriate. This may be because the test is not covered by tables or accessible software, or because there is no agreed-upon method of calculating power for that test. One example of the second situation is nonparametric tests, where the distribution of the processes producing the data are not fully specified so their distribution under the alternative hypothesis is unknown (see the following example). Specific examples in ecology include analyzing multisite trends (Gibbs and Melvin 1997), modeling predator functional responses (Marshal and Boutin 1999), and assessing trends in fish populations (Peterman and Bradford 1987).

   In these situations, power analyses can be performed using stochastic (Monte Carlo) simulations. The approach is simple. First, write a computer routine that mimics the actual experiment, including the analysis. In the plant biomass experiment, for example, the program would use a pseudorandom number generator to create 10 biomass measurements from a normal distribution with mean μ_C and standard deviation σ for controls, and 10 measurements from a normal distribution with mean μ_T and standard deviation σ for treatments. The routine then would analyze these data using a *t*-test. Second, for each level of the input parameters (in this case μ_C, μ_T, and σ), program the routine to run many times (see subsequent discussion), and tally whether the results were statistically significant for each run. Finally, calculate the proportion of results that were significant. If the computer model is an accurate representation of the real experiment, then the probability of getting a statistically significant result from the model is equal to the probability of getting a statistically significant result in the real experiment, in other words, the statistical power. Hence, the proportion of significant results from the simulation runs is an estimate of power.

The number of simulation runs required depends on the desired precision of the power estimate (table 2.2 and appendix). For precision to one decimal place, 1,000 runs should suffice, whereas for precision to two decimal places, 100,000 runs are necessary.

Example 4. Power analyses by simulation for a nonparametric test. In an experiment investigating the effect of prey size on predation rates, several replicate groups of six large and six small juvenile fish were exposed to a predatory fish; the number of small and large fish depredated was recorded for each group. A Wilcoxon signed-ranks test (a nonparametric equivalent of the one-sample *t*-test) was used to test the null hypothesis that the median difference in the number killed between size classes was zero. Thomas and Juanes (1996) explored the power of this experiment using simulations. They assumed that, within a group, the number of fish killed in each size class was a binomial random variable, and they varied the number of replicate groups (the sample size) and the difference in probability of predation between large and small fish. Their results (figure 2.2) suggested that at least 14 groups were necessary to achieve power of 0.8 given a difference in survival between size classes (effect size) of 0.23.

To simulate experiments analyzed using nonparametric tests, such as the previous one, we must specify fully the data-generating process. In these situations, simulations allow us to explore the power of the experiment under a range of different assumptions. In the example, the probability that a fish is depredated was assumed to be constant within each size class. We could arguably make the model more realistic by allowing probability of predation to vary within groups according to some distribution (for example, the beta distribution). However,

Table 2.2 Dependence of the precision of power estimates from Monte Carlo simulations on the number of simulation runs[a]

Number of simulations	SE ($\hat{\beta}$)	99% CI
100	0.050	0.371–0.629
1 000	0.016	0.460–0.541
10 000	0.005	0.487–0.513
100 000	0.002	0.496–0.504

[a]Calculations are performed at a true power (β) of 0.5 and therefore represent minimum levels of precision (see appendix).

there is always a trade-off between simplicity and realism, and we should be content to stop adding complexity to models when they adequately mimic the features of the experiment and subsequent data that are of particular interest. In the example, the variance of the data generated by the model was similar to that of the experimental data, providing a degree of confidence in the model.

Another related approach is the use of bootstrap resampling (chapter 14) to obtain retrospective power estimates from experimental data. In this approach, many bootstrap data sets are generated from the original data, and the same statistical test is performed on each one. The proportion yielding statistically significant results is an estimate of the power of the test for the given experiment. Unless modified in some way, this approach will estimate power at the observed effect size, which is not useful (section 2.3.2). Therefore, power must be estimated over a range of effect sizes, by adding and subtracting effects to the observed data (e.g., Hannon et al. 1993).

2.5   Related Issues and Techniques

2.5.1 Bioequivalence Testing

There have been numerous criticisms of the hypothesis-testing approach (e.g., Yoccoz 1991; Nester 1996; Johnson 1999; references in Harlow et al. 1997 and Chow 1998). One criticism is that the null hypothesis can virtually never be true and therefore is of no interest. For example, no matter how homogeneous a population, no two samples drawn from the population will be identical if measured finely enough. Consequently, bioequivalence testing was developed in part to counter this criticism (Chow and Liu 1999) and is used commonly in pharmaceutical studies and increasingly in ecological studies (e.g., Dixon and Garrett 1994).

Bioequivalence testing reverses the usual burden of proof, so that a treatment is considered biologically important until evidence suggests otherwise. This is achieved by switching the roles of the null and alternative hypotheses. First, a minimum effect size that is considered biologically important is defined (say, Δ_{crit}). Next, the null hypothesis in stated such that the true effect size is greater than or equal to Δ_{crit}. Finally, the alternative hypothesis is stated such that true effect size is less than Δ_{crit}. The plant biomass experiment discussed previously, for example, could be phrased as:

H_0: $|\mu_T - \mu_C| \geq \Delta_{crit}$, which represents the case where a biologically important effect exists

H_a: $|\mu_T - \mu_C| < \Delta_{crit}$, which represents the case where no biologically important effect exists

In this context, a Type I error occurs when the researcher concludes incorrectly that no biologically important difference exists when one does. This is the type of error that is addressed by power analysis within the standard hypothesis-testing framework; in bioequivalence testing, this error rate is controlled a priori by setting the α-level of the test. Some have argued that this approach is preferable and eliminates the need for retrospective power analysis (Hoenig and Heisey 2001). However, Type II errors still exist within this framework when the researcher concludes incorrectly that an important difference exists when one does not. If this type of error is a concern, then a retrospective investigation will still be necessary when the null hypothesis is not rejected.

2.5.2 Estimating Effect Sizes and Confidence Intervals

Another criticism of hypothesis testing is that the statistical significance of a test does not reflect the biological importance of the result, because any two samples will differ significantly if measured finely enough. For example, a statistically significant result can be found for a biologically trivial effect size when sample sizes are large enough or variance small enough. Conversely, a statistically insignificant result can be found either because the effect is not biologically important or because the sample size is small or the variance large. These scenarios can be distinguished by reporting an estimate of the effect size and its associated confidence interval, rather than simply reporting a P-value.

Confidence intervals can function to test the null hypothesis. When estimated for an observed effect size, a confidence interval represents the likely range of numbers generated from the data that cannot be excluded as possible values of the true effect size with probability $1 - \alpha$. If the $100(1 - \alpha)\%$ confidence interval for the observed effect does not include the value established by the null hypothesis, you can conclude with $100(1 - \alpha)\%$ confidence that a hypothesis test would be statistically significant at level α. In addition, however, confidence intervals provide more information than hypothesis tests because they establish approximate bounds on the likely value of the true effect size. More precisely, on average, $100(1 - \alpha)\%$ confidence intervals will contain the true value of the estimated parameter $100(1 - \alpha)\%$ of the time. Therefore, in situations where the null hypothesis would not be rejected by a hypothesis test, we can use the confidence interval to assess whether a biologically important effect is plausible (figure 2.3). If the confidence interval does not include a value large enough to be considered biologically important, then we can conclude with $100(1 - \alpha)\%$ confidence that no biologically important effect occurred. Conversely, if the interval does include biologically important values, then results are inconclusive. This effectively answers the question posed by retrospective power analysis, making such analyses unnecessary (Goodman and Berlin 1994; Thomas 1997; Steidl et al. 1997; Gerard et al. 1998).

Confidence interval estimation and retrospective power analysis are related but not identical. In the estimation approach, the focus is on establishing plausible bounds on the true effect size and determining whether biologically important effect sizes are contained within these bounds. In power analysis, the focus is on the probability of obtaining a statistically significant result if the effect size were truly biologically important. Despite these differences, the conclusions drawn

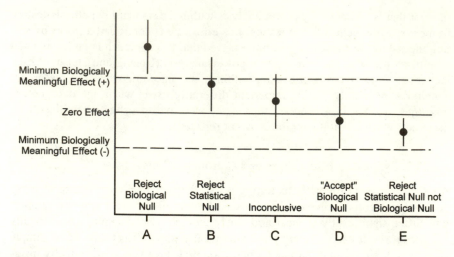

Figure 2.3 Hypothetical observed effects (circles) and their associated $100(1 - \alpha)\%$ confidence intervals. The solid line represents zero effect, and dashed lines represent minimum biologically important effects. In case A, the confidence interval for the estimated effect excludes zero effect and includes only biologically important effects, so the study is both statistically and biologically important. In case B, the confidence interval excludes zero effect, so the study is statistically significant; however, the confidence interval also includes values below those thought to be biologically important, so the study is inconclusive biologically. In case C, the confidence interval includes zero effect and biologically important effects, so the study is both statistically and biologically inconclusive. In case D, the confidence interval includes zero effect but excludes all effects considered biologically important, so the "practical" null hypothesis of no biologically important effect can be accepted with $100(1 - \alpha)\%$ confidence. In case E, the confidence interval excludes zero effect but does not include effects considered biologically important, so the study is statistically but not biologically important.

from both approaches are often similar. Nevertheless, we prefer the confidence interval approach because interpretation of results is straightforward, more informative, and viewed from a biological rather than probabilistic context.

Example 5. Confidence intervals in lieu of retrospective power. In the osprey eggshell study from example 3, the mean difference in eggshell thickness between regions (the observed absolute effect size) was estimated to be 0.022 mm with a standard error of 0.0169. In the hypothesis-testing approach (example 3), assume we established α at 0.05; we would then use a $100(1 - \alpha) = 95\%$ confidence interval. The 95% confidence interval on this observed effect size (mean difference) ranges from -0.012 to 0.056 mm. This interval contains the value of 0 predicted by the null hypothesis, so we know the statistical test would not be rejected at $\alpha = 0.05$, as we showed previously ($P = 0.19$). However, our conclusion about the results of this study will depend on the effect size we consider biologically important. If we consider a relative difference of 10% (0.048 mm) or greater between colonies to be important, then we can consider the results to be inconclusive because the confidence interval includes this value (figure 2.3). If instead we consider a relative difference of $\geq 20\%$ (0.096 mm) to be important, then we can conclude with 95% confidence that the study showed no important effect because this value is excluded by the confidence interval.

2.5.3 Design Using Confidence Intervals

If the results of a study are to be evaluated using confidence intervals about effect size, then we might design the study to achieve a specified level of precision, or equivalently, a confidence interval of specified width, rather than a desired level of power. For example, we could plan a study to attain a confidence interval that is narrow enough to exclude the null effect size if the true effect size is that which we establish as the minimum to be biologically important. The confidence interval width that we determine is a random variable, however, so there is a 50% chance that it will be wider or narrower than the planned width. Therefore, a conservative approach to design in an estimation context is important (as it is in all aspects of design), and power analysis is a useful tool for this approach (Greenland 1988; Borenstein 1994; Goodman and Berlin 1994).

As mentioned previously, when the realized $100(1 - \alpha)\%$ confidence interval excludes the null effect size, this is equivalent to rejecting the null hypothesis at level α. Therefore, the probability that the confidence interval excludes the null effect size, given some specified true effect size, is equal to the power of the test. So, to have a $(1 - \beta)$ probability of achieving $100(1 - \alpha)\%$ confidence intervals narrow enough to exclude the null hypothesis at a specified true effect size, we must have $(1 - \beta)$ power at that effect size.

Example 6. Prospective power analysis for prespecified confidence interval width.   We are planning to evaluate the results of the next plant biomass experiment using confidence intervals. As in example 2, we will assume that the variance is not known. For planning purposes, we will base our calculations on a previous study where the estimated standard deviation was 16 with sample size 20. Assume that we wish to have a 90% chance of obtaining 90% confidence limits large enough to exclude zero difference should the true difference be 20% or greater (i.e., ≥ 20.6 kg/ha). Because the confidence limits are symmetric, the desired confidence interval width is therefore $2 \times 20.6 = 41.2$ kg/ha.

This scenario leads to exactly the same power analysis as in example 2: the estimated sample size required is 24, but when we incorporate our uncertainty about the variance estimate, the sample size required is between 14 and 44. Further, calculating the expected confidence interval widths, given the expected variance and sample size, is instructive. With a sample size of 24 and standard deviation of 16, the expected confidence interval width is 13.5 kg/ha. So, we can be 90% sure of achieving a confidence interval width of less than 41.2 kg/ha, but 50% sure that the width will be less than 13.5 kg/ha. As with all prospective design tools, figures displaying how these values change as other factors in the design change prove extremely useful in research planning.

2.5.4   Consequences and Considerations for Establishing α and β

Results of all prospective and retrospective power analyses depend on the levels at which α and β are established. In prospective power analyses, for example, decreasing α (say from 0.10 to 0.05) or increasing the target level of power (say, from 0.7 to 0.9) will always increase the sample sizes necessary to detect a given effect size. As with establishing meaningful effect sizes, choosing these error rates will forever be a challenge.

In general, establishing α and β requires balancing the costs and consequences of Type I and Type II errors (Shrader-Frechette and McCoy 1992; table 2.1). Traditionally, scientists have focused only on Type I errors (hence the impetus for this chapter). However, when there are considerable risks associated with decisions based on the results of hypothesis tests that are not rejected, the consequences of Type II errors often can exceed those of Type I errors (Hayes 1987; Peterman 1990; Steidl et al. 1997). Decisions resulting from hypothesis tests that were not rejected have an underlying, often unrecognized, assumption about the relative costs of Type I and Type II errors that is independent of their true costs (Toft and Shea 1983; Cohen 1988; Peterman 1990). In particular, when $\beta = \alpha$, scientists have decided, perhaps unknowingly, that the costs of Type I errors exceed those of Type II errors when their recommendations assume that a null hypothesis that was not rejected was actually true (i.e., when the null hypothesis was inappropriately accepted). Some have suggested that Type II errors be considered paramount when a decision would result in the loss of unique habitats or species (Toft and Shea 1983; Shrader-Frechette and McCoy 1992). Other approaches have been suggested to balance Type I and Type II error rates based on their relative costs (Osenberg et al. 1994).

As we discussed previously (section 2.5), hypothesis testing has been misused by scientists too often (see also Salsburg 1985, Yoccoz 1991), especially in the context of environmental decision making. Hypothesis tests assess only "statistical significance." The issue of "practical or biological importance" may be better evaluated using confidence intervals (section 2.5.2, although we must still choose the level of confidence to use). We suggest that the reliance on hypothesis testing in decision-making circumstances be decreased in favor of more informative methods that better evaluate available information, including confidence intervals (section 2.5.2), bioequivalence testing (section 2.5.1), Bayesian methods (chapter 17), and decision theory, an extension of Bayesian methods that incorporates the "cost" of making right and wrong decisions (Barnett 1999). Nearly all resource-based decisions are complex, and reducing that complexity to a dichotomous yes–no decision is naïve. Typically, the relevant issue is a not whether a particular effect or phenomenon exists, but whether the magnitude of the effect is biologically consequential. Hypothesis testing should not be the only tool used in decision making, especially when the risks associated with an incorrect decision are considerable. In these instances, knowledge of the potential risks and available evidence for each decision should guide the decision-making process.

2.6   Conclusions

We recommend that researchers evaluate their study design critically before committing to a particular scheme. Only before serious data collection has begun and considerable investment been made can research goals be evaluated freely and details of the experimental design be changed to improve efficiency. We recommend power analysis as a tool for evaluating alternatives in design. This technique forces us to explicitly state our goals (including effect sizes considered biologically important and tolerable levels of error) and make a plan for the analysis of the data—something done far too rarely in practice. In many cases, the power analysis will force us to be more realistic about our goals and perhaps convince us of the need to consult a statistician, either to help us with the power analysis or to help outline the options for study design and analyses. No matter how harsh the realism, the insights gained will save much time and effort. As Sir Ronald Fisher once said, perhaps the most a statistician can do after data have been collected is pronounce a postmortem on the effort.

Acknowledgments   Thanks to John Hoenig for sharing his unpublished manuscript and to Ken Burnham, John Hayes, John Hoenig, and Eric Schauber for many insightful discussions. We also thank Sam Scheiner, Jessica Gurevitch, and two anonymous reviewers for their helpful comments on the first draft of this chapter.

Appendix

Here we present formulae and sample calculations for Examples 1–3. SAS code for these examples is available at http://www.oup-usa.org/sc/0195131878/. This appendix also contains the formula for calculating confidence limits for an estimated standard deviation. Confidence intervals on power, required sample size, or minimum detectable effect size can be calculated by substituting the upper and lower confidence limits for standard deviation into the power formulas in the SAS code supplied. Finally, we outline the method used to calculate the precision of power estimates from Monte-Carlo simulation (Table 2.2).

All of the examples in this chapter are of two-tailed two-sample tests, where the test statistic is a Z or t value. In these cases, power is calculated as the probability that the test statistic is greater than or equal to the upper critical value of the appropriate distribution, plus the probability that the test statistic is less than or equal to the lower critical value of the distribution. For a Z-test, power $(1 - \beta)$ is:

$$1 - \beta = (1 - F_z(Z_{1-\alpha/2} - Z_{hyp})) + F_z(Z_{\alpha/2} - Z_{hyp}) \qquad (1)$$

where $F_z(x)$ is the cumulative distribution function of the Z distribution at x, and Z_{hyp} is the $100p$ percentile from the standard normal distribution, calculated as:

$$Z_{hyp} = \frac{\Delta}{\sigma} \sqrt{\frac{n_1 n_2}{n_1 + n_2}} \qquad (2)$$

where Δ is the specified difference between group means, σ is the pooled standard deviation, and n_1 and n_2 are samples sizes for each group. For a t-test, power is:

$$1 - \beta = (1 - F_t(t_{1-\alpha/2,v} \mid v,\delta)) + F_t(t_{\alpha/2,v} \mid v,\delta) \qquad (3)$$

where $F_t (x \mid v,\delta)$ is the cumulative density function of the noncentral t distribution with v degrees of freedom and noncentrality parameter δ, evaluated at x, and $t_{p,v}$ is the $100p$ percentile from the central t distribution with v degrees of freedom. The noncentrality parameter δ is calculated as:

$$\delta = \frac{\Delta}{\sigma} \sqrt{\frac{n_1 n_2}{n_1 + n_2}} \qquad (4)$$

Example 1. Sample sizes necessary to achieve a specified level of power, where population variance is known. Estimate sample sizes ($n = n_1 + n_2$; $n_1 = n_2$) necessary to achieve a specified level of power (1- β) to detect a minimum biologically important difference (Δ) between means of two groups, given α, and the known, pooled standard deviation, σ. Formulas (1) and (2) could be used iteratively using different values of n_1 and n_2, however Zar (1996) provides a more direct formula:

$$n_1 = n_2 = \frac{2\sigma^2 (Z_{\alpha/2} + Z_\beta)^2}{\Delta^2}. \qquad (5)$$

In this example, $\alpha = \beta = 0.1$, $\Delta = 20.6$ kg/ha, $\sigma = 16$ kg/ha; $Z_{0.1/2} = 1.64$ and $Z_{0.1} = 1.28$. Therefore,

$$n_1 = n_2 = \frac{2 \cdot 16^2 (1.64 + 1.28)^2}{20.6^2} = 10.3$$

which indicates than 11 samples (rounding up) would be necessary for each group to meet the specified level of power, yielding a total $n = 22$.

Example 2. Sample sizes necessary to achieve a specified level of power, where a previous estimate of population variance used. Because the pooled standard deviation is estimated rather than known, the t-test rather than the Z-test is appropriate, and (5) cannot be used. Instead, we provide an initial estimate of n required in (4), substitute the calculated noncentrality (δ) into (3), and calculate power. We then continue to adjust our estimate of n until we equal or exceed the level of power specified. For example, beginning with an estimated $n_1 = n_2 = 11$ or $n = 22$ (from example 1), calculate noncentrality using equation (4):

$$\delta = \frac{20.6}{16} \sqrt{\frac{11 \cdot 11}{11 + 11}} = 3.019.$$

To estimate power using equation (3), first calculate degrees of freedom $v = n_1 + n_2 - 2 = 20$, $t_{0.95, 20} = 1.725$, and $t_{0.05, 20} = -1.725$. Power is then:

$$1 - \beta = (1 - F_t(1.725|20,3.02)) + F_t(-1.725|20,3.02)$$
$$= 1 - 0.102030 + 0.000002 = 0.897$$

This is slightly less than power of 0.9 specified, so increase the estimate of n to $n_1 = n_2 = 12$, which yields $\delta = 3.154$, $v = 22$, $t_{0.95, 22} = 1.717$, $t_{0.05, 22} = -1.717$, so power is:

$$1 - \beta = (1 - F_t(1.717|22,3.15)) + F_t(-1.717|22,3.15)$$
$$= 1 - 0.079325 + 0.000001 = 0.921.$$

Therefore, 12 samples from each group are necessary to meet the specified level of power, yielding a total $n = 24$.

Example 3. Retrospective power analysis. First, estimate power for a hypothesis test already conducted that was not rejected given a minimum biologically important difference in means between two groups ($\Delta = |\mu_1 - \mu_2|$), sample size, α, and an estimate (s) of the pooled standard deviation (σ). In this example, $\Delta = |\mu_1 - \mu_2| = 0.024$ mm, $s = 0.048$, $n_1 = 10$, $n_2 = 41$, and $\alpha = 0.05$.

Calculate an estimate of noncentrality appropriate for the two-tailed, two-sample t-test (4):

$$\hat{\delta} = \frac{0.024}{0.048} \sqrt{\frac{10 \cdot 41}{10 + 41}} = 1.418$$

Calculate degrees of freedom $v = n_1 + n_2 - 2 = 49$, $t_{0.975, 49} = 2.010$, and $t_{0.025, 49} = -2.010$, and estimate power using equation (3):

$$1 - \beta = (1 - F_t(2.010|49,1.418)) + F_t(-2.010|49,1.418)$$
$$= 1 - 0.71567 + 0.00040 = 0.285.$$

Minimum detectable effect size (Δ_{mde}) is estimated iteratively. Begin with an arbitrary estimate of detectable effect size, calculate power, then adjust Δ_{mde} until the specified level of power is obtained. For example, begin with an estmate of $\Delta_{mde} = 0.030$ in (4):

$$\hat{\delta} = \frac{0.030}{0.048} \sqrt{\frac{10 \cdot 41}{10 + 41}} = 1.772$$

Calculate degrees of freedom $v = n_1 + n_2 - 2 = 49$, $t_{0.975, 49} = 2.010$, and $t_{0.025, 49} = -2.010$, and substitute the estimate of δ from above into (3) to calculate power:

$$1 - \beta = (1 - F_t(2.010|49,1.772)) + F_t(-2.010|49,1.772)$$
$$= -0.58808 + 0.00011 = 0.412$$

which is below the power of 0.9 specified. Therefore, increase the estimate of Δ_{mde} until you determine that the minimum effect size that could have detected was 0.0484.

Finally, the sample size that would have been necessary to detect the observed effect size ($\hat{\Delta}$) at the specified level of power (0.80) is also calculated iteratively. Begin with an arbitrary estimate of sample size, calculate power, then adjust the estimated sample size until the specified level of power is obtained. For example, begin with $n_1 = n_2 = 30$ or $n = 60$ in (4):

$$\hat{\delta} = \frac{0.024}{0.048} \sqrt{\frac{30 \cdot 30}{30 + 30}} = 1.936.$$

Calculate degrees of freedom $v = n_1 + n_2 - 2 = 58$, $t_{0.975, 58} = 2.001$, and $t_{0.025, 58} = -2.001$, and substitute the estimate of δ from above into (3) to calculate power:

$$1 - \beta = (1 - F_t(2.001 \,|\, 58, 1.936)) + F_t(-2.001 \,|\, 58, 1.936)$$
$$= 1 - 0.52216 + 0.00006 = 0.478$$

which is well below the power of 0.9 specified, so we increase the estimate of n until we determine that $n = 128$ (64 per group) were necessary to detect the observed effect size at the level of power specified.

Confidence limits for population standard deviation

The $(1 - \alpha)$ confidence limits for the population standard deviation, based on an estimated standard deviation, s, are given by:

$$\sqrt{\frac{vs^2}{\chi^2_{(\alpha/2),v}}} \text{ and } \sqrt{\frac{vs^2}{\chi^2_{(1-\alpha/2),v}}}$$

where v is the degrees of freedom ($n - 2$ for the examples in this chapter) and $\chi^2_{p,v}$ is the $100p$ percentile from a χ^2 distribution with v degrees of freedom (e.g., Zar 1996, p. 113–115).

Precision of power estimates from Monte-Carlo simulations

Each simulation is assumed to be an independent Bernoulli trial with probability of success, β, equal to the true power of the test. Under these conditions, $SE(\hat{\beta}) = \beta(1 - \beta)/n$, where n is the number of simulations. $SE(\hat{\beta})$ will be at its maximum (and so precision at its minimum) when $\beta = 0.5$.

3

Exploratory Data Analysis and Graphic Display

AARON M. ELLISON

3.1 Introduction

You have designed your experiment, collected the data, and are now confronted with a tangled mass of information that must be analyzed, presented, and published. Turning this heap of raw spaghetti into an elegant fettucine alfredo will be immensely easier if you can visualize the message buried in your data. Data graphics, the visual "display [of] measured quantities by means of the combined use of points, lines, a coordinate system, numbers, symbols, words, shading, and color" (Tufte 1983, p. 9) provide the means for this visualization.

Graphics serve two general functions in the context of data analysis. First, graphics are a tool used to explore patterns in data before the formal statistical analysis (Exploratory Data Analysis, or EDA, Tukey 1977). Second, graphics communicate large amounts of information clearly, concisely, and rapidly, and illuminate complex relationships within data sets.

Graphic EDA yields rough sketches to help guide you to appropriate, often counterintuitive, formal statistical analyses. In contrast to EDA, presentation graphics are final illustrations suitable for publication. Presentation graphics of high quality can leave a lasting impression on readers or audiences, whereas vague, sloppy, or overdone graphics easily can obscure valuable information and engender confusion. Ecological researchers should view EDA and sound presentation graphic techniques as essential components of data analysis, presentation, and publication.

This chapter provides an introduction to graphic EDA, and some guidelines for clear presentation graphics. More detailed discussions of these and related topics can be found in texts by Tukey (1977), Tufte (1983, 1990), and Cleveland (1985). These techniques are illustrated for univariate, bivariate, multivariate, and classified quantitative (ANOVA) data sets that exemplify the types of data sets encountered commonly in ecological research. Sample data sets are described briefly in section 3.3; formal analyses of three of the illustrated data sets can be found in chapters 14 (univariate data set) and 10 (predator–prey data set), and Potvin (1993, chapter 3, ANOVA data set). You may find some of the graphics types presented unfamiliar or puzzling, but consider them seriously as alternatives to the more common bar charts, histograms, and pie charts. The majority of these graphs can be produced by readily available Windows-based software (Kardia 1998). I used S-Plus (MathSoft, Inc.) and SYSTAT (SPSS, Inc.) to construct the figures in this chapter.

Guiding Principles.   The question or hypothesis guiding the experimental design also should guide the decision as to which graphics are appropriate for exploring or illustrating the data set. Sketching a mock graph, without data points, *before* beginning the experiment usually will clarify experimental design and alternative outcomes. This procedure also clarifies a priori hypotheses that will prevent inappropriately considering a posteriori hypotheses (suggested by EDA) as a priori. Often, the simplest graph, without frills, is the best. However, graphs do not have to be simple-minded, conveying only a single type of information, and they need not be assimilated in a single glance. Tufte (1983) and Cleveland (1985) provide numerous examples of graphs that require detailed inspection before they reveal their messages. Besides the aesthetic and cognitive interest they provoke, complex graphs that are information-rich can save publication costs and time in presentations.

Regardless of the complexity of your illustrations, you should adhere to the following four guidelines in EDA and production graphics:

1. Underlying patterns of interest should be illuminated, while not compromising the integrity of the data.
2. The data structure should be maintained, so that readers can reconstruct the data from the figure.
3. Figures should have a high data-to-ink ratio and no "chartjunk"—"graphical paraphernalia routinely added to every display" (Tufte 1983, p. 107), including excessive shading, grid lines, ticks, special effects, and unnecessary three-dimensionality.
4. Figures should not distort, exaggerate, or censor the data.

With the increasing availability of hardware and software capable of digitizing information directly from published sources, adherence to these guidelines has become increasingly important. Gurevitch (chapter 18; Gurevitch et al. 1992), for example, relied extensively on information gleaned by digitizing data from many different published figures to explore common ecological effects across many experiments via meta-analysis. Readers will be better able to compare published data sets that are represented clearly and accurately.

3.2 Graphic Approaches

3.2.1 Exploratory Data Analysis (EDA)

Tukey (1977) established many of the principles of EDA, and his book is an indispensable guide to EDA techniques. You should view EDA as a first pass through your data set prior to formal statistical analysis. EDA is particularly appropriate when there is a large amount of variability in the data (low signal-to-noise ratio) and when treatment effects are not immediately apparent. You can then proceed to explore, through formal analysis, the patterns illuminated by graphic EDA.

Since EDA is designed to illuminate underlying patterns in noisy data, it is imperative that the underlying data structure not be obscured or hidden completely in the process. Also, because EDA is the predecessor to formal analysis, it should not be time-consuming. Personal computer-based packages permit rapid, interactive graphic construction with little of the effort necessary in formal analysis. Finally, EDA should lead you to appropriate formal analyses and models. A common use of EDA is to determine whether the raw data satisfy the assumptions of the statistical tests suggested by the experimental design (see sections 3.3.1 and 3.3.4). Violation of assumptions revealed by EDA may lead you to use different statistical models from those you had intended to employ a priori. For example, Antonovics and Fowler (1985) found unanticipated effects of planting position in their studies of plant competitive interactions in hexagonal planting arrays. These results led to a new appreciation for neighborhood interactions in plant assemblages (e.g., Czárán and Bartha 1992).

3.2.2 Production Graphics

Graphics are an essential medium of communication in scientific literature and at seminars and meetings. In a small amount of space or time, it is imperative to deliver the message and fix it clearly and memorably in the audience's mind. Numerous authors have investigated and analyzed how individuals perceive different types of graphs, and what makes a "good" and "bad" graph from a cognitive perspective (reviewed concisely by Wilkinson 1990 and in depth by Cleveland 1985). It is not my intention to review this material; rather, through example, I hope to change the way we as ecologists display our data to maximize the amount of information communicated while minimizing distraction.

Cleveland (1985) presented a hierarchy of graphic elements used to construct data graphics that satisfy the guidelines suggested in section 3.1 (figure 3.1). Although there is no simple way to distinguish good graphics from bad graphics, we can derive general principles from Cleveland's ranking. First, color, shading, and other chartjunk effects do not as a rule enhance the information content of graphs. They may look snazzy in a seminar, but they lack substance and use a lot of ink. Second, three-dimensional graphs that are mere extensions of two-dimensional graphs (e.g., ribbon charts, three-dimensional histograms, or pie

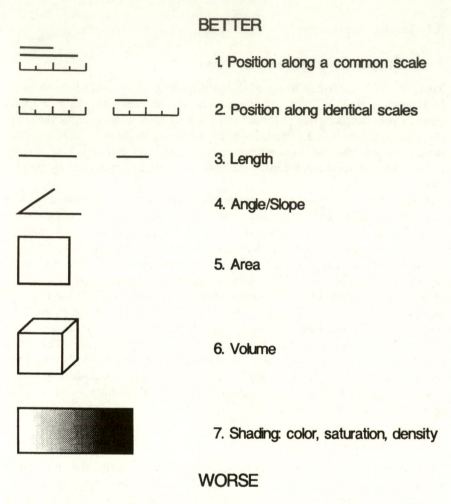

Figure 3.1 Ordering of graphic features according to their relative accuracy in representing quantitative variation (after Cleveland 1985).

charts) not only do not increase the information content available, but often obscure the message (a dramatic, if unfortunate, set of examples can be found in Benditt 1992). These graphics, common in business presentations and increasingly rife at scientific meetings, violate all of the suggested guidelines. Finally, more dimensions often are used than are necessary, for example, "areas" and lines where a point would do. Simken and Hastie (1987) discuss exceptions to Cleveland's graphic hierarchy. In general, when designing graphics, adhere to the Shaker maxim: form follows function.

High-quality graphical elements can be assembled into effective graphic displays of data (Cleveland 1985). First, emphasize the data. Lines drawn through data points should not hide the points themselves. Second, data points should never lie on axes themselves, as the axes can obscure data points. If, for example, there are many points that would fall along a zero line, then extend that axis beyond zero (figure 3.6). Third, reference lines, if needed (which they rarely are), should be deemphasized relative to the data. This can be accomplished with different line types (variable thicknesses; dotted, dashed, or solid) or shading. Fourth, overlapping data symbols or data sets should be clearly distinguishable. You can increase data visibility and minimize overlap by varying symbol size or position, separating data sets to be compared into multiple plots, or changing from arithmetic to logarithmic scales. Exemplars include the jitter plot, which avoids overlap of identical values (figure 3.3B) and spreading of responses to categories across an axis (figure 3.12D). Fifth, the plot must be easily readable following reduction for publication or when projected as a slide to a seminar audience. Finally, Cleveland recommends using a full rectangular plot frame, not the more common bottom axis/left axis only combination seen in many articles. This, together with tick marks *outside* the plot frame (1) emphasize the data and (2) help the reader accurately place individual data points. Tufte (1983) disagrees, as the extra axes are an excessive use of ink and convey no information. Examples in this chapter illustrate most of these possibilities. In the final analysis, many of these rules reflect not only insight into cognitive perception, but also aesthetic judgments by you, the author.

From this discussion, we could ask, Isn't all this too much trouble? Should we dispense with graphs altogether in favor of tables? Because of their conciseness, graphics are almost always preferable in oral presentations. Graphs illustrate more clearly relationships among variables and are a quick means of displaying multivariate information. However, where exact values are important (as in final publications), tables are more precise. Although the need for precise tables has been obviated by the increasing availability of digitizing software and on-line data archives, data presented graphically must be unbiased and uncensored. A discussion of what data should be provided, in either graphs or tables, follows in section 3.4.

3.3 Examples

3.3.1 Univariate Data: Frequency (Density) Distributions

Distributions of height, biomass, or other size metrics are often the primary descriptor of populations or communities. As an example of size distributions, I use a data set containing the number of leaf nodes of 75 *Ailanthus altissima* plants. The experimental design and formal analysis of these data are given in chapter 14.

With univariate data, two questions are paramount: (1) How are the data distributed (including summary statistics such as the mean, variance, and median)? and (2) Are the data normally distributed or can they be transformed to make them amenable to parametric analyses? Investigators often explore these questions via histograms or normality plots.

A histogram is an example of a *density* plot; that is, each bar illustrates the frequency, or density, of the values occurring in the data set between the lower bound and the upper bound of each bar. Histograms are commonly confused with bar charts (see section 3.3.4). The latter are used to illustrate some summary measure (often the mean, sum, or percentage) of all the values within a given treatment category. Histograms of the *Ailanthus* data are shown in figure 3.2.

A histogram is not the best method for answering the two questions posed previously, for three reasons. First, the raw data are hidden. In this example, there are 75 plants, which have been divided into 12 biomass groups, or *bins* (figure 3.2A). It is impossible to know, for example, if the third bar (range 12–14 nodes) contains 10 observations of 12 nodes, 10 observations of 14 nodes, or any other of the possible combinations of 12–14 nodes in 10 observations. Second, the division into 12 bins is arbitrary; it was the default of the graphics program. We could just as easily use 24 or 6 bins, both of which change the apparent shape of the distribution (figures 3.2B,C) without conveying additional information. Third, summary statistics cannot be computed from the data illustrated in the histogram. Thus, a histogram does not enable us to answer key questions about univariate data. In addition, histograms fall low on Cleveland's hierarchy of graphic primitives. Bars in a histogram use vertical lines, horizontal lines, and shading in concert to present information embodied in the single point indicated by the top of the bar.

Tukey (1977) introduced the stem-and-leaf diagram as the simplest alternative to the histogram (figure 3.3A). The main advantage of the stem-and-leaf diagram is that the raw data are presented in toto. Summary statistics can be derived easily from or incorporated into the figure. Nevertheless, stem-and-leaf diagrams suffer visually from one of the same drawbacks as histograms: the number of bins is arbitrary. Two other alternatives to histograms are jitter plots (figure 3.3B) and dit plots (figure 3.3C). These two figures preserve the underlying data structure (all values are presented), do not use arbitrary bins, and can be constructed quickly without additional preparation (e.g., sorting) of the data set. Both plots permit rapid assessment of density patterns and are simple to understand.

Stem-and-leaf plots and the density diagrams presented in figure 3.3 can be used as simple alternatives to histograms. However, these plots do not clearly convey some of the information that ecologists may want to communicate, and it is difficult to compare the information in two or more of these plots. I suggest the box-and-whisker plot (Tukey 1977), often called simply a box plot, as a presentation alternative to the univariate histogram (figures 3.2 and 3.4A). An advantage of the box plot is that it provides more summary statistical information than a histogram—it includes medians, quartiles, ranges, and outliers (extreme variates)—in much less space and with much less ink. Box plot construction is not dependent on arbitrary bins, so these plots do not exaggerate or distort the data distribution. By notching the box plot (figure 3.12E), you can easily add confidence intervals so that plots of several distributions can be compared easily.

Wilkinson (1990; Haber and Wilkinson 1982) developed the fuzzygram (figure 3.4B), another alternative to the histogram. Fuzzygrams are histograms with probability distributions superimposed on each bar. Consequently, fuzzygrams present

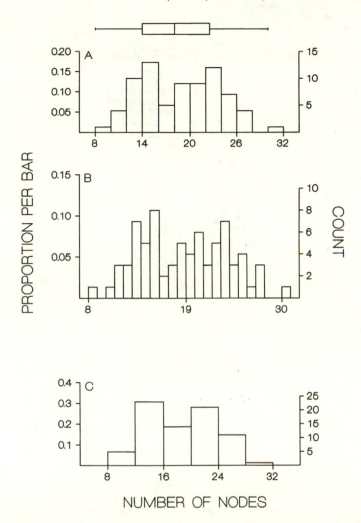

Figure 3.2 Histograms of the number of nodes per plant of 75 surviving *Ailanthus altissima* individuals grown in a 5 × 20 plant rectangular array. Each bar represents the frequency or count (right axis) of observations within the bounds indicated by the ticks on the *x*-axis, and the proportion of the total sample (left axis) represented by each bar. The three plots illustrate the variation in histogram presentation obtained by changing the bin width: (A) default (bin width = 4); (B) bin width = 2; (C) bin width = 8. At the top of the figure, a box plot (see figure 3.4 for construction details) illustrates summary statistics and is a better indication of the true data distribution.

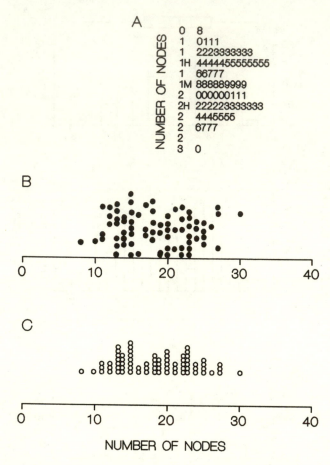

Figure 3.3 Alternative density plots that convey more information than a histogram. (A) A stem-and-leaf plot. In this plot, each line is a *stem*, and each datum on a stem is a *leaf*. The label for the stem is the first digit (*starting part*) of the number, followed by the value of the leaf. On the first line, the starting part is 0 and the only leaf is 8, indicating a value of 08 nodes. On the second line, the starting part is 1, and there are four leaves, indicating four data points: 10, 11, 11, and 11 nodes. The location of the sample median (M) and upper and lower quartiles (H) are also marked on this plot. (B) A jittered density plot. Each point is placed along the horizontal scale at the exact location of its value. To keep points of equal value from overlapping, they are located at random heights above the *x*-axis. (C) A dit plot. Each point indicates an individual observation, stacked along the *y*-axis at its location along the *x*-axis. In essence, a dit plot is a stem-and-leaf plot with symbols substituted for leaves.

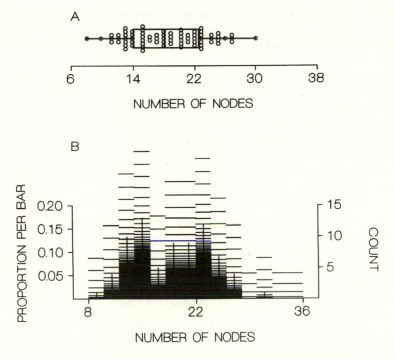

Figure 3.4 Information-rich production alternatives to histograms. (A) A box-and-whisker plot. The vertical line in the center of the box plot indicates the *sample median*. The left and right vertical sides of the box indicate, respectively, the location of the 25th and 75th percentile of the data (*lower* and *upper quartiles*, or *hinges*). The absolute value of the distance between the hinges (obtained by subtracting the value of the lower quartile from the value of the upper quartile) is the *hspread*. The whiskers of the box extend to the last point occurring between each hinge and its *inner fence*, a distance 1.5 *hspreads* from the hinge. Two kinds of outliers can be distinguished on a box plot. Points occurring between 1.5 hspreads and 3 hspreads (the outer fence) are indicated by an asterisk (see figure 3.12E). Points occurring beyond the outer fence are indicated by an open circle. The various summary statistics are clearly seen in relation to the raw data, which are overlain on this box plot as a symmetric dit plot. The distance encompassed by the whiskers includes ≈90% of the data (Norusis 1990). (B) A fuzzygram (Wilkinson 1990). This plot is a standard histogram (counts and proportions of each bin indicated by the height of the vertical line), with a probability distribution superimposed on each bar. The shading of the bars is based on a gray-scale distribution according to the probability that the ith observation will occur in that region: $P_i = P(p_i > \pi_i)$, where $p_i = n_i/n$ is the sample estimate of π_i (the expected proportion of a sample of n values from a continuous distribution to fall in the ith bin of the histogram). The more likely that $p_i > \pi_i$, the lighter the bar. Consequently, for large sample sizes, the bars will appear in sharp focus, whereas for small counts, the bars will be fuzzy. See Haber and Wilkinson (1982) for a discussion of the cognitive perception of fuzzygrams.

not only the data, but also some estimation of how realistically the data represent the actual population distribution. Such a presentation is particularly useful in concert with results derived from sensitivity analyses (Ellison and Bedford 1991) or resampling methods (Efron 1982; chapters 7 and 14). Haber and Wilkinson (1982) discuss, from a cognitive perspective, the merits of fuzzygrams and other density plots relative to traditional histograms. Histograms (figure 3.2), stem-and-leaf plots (figure 3.3A), dit plots (figure 3.3C), and fuzzygrams (figure 3.4B) can indicate possible bimodality in the data. Bimodal data, observed commonly in plant ecology, are obscured by box plots and jittered density diagrams.

Probability plots are common features of most statistical packages, and they provide a visual estimate of whether the data fit a given distribution. The most common probability plot is the normal probability plot (figure 3.5A). Here, the observed values are plotted against their expected values if the data are from a normal distribution; if the data are derived from an approximately normal distribution, the points will fall along a relatively straight diagonal line. There are also numerical statistical tests for normality (e.g., Sokal and Rohlf 1995; Zar 1996). If, for biological reasons, the investigator believes the data come from a population with a known distribution different from a normal one, it is similarly possible to construct probability plots for other distribution functions (figure 3.5B).

3.3.2 Bivariate Data: Examining Relationships Between Variables

Ecological experiments often explore relationships between two or more continuous variables. Two general questions related to bivariate data can be addressed with graphical EDA: (1) What is the general relationship between the two variables? and (2) Are there any outliers—points that disproportionately affect the apparent relationship between the two variables? The answers to these questions lead, in formal analyses, to investigations of the strength and significance of the relationship (chapters 6, 9, and 10). Scatterplots and generalized smoothing routines are illustrated here for exploring and presenting bivariate data. Extensions of these techniques to multivariate data are presented in section 3.3.3.

Bivariate data sets can be grouped into two types: (1) those for which we have a priori knowledge about which variable ought to be considered independent, leading us to consider formal regression models (chapters 8 and 10), and (2) those for which such a priori knowledge is lacking, leading us to examine correlation coefficients and subsequent a posteriori analyses. The functional response of *Notonecta glauca*, a predatory aquatic hemipteran, presented experimentally with varying numbers of the isopod *Asellus aquaticus* is used to illustrate the first type of data set; these data are described in detail in chapter 10. For the latter type of data, I use a data set consisting of the height (diameter at breast height, dbh) and distance to nearest neighbor of 41 trees in a 625-m^2 plot within an approximately 75-year-old mixed hardwood stand in South Hadley, Massachusetts (A. M. Ellison, unpubl. data, 1993). Data sets of this type are commonly used to construct forestry yield tables (e.g., Tritton and Hornbeck 1982) and have been used to infer

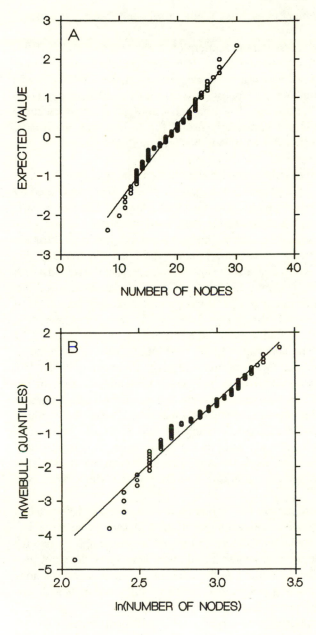

Figure 3.5 Probability plots of the *Ailanthus* data. (A) A normal probability plot. (B) A probability plot with the predicted values coming from a Weibull distribution: $f(y) = 1 - \exp[(-y/s)^t]$, where s is a spread parameter and t is a shape parameter. In this probability plot, the slope of the line is an estimate of $1/t$, and the intercept is an estimate of $\ln(s)$. See Gnanadesikan (1977) for a general discussion of probability plots.

competitive interactions among trees (e.g., Weller 1987) and forest successional dynamics (e.g., Horn et al. 1989).

For both exploration and presentation, scatterplots are the most straightforward way of displaying bivariate data (figure 3.6A). However, because scatterplots are merely a display, they do not necessarily reveal pattern. Figure 3.6A illustrates clearly this idea. Three functional response curves (Holling 1966; chapter 10) could be fit to these data, but it is not clear from the scatterplot itself which curve would best fit the data. EDA is particularly useful for dealing with these data, which show high variability and no obvious best relationship between the two variables.

Recent computer-intensive innovations in smoothing techniques (reviewed by Efron and Tibshirani 1991) have expanded the palette of smoothers developed by Tukey (1977). Basically, to construct a smoothed curve through the data, a best-fit line is constructed through a subset of the data, local to each point along the x-axis. This process is repeated for each point, and a smooth line is constructed by connecting the intersections of each local regression line. The result of this process, using LOWESS (robust LOcally WEighted regrESSion: Cleveland 1979; Efron and Tibshirani 1991), is shown for the predator–prey data in figure 3.6B. In this case, 50% of the data were used to construct each segment of the smoothed curve. That is, to construct the first segment, the response data from $0 \leq N_0 \leq 50$ were used; to construct the second segment, the response data from $1 \leq N_0 \leq 51$ were used, and so forth. The apparent type III functional response observed in the smoothed curve is supported by the formal analysis of these data (chapter 10). The lack of underlying assumptions about the distribution and variance of the data and the ability to elucidate patterns very noisy data are two advantages of smoothing over traditional regression techniques. One disadvantage of smoothing is that relative weighting of data used for each segment must be specified in advance, usually with little or no rational basis for the decision. Moreover, statistical comparison of different smoothed curves is virtually impossible. Most statistical software packages compute a variety of smoothers (see reviews by Ellison 1992; Kardia 1998).

Smoothers are used appropriately only when there is clear a priori knowledge of an independent variable and a corresponding dependent variable or variables. When this is not the case, other exploratory techniques are more appropriate for examining relationships between variables. In addition, smoothing does not provide information about potential outliers in the data set. To examine correlations between variables and to search a posteriori for outliers, influence plots and convex hulls are useful exploratory tools.

A scatterplot of the relationship between tree height and stem diameter (A. M. Ellison, unpubl. data, 1993) is illustrated in figure 3.7A. The raw data are shown, and there appears to be an apparent outlier (a 30-m-tall tree with a dbh > 70 cm). In an influence plot of these data (figure 3.7B), the size of each point becomes directly proportional to the magnitude of the change its removal would have on the Pearson correlation coefficient (r) between the two variables. By overlaying a bivariate 50% confidence ellipse, it becomes obvious that outlying points have greater influence on r than do points within the ellipse.

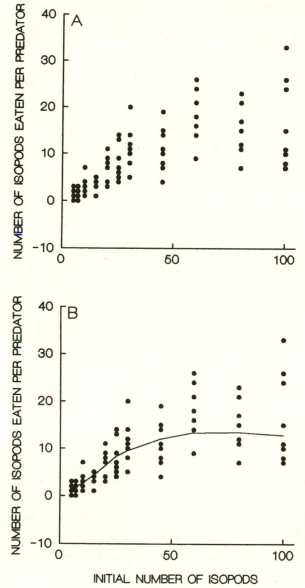

Figure 3.6 Scatterplots of the functional response of *Notonecta* to varying levels of *Asellus*. (A) A simple scatterplot showing the raw data. (B) A scatterplot with a lowess smooth fitted to the data. Note the apparent type III functional response revealed by the smoother (see chapter 10).

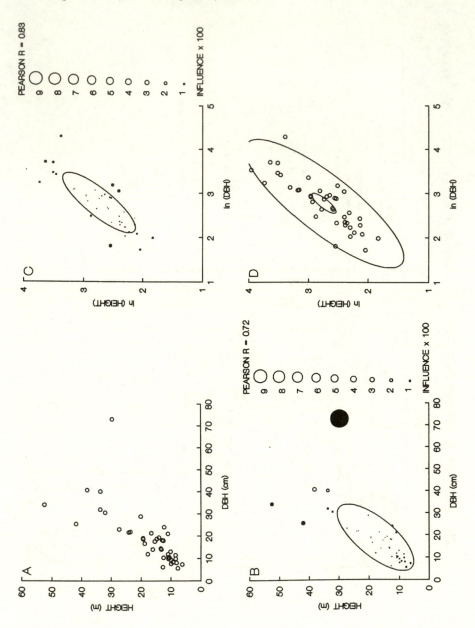

In an influence plot of the logarithmically transformed data (figure 3.7C), the apparent outliers have all but disappeared (the large outlier in figure 3.7B now has an influence on r of only .01), and the data are better distributed for formal analysis. Figure 3.7D supports this notion. The outer ellipse is a 95% confidence ellipse centered on the sample (dbh and height) means, with the ellipses' major and minor axes equal in length to the unbiased *sample* standard deviations of height and dbh, respectively. The orientation of the ellipse is determined by the sample covariance. All of the points, expect the apparent outlier, fall within this confidence ellipse. For comparison, the inner ellipse is a 95% confidence ellipse with axes computed from the standard errors of the means of each variable and centered on the sample centroid—a graphic illustration of the real difference between the standard deviation and the standard error (see section 3.4).

Convex hulls and subsequent peeled convex hulls (Barnett 1976) are useful exploratory tools when the distribution underlying the data is not normal or not known. Convex hulls illustrate order in bivariate or multivariate data, and they are used to distinguish distinct groups, outliers, and general shapes of multivariate distributions (for a detailed discussion, see Barnett 1976). Peeled convex hulls are essentially bivariate smoothers. Figure 3.8 illustrates a convex hull and a subsequent peel around the same data set illustrated in figure 3.7. The initial hull (figure 3.8A) describes the boundaries of the data—it encompasses the full range of variation in the data set. The peeled hull, referred to as "peeled to depth 2" (figure 3.8B), includes all but the most extreme values of the data set (compare the points outside the peeled hull of figure 3.8B to the points with strong influence on r in figure 3.7B). This process can be repeated ad infinitum, but normally does not proceed beyond depth 3. This is analogous to Tukey's (1977) running median (3R) smoother, extended in two dimensions. Like smoothers, convex hulls are constructed most easily with pencil and paper, or fast, interactive computer software (S-Plus). Convex hulls are useful for highlighting patterns within noisy data; they make no assumptions about the underlying distribution of the data.

Bivariate plots suitable for EDA are also suitable for final presentation. In preparing these plots for publication, however, several conventions often observed in the literature should be dropped in favor of clarity of presentation. First, it is common in scatterplots to always start each axis at the origin (0, 0). In fact, closely adhering to the actual range of the data when scaling axes is far more

---

Figure 3.7 Scatterplots of tree diameter versus tree height for 41 trees in a mixed hardwood stand. (A) Raw data. (B) An influence plot, where the size of each point is directly proportional to the magnitude of its influence on r. Shading of the points indicates the direction of the influence (open circles have a positive influence on r, solid circles a negative influence). In this case, the putative outlier is shown as a large solid point (influence $\times$ 100 = 11). Removal of this point alone, therefore, would increase the value of r from 0.72 to 0.83. A 50% bivariate confidence ellipse is overlain on the figure. (C) An influence plot of the data following log transformation. (D) Two different 95% confidence ellipses, the outer constructed based on the variables' standard deviations, and the inner constructed based on the standard errors of the means of the variables.

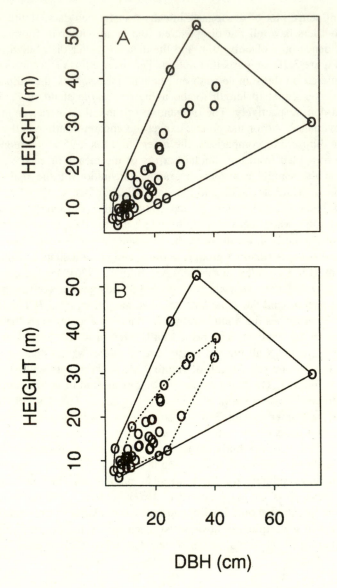

Figure 3.8 A convex hull (A and B, solid line) and a depth-2 peel (B, dotted line) around the tree size data. The hull is constructed by determining which points are farthest from the centroid of the data and by joining those points to form a polygon that envelopes the other points. To peel the hull, all the points that lie on the initial convex hull are deleted, and a new convex hull is constructed for the remaining points.

useful and informative than always including 0, especially if the extreme value of either variable is $\ll 0$ or $\gg 0$. Restricting the values on the axes to just beyond the extreme values of the data improves clarity and highlights pattern. Axis breaks do not always help, and changing the relative scaling after an axis break usually hinders accurate perception of the data and can stymie future digitizers.

3.3.3 Extensions of Bivariate Techniques to Multivariate Data Sets

For data sets that include a number of continuous variables, it may not be clear which, if any, pair(s) of variables should be subjected to bivariate correlation or regression analysis, or whether you should resort to multivariate techniques (chapter 7). Three-dimensional plots (e.g., figure 3.11A) are often used to examine and illustrate higher dimensional data. Although these graphs are aesthetically pleasing and easy to produce with current graphic software, accurate interpretation and digitizing depend on the perspective and orientation of the plot.

The scatterplot matrix, whose origins are shrouded in mystery, provides an alternative exploratory and presentation tool for higher dimensional data. A symmetrical scatterplot matrix of the tree data is shown in figure 3.9. This is simply a plot of all possible bivariate combinations of the variables in the data set. Plots above the diagonal have x- and y-axes transposed relative to those below the diagonal, which frees the investigator from preconceived notions of dependent and independent variables. We can, of course, apply the bivariate exploratory techniques described previously to each of the scatterplots within the matrix. The possible addition of density plots of each variable along the diagonal gives the investigator a simultaneous feel for the distribution of individual variables (Ellison and Bedford 1991). The final construction provides an information-rich, but rapidly comprehensible, picture of the overall data set. Advanced, interactive data exploration and visualization techniques have been extended to n-dimensional data by Cook et al. (1995) and Buja et al. (1996).

3.3.4 Classified Quantitative Data: Alternatives to Bars and Pies

Classified quantitative data are common in many experimental situations. This type of data set consists of responses of a given parameter to discrete treatments. Such experiments may be analyzed by ANOVA (chapters 4 and 5), and the results expressed in terms of the significance of treatment effects or interaction effects. Data from these types of experiments often are not explored before the formal analysis, although the univariate techniques described in section 3.3.1 are appropriate for examining the data structure of individual treatment groups. The exception to this generalization is common tests of the critical assumptions of ANOVA: homoscedasticity (variances among treatment groups are equal) and normal distribution of residuals within treatment groups. In particular, failure to test for homoscedasticity is one of the most common statistical errors (Fowler 1990). Hetero-

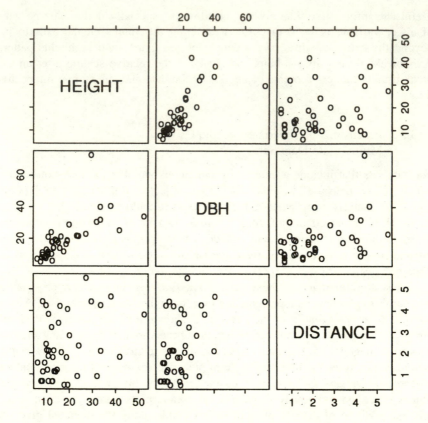

Figure 3.9 A scatterplot matrix of the tree size data. This plot illustrates bivariate relationships between all possible combinations of variables in a multivariate dataset. The variable name in the boxes along the diagonal corresponds to x-axis variables of plots below the diagonal and y-axis variables above the diagonal.

scedastic (unequal variances) data can complicate or compromise results obtained from ANOVA (Sokal and Rohlf 1995).

To illustrate EDA and graphical presentation of classified quantitative data, I used two data sets from Potvin (1993, tables 4.2 and 4.3; see also [http://www.oup-usa.org/sc/0195131878/]). In one data set, the effects of genotype (the classifying variable) on fresh mass of *Plantago major*, were examined. The second set comprised data on the interaction effects of bench position and genotype on stem dry mass of *Helianthus annuus* grown in a Latin square design. In each of these data sets, there is only one response variable: plant mass. More complex data sets include responses of several variables to multiple levels of a given treatment. As an example of this latter type of data set, I use data from Ellison et al. (1993). We measured a number of growth and morphological characteristics of *Nepsera aquatica* (an herbaceous species of disturbed areas in tropical wet forests) in response to varying light levels (2%, 20%, and 40% of full sunlight).

Spread (some measure of variance) versus level (mean, median) plots (Norusis 1990) are a rapid, graphic way to examine the within- and between-treatment group variances, as well as to provide clues as to appropriate data transformations to bring heteroscedastic data into line. Norusis (1990), modifying the technique of Box et al. (1978), suggests plotting the natural log of the interquartile distance (i.e., the hspread; fig. 3.4A) versus the natural log of the median for *each* treatment group. An appropriate transformation of the data to remove dependency of the spread on the level is then given as 1 minus the slope of the linear regression line fit to the spread versus level plot. Figure 3.10A illustrates a spread versus level plot for Potvin's *Plantago* data. Note that the raw data are not homoscedastic; the variance increases with the mean. Following Norusis (1990) and Box et al. (1978), the slope of the regression line for this plot is 1.71, suggesting that the data be transformed by raising each observation to the −0.71 power. After such a transformation, the spread versus level plot (figure 3.10B) illustrates that the strict dependency of spread on level no longer exists, and the data are somewhat more suitable for ANOVA (the variances are no longer correlated with the mean, although they are still not equalized). Plant size data are often subject to logarithmic transformations to equalize variances within treatment groups. A log transformation of these data is almost as good as the negative exponential transformation in equalizing these variances (table 3.1). Box and Cox (1964) and Zar (1996) provide detailed methods on determining the "best" transformation to be used on heteroscedastic data. Such transformations may not make biological sense, but keep in mind that the role of transformation is to bring your data in line with the assumptions and requirements of the statistical model(s) you are testing.

Graphic EDA can also be used to examine interaction effects in data. An example is illustrated in figure 3.11 for Potvin's *Helianthus* data. In this experiment, Potvin illustrates how position on a greenhouse bench interacts with genotype to determine plant mass. The top figure illustrates the relatively small size of genotype A and the relatively large size of genotype E. Although a scatterplot matrix might have made this pattern clearer, there is no real reason to plot row × column, or row × genotype, or column × genotype when the point is to illustrate the row × column *interaction* effect on genotype. The lower figure, a contour plot of the top figure, illustrates the clear "hot spot" in the upper left corner of the bench. Because interaction effects often involve visualizing data in more than two dimensions, you can use many of the techniques normally applied to multivariate data in the exploration of interactions.

Classified quantitative data are presented poorly in the ecological literature. These problems are illustrated with the data of Ellison et al. (1993) on resource allocation and morphological responses to light by *Nepsera* (figure 3.12). The most common ways of presenting classified quantitative data are bar charts, separated or stacked (figures 3.12A,B), and pie charts (figure 3.12C). Separated bar charts (figure 3.12A), where a single bar represents the results of a single treatment, suffer from the same problems as histograms. The bars themselves use a lot of ink—horizontal lines, vertical lines, shading of bars of arbitrary width—to convey information about only a single point at the top of the bar (compare figure

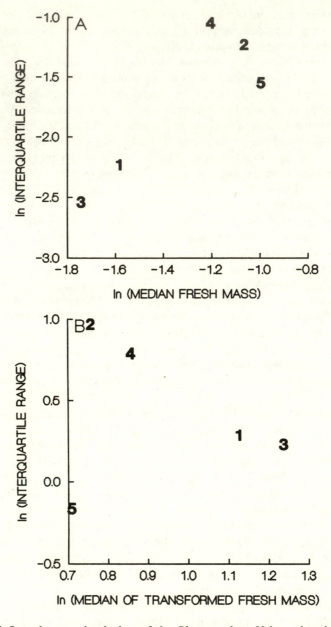

Figure 3.10 Spread versus level plots of the *Plantago* data. Values plotted on (A) are ln(interquartile distance) on the *y*-axis versus ln(median plant mass) on the *x*-axis of seven replicate individuals of each of five genotypes. Genotype number is indicated on the plot. (B) Spread versus level plot of data following a negative exponential transformation (Norusis 1990). See text and table 3.1 for further explanation.

Table 3.1 Variance (s^2) of $n = 7$ observations per genotype of *Plantago* fresh mass[a]

		Variance		
Genotype	Mean	Untransformed	Log transformation	Negative exponential transformation ($y^{-0.71}$)
1	0.198	0.006	0.179	1.245
2	0.309	0.034	0.440	1.798
3	0.109	0.008	0.151	0.710
4	0.298	0.029	0.354	1.302
5	0.412	0.039	0.196	0.392

[a]Variances are shown (1) before transformation, (2) after transformation by natural logarithms, and (3) after transformation by the negative exponential suggested by the spread versus level plot (figure 3.10).

3.12A with 3.12D). Stacked bar charts (figure 3.12B), where treatment groups are divided into subsets and groups are compared against one another, are virtually unintelligible and never should be used. In this example, the percent allocation to leaves, roots, and stems sums to roughly 100% (allowing for error and missing values). Figure 3.12A (bars side by side) at least clearly illustrates the relative allocation to each part. It is not so simple, on the other hand, to determine the relative allocation in figure 3.12B.

Because we use 0 as our reference point, the first guess would be that the allocation to roots in 2% light is approximately 70% and that to stems is 100%, when clearly this cannot be true. However, it is difficult to determine visually the beginning point of any of the stacked segments beyond the lowest one. Although measures of variance can be placed clearly on side-by-side bar charts, error bars cannot be placed on stacked bar charts (see section 3.4). Shading, hatching, and other chartjunk used in bar charts also can interfere with accurate perception of the data and decrease the data-to-ink ratio. Pies share all of the problems of stacked bar charts, and none of the advantages of side-by-side bar charts. I can think of no cases in which a pie chart should be used.

There are several alternatives to bar charts and pie charts. Plots in which the mean value of the response variable is plotted as a single point, along with some measure of error, clearly illustrate the same data as in a bar chart with greater clarity and less ink (figure 3.12D). Sets of box plots better illustrate the underlying data structure and convey more information with less ink and confusion (figure 3.12E). These box plots have been "notched" (McGill et al. 1978) to show 95% confidence intervals. Polar category plots (with or without error bars; the latter are shown in figure 3.12F) are the minimalist alternative to bar charts and are a visually comparable substitute for pie charts. These polar category plots illustrate the response of eight measured variables to the three light environments and clearly convey overall differences between treatment groups.

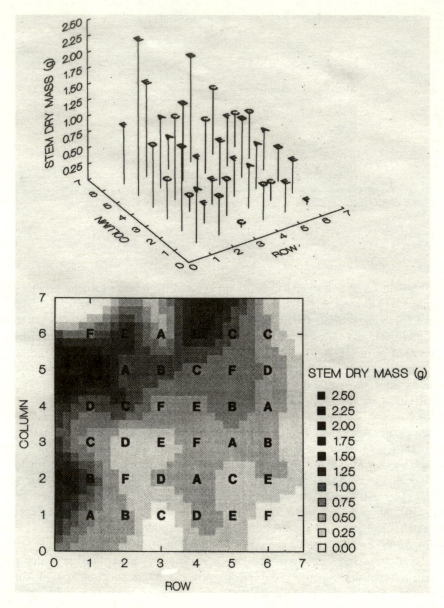

Figure 3.11 Two ways of visualizing the effect of bench position and genotype on stem dry weight of *Helianthus*. The top figure is a three-dimensional scatterplot, with genotype letter (A–F) as the plotting symbol. The addition of sticks connecting each point to its position on the x–y plane permits more accurate perception of the true height along the z-axis of each point. The lower figure is a contour plot, with intensity of shading indicating the biomass at a particular row × column location on the bench. These contours were determined by a negative exponential smoothing routine, where the influence of neighboring values decreases exponentially with distance. Shading density increases with biomass.

3.4 A Word About Error Bars

Any reported parameter must include a measure of the reliability of that parameter, as well as the sample size. For example, sample means, whether reported graphically or in tables, must be accompanied by the sample size and some estimator of the variance. Error bars on graphs must be correctly identified. Three kinds of error bars are seen commonly in the ecological literature: standard deviations, standard errors, and *n%* confidence intervals. Strictly speaking, the first is the *sample standard deviation*. The second, more properly referred to as the *standard error of the mean*, is an estimate of the accuracy of the estimate of the mean. We compute it as the standard deviation of a distribution of means of samples of identical sizes from the underlying population (see Zar 1996, section 6.3 for a complete description). Thus, calling error bars simply *standard deviation* bars confounds the two. Measures of error are used to calculate *n%* confidence intervals. We can easily compute confidence intervals of normally distributed data from the standard error of the mean (Sokal and Rohlf 1995). For other distributions, approximations of confidence intervals can be computed using bootstraps, jackknifes, or other resampling techniques (Efron 1982; chapter 13). All of these measures require information about sample size, which must be reported to ensure accurate interpretation of results.

In general, error bars are useful only when they convey information about confidence intervals. Typically, in the ecological literature, means are plotted along with error bars illustrating 1 standard error of the mean. For suitably large *n*, or for samples from a normal distribution, 1 standard error bar approximates a 68% confidence interval. This conveys little information of interest, since we are accustomed to thinking in terms of 50%, 90%, 95%, or 99% confidence intervals. Further, most ecological samples are small, or the underlying data distributions are unknown. In those cases, error bars representing 1 standard error of the mean convey no useful information at all. In keeping with the guidelines for graphical display presented at the beginning of the chapter, I suggest that sample standard deviations or 95% confidence intervals be the error bars of choice. Two-tiered error bars (Cleveland 1985) that display both quantities are an excellent compromise. Meta-analysis (chapter 18) requires sample standard deviations, and if reported together with sample size, they permit rapid calculation of confidence intervals, standard errors, or most other measures of variation. In the end, the choice of error bar lies with you. It is most important that they be identified accurately.

If you transformed the data before analysis, your calculated standard deviation will be symmetrical only with respect to the transformed mean. If you present the results back-transformed (as is common practice), the error bars may be asymmetric.

3.5 Conclusion

Ecologists traditionally have used a limited palette of graphic elements and techniques for exploring and presenting data. We must refocus our vision to grasp

Figure 3.12 Six alternatives for presenting classified quantitative data. Data are from an experiment examining the effect of three different light levels (2%, 20%, and 40% of full sun) on growth, resource allocation, and morphology of *Nepsera aquatica*. Each treatment consisted of 20 individually potted plants, harvested after 6 months of growth (Ellison et al. 1993). (A) A side-by-side bar chart illustrating percent allocation to leaves, roots, and stems by plants in each light treatment. Height of the bar indicates mean percent allocation, and error bars indicate 1 standard deviation of the mean. (B) A stacked bar chart illustrating the same data. (C) Pie charts illustrating the relative resource allocation in the three light environments (dark shading: 2% light; intermediate shading: 20% light; no shading: 40% light). Note that it is not possible to place error bars on stacked bar charts or pie charts. (D) Simple category plot of the data illustrated in figure 3.12A. Each point represents the mean percent allocation to leaves (circles), roots (squares), and stems (triangles); error bars are 1 standard deviation. (E) Notched box plots of the data. Box plot construction as in figure 3.4A. Plots are "notched" to illustrate 95% confidence intervals. Where the box reaches full width on either side of the median indicates the limits of the

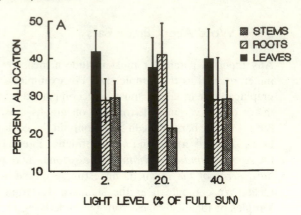

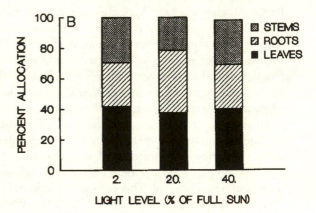

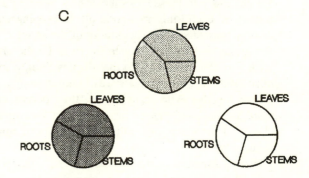

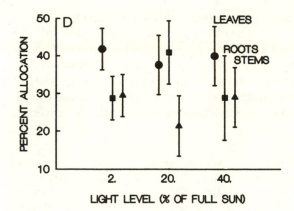

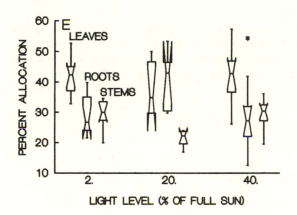

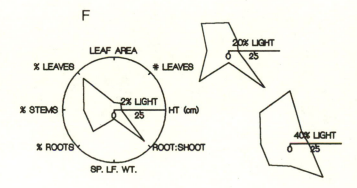

confidence interval. (F) Polar projections of category plots (also known as star plots) of the response of eight measured parameters to the three light treatments. The radius of the circle is equivalent to the *y*-axis of a rectangular plot; the distance from the center of the circle to each vertex of the polygon is the mean response of each variable to the treatment. Variables are arranged equidistantly around the perimeter of the circle (equivalent to the *x*-axis of a rectangular plot). One obtains a picture of the overall response of the plant to each light treatment by constructing a polygon whose vertices are equal to the value of the response variable. Different shapes in the different light treatments indicate overall treatment effects. For this type of plot to be effective, all data must be similarly scaled; for this plot, root-to-shoot ratio (g g^{-1}) was multiplied by 10^2, and specific leaf weight (g cm^{-2}) was multiplied by 10^4. Leaf area (cm^2), is a measure of total leaf area per plant.

new or unfamiliar graphic elements and techniques that will permit clear communication of our data. We can now use available computer hardware and software with expanded EDA and presentation capabilities to display our results accurately, concisely, and in aesthetically pleasing ways (Ellison 1992; Kardia 1998). We can improve our comprehension and appreciation of data by using many of the graphic techniques presented in this chapter, just as we can increase our appreciation of the diversity of pasta entrées with a trip to a fine Italian restaurant.

Acknowledgments   I am grateful to the late Deborah Rabinowitz for introducing me to EDA and data-rich graphic techniques. Philip Dixon, Steve Juliano, and Catherine Potvin generously shared data from their respective chapters. The data on tree size was collected by the 1992 population ecology class at Mount Holyoke College. The work on *Nepsera* was supported by NSF Grant BSR-8605106 to Julie Denslow. Technical support personnel at Systat, Inc (now SPSS, Inc.). and Statistical Sciences, Inc. (now MathSoft, Inc.) helped immensely with final graphics production. Philip Dixon, Elizabeth Farnsworth, Jessica Gurevitch, Catherine Potvin, Sam Scheiner, and one anonymous reviewer provided constructive reviews of early drafts of this chapter that resulted in a much-improved final version. Hardware for graphics production was provided by the BioCIS grant from IBM Corporation. Additional support was provided by NSF Grant BSR-9107195 and the Internet.

4

ANOVA

Experimental Layout and Analysis

CATHERINE POTVIN

4.1 Ecological Issues

When the understanding of ecological questions necessitates partitioning the effects of environmental factors, ecologists rely on experiments. Manipulative experiments, either in the field or in controlled environments, enable ecologists to vary single factors to isolate their effects. For example, growth cabinets make it possible to raise organisms at identical temperatures but different photoperiods, or at identical light intensities but different temperatures. In manipulative experiments, it is often desirable that environmental "background," that is, all the factors that are not altered voluntarily, be controlled precisely. This ensures that responses observed when varying a target factor are not confounded with uncontrolled sources of variation. Thus, controlled environments, mainly growth chambers and greenhouses, are a frequent tool in plant ecology, as are growth cabinets and aquariums in animal ecology.

In the first section of this chapter, I present analysis of variance (ANOVA) as a fundamental tool in experimental ecology. The core of the chapter will address the design of experiments. Although growth chambers are often perceived as uniform environments, environmental heterogeneity exists within a single growth unit, as well as between units (Lee and Rawlings 1982; Potvin et al. 1990a). Experimental designs that can adequately account for environmental heterogeneity will be examined. Although my emphasis will be on growth chambers, the same principles prevail for research in other types of controlled or field environments (chapters 5, 15, and 16). I also explore the cost of erroneous designs. This chapter should be viewed as a point of departure to illustrate the considerations that come into play when it is time to design an experiment. Experimentalists will often have more complex designs than those that are presented here. Once the underlying principles are understood, it becomes relatively simple to elaborate the appropriate experimental designs. Sources for more details include Cochran and Cox (1957) and Winer et al. (1991).

4.2 Statistical Issues: Environmental Variation and Statistical Analysis

As suggested by Underwood (1997), the first step in the design of an ecological experiment is formulating a linear model that allows the researcher to isolate the effect(s) of interest. Formulating a linear model depends on the factors studied and on a specific design, because experimental design dictates error terms. It is essential, at the onset of any experiment, to examine patterns of spatial and temporal variability. Experimental designs provide a way to account for these otherwise uncontrolled sources of variability. Thus, a good experimental design will reduce the size of the experimental error. Examining alternative designs will facilitate the choice of an appropriate experimental design and will clarify the degrees of freedom associated with each source of variation. Thus, the choice of an appropriate experimental design is essential to avoid problems of pseudoreplication and confounding (Hurlbert 1984). The proper testing of effects depends on an appropriate choice of error terms. This discussion assumes a basic knowledge of ANOVA, hence I will focus on less familiar aspects. Details of the statistical treatment of ANOVA can be found in Searle (1971).

Analysis of variance (ANOVA) uses sampled data to test hypotheses regarding a population. ANOVAs are based on the specification of linear models used to partition the variance attributable to factor(s) (often treatments). A factor may be represented by any number of levels, that is, categories into which a factor is divided (Searle 1971). The parameters of the linear model describing the data can be estimated, among other techniques, by least-square or maximum-likelihood methods. Least-square estimators, traditionally used in ANOVA, minimize the sum of squares of deviations of the observed data from their expected values (Searle 1971). In least-squares analysis, if the data set is balanced (i.e., with an equal number of observations per cell), the total sum of squared deviations can be readily partitioned in sums of squares (SS) that contribute to each factor in the design. As a reminder, a deviate is the difference between an observation and the mean. The result is minimum-variance, unbiased estimators, which are desirable properties for estimators (Winer et al. 1991). Mean squares (MS), a measure of the average variation for each degree of freedom, are then derived by dividing every SS by its degrees of freedom (SS/df). In that sense, a MS is equivalent to a statistical variance. An expected value is associated with each calculated MS. Table 4.1 shows that the expected value of a MS is a linear combination of the components of the variance. The statistic, F, on which ANOVA relies for hypothesis testing, is obtained from the ratios of two MSs, the treatment MS and the

Table 4.1 Expected mean squares and F-ratios for a two-way ANOVA[a]

Effect	Expected mean square	F-ratio
A		
A_i	$\sigma_e^2 + \dfrac{nb}{a-1}\sum\limits_{i=1}^{a}\alpha^2$	MS_A/MS_e
B_j	$\sigma_e^2 + \dfrac{na}{b-1}\sum\limits_{j=1}^{b}\beta^2$	MS_B/MS_e
AB_{ij}	$\sigma_e^2 + \dfrac{n}{(a-1)(b-1)}\sum\limits_{i=1}^{a}\sum\limits_{j=1}^{b}(\alpha\beta)^2$	MS_{AB}/MS_e
Residual (error)	σ_e^2	
B		
A_i	$\sigma_e^2 + n\sigma_{AB}^2 + nb\,\sigma_A^2$	MS_A/MS_{AB}
B_j	$\sigma_e^2 + n\sigma_{AB}^2 + na\sigma_B^2$	MS_B/MS_{AB}
AB_{ij}	$\sigma_e^2 + n\sigma_{AB}^2$	MS_{AB}/MS_e
Residual (error)	σ_e^2	

[a]The analysis is for (A) fixed effects and (B) random effects models.

error MS. Thus, the probability of detecting the effect of the factor of interest hinges on the use of an appropriate error term.

The importance of error terms can be illustrated by the following example. The theory underlying ANOVA recognizes an essential difference between two types of effects: random and fixed. We can view levels of a random factor as being randomly drawn from a larger defined set and levels of a fixed factor as being deliberately chosen by the experimenter. Biologically, the inferential aspect of whether an effect is fixed or random is crucial. If an effect is considered fixed, the conclusions cannot be generalized or extended beyond the levels under study because the levels of the factor have been deliberately chosen. To infer to other levels of a treatment factor, the effect of that factor has to be considered random. Study of increased atmospheric CO_2 concentration provides a clear example of a fixed-factor effect. Researchers commonly compare the effect of current CO_2 level (350 ml/l) with the doubling predicted by the middle of the twenty-first century (650 ml/l). In these experiments, no attempt is made to generalize to other CO_2 levels. However, if the experiment focuses on the response of various genotypes of *Arabidopsis* to elevated CO_2, it is most likely that genotypes are selected randomly to represent the general population of *Arabidopsis* genotypes. The genotype effect would then be random and the findings could be extended to all *Arabidopsis* genotypes.

When the data set is balanced, that is, it has an equal sample size in each cell of the analysis, computation of the SSs and of the MSs is identical whether a factor is fixed or random (Herr 1986). However, the expected MSs differ, which is important since the F-ratios are dictated by expected MSs. The simplest case, the two-way model, is illustrated in table 4.1. Computer codes for the analysis in SAS (SAS Institute Inc., 1989a,b) are given in appendix 4.1. In fixed models, the expected MS for each factor is the sum of the error variance and the constant effect of that factor. Therefore, the appropriate MS to be used as the denominator is always the error MS. In the random model, the expected MS for each main effect is the sum of the error variance, the interaction variance, and the variance due to the effect tested. Thus, F-tests of the main factors use the interaction MS as denominator, but the interaction is still tested over the error MS. In three-way (and higher order) random or mixed factorial models, the appropriate denominator is often a combination of MS terms (Winer et al. 1991). I emphasize the central role of the expected MS in determining the appropriate tests of significance because practitioners frequently use statistical packages that test all factors over the error MS as a default. This discussion indicates that the resulting analysis is valid only for a fixed model, regardless whether this is appropriate. The remaining sections of the chapter present different experimental designs and point out appropriate error terms, emphasizing the distortion that can result from choosing the wrong error term.

4.3 The Statistical Solution: Designing an Experiment

Data analysis depends on the design of the experiment itself, how the levels of the factors of interest are assigned to the experimental units. In general, the smaller the experimental error, the more efficient the design. Designing an experiment also involves choosing the sample size and the physical and temporal layout of the experiment. A variety of standard experimental designs are available, each associated with a mathematical model and analysis. Here I examine two such designs and how each of them can account for specific patterns of variability. These designs are typical of ecological experiments. Alternative designs that address various specific problems can be found in Cochran and Cox (1957), Winer et al. (1991), and Underwood (1997).

In manipulative experiments, different experimental units receive different levels of the treatment factors. It is then assumed that differences between experimental units represent differences between these levels (Hurlbert 1984). Randomization of the levels of treatment factors over the experimental units and treatment replication are the central tenets of a good experimental design. Sir Ronald Fisher (1935) was a strong advocate of randomization. He argued convincingly that randomization was a warranty against confounding sources of variation.

Suppose that we are interested in comparing the photosynthetic performance of three different species and that sampling takes place over three 2-hour periods between 10:00 A.M. and 4:00 P.M. A good design would randomly assign each species to each time period on every sampling day. An erroneous design would systematically measure species A in the morning, species B at noon, and species C in the afternoon. In the latter design, species photosynthetic rates would be confounded with the time of day at which they were sampled. It would be impossible to know from statistical inference whether differences were due to differential species performance or to differential time of day.

The second most important tenet of good experimental design is replication. According to Fisher (1971), replication has two main purposes: "whereas replication of the experimental varieties or treatments on different plots is of value as one of the means of increasing the accuracy of the experimental comparison, its main purpose, which there is no alternative method of achieving, is to supply an estimate of error by which the significance of these comparisons is to be judged." Hurlbert (1984) introduced the term pseudoreplication as "the use of inferential statistics to test for treatment effects (factors) with data from experiments where either treatments are not replicated or replicates are not statistically independent." The core of Hurlbert's article addresses pseudoreplication in experimental layout. Often, however, experiments are well laid out but problems arise during data analysis because the investigator failed to identify the actual experimental units or replicates and, hence, the appropriate error terms.

4.3.1 Blocking

Blocking, the grouping of like experimental units, can accommodate environmental heterogeneity and improve statistical power. In keeping with the idea of randomization (Fisher 1971), each level of the treatment factor is randomly applied to a different experimental unit within each block. In a randomized block design, experimental units are grouped into blocks within which the environment is relatively constant. Differences between experimental units within a block provide a measure of the treatment effect, whereas repetition of the blocks provides replication of the treatments. This design enables us to partition the random deviation into that due to the treatment factor of interest, the experimental error, and the undesirable environmental (block) effect. The resulting experimental error term will be smaller, and the design will be more powerful than in a completely randomized design.

In the classic randomized block design, each level of the treatment factor will be applied at random to one replicate per block. The number of experimental units within each block will therefore be equal to the levels of the factor under study. This design can thus be viewed as a special case of ANOVA with no replication per cell. Consequently, the model does not include an interaction term. A randomized block design is described by the following linear model:

$$X_{ijk} = \mu + \tau_i + \beta_j + \varepsilon_{ijk} \tag{4.1}$$

where X_{ijk} is the response of the jth experimental unit under the ith level of treatment factor τ, μ is the population mean for the response, τ_i is the effect of the ith level of treatment τ, β_j is the effect of the jth block, and ε_{ijk} is the random deviation or error. The SAS commands for this design are given in appendix 4.2. Note that the MS error corresponds to the interaction between block and treatment (Sokal and Rohlf 1995, pp. 328, 347). The expected error mean square of the randomized block design ($\sigma_e^2 + \sigma_{AB}^2$) corresponds to the expected interaction mean square of a two-way ANOVA.

The analogous completely randomized design is described by

$$X_{ijk} = \mu + \tau_i + \varepsilon_{ijk} \tag{4.2}$$

Comparing equations 4.1 and 4.2 further shows that if blocking is done appropriately and if the blocks correspond to the various environmental conditions, the β_j

term will remove from the error term variability due to environmental heterogeneity. As a consequence, the error term will be reduced, and the randomized block design will be more likely to detect a significant treatment effect than the completely randomized design.

The classic model, equation 4.1, implicitly assumes the absence of an interaction between the treatment and block effects. Both Underwood (1997) and Newman et al. (1997) are critical of this assumption for field experiments. They argue that the presence of an interaction invalidates the test of the treatment effect. This argument is not a new one. Kempthorne (1975) argued that in the presence of an interaction, an overall statement about a main factor would have little meaning. However, as clearly illustrated by Sokal and Rohlf (1995, p. 336) in many cases, it may be important to test for the overall significance of a main effect despite the presence of an interaction. It is worth going back to Sheffé (1959) for clarification. According to this author, a case of no interaction is a case of additivity with a simple interpretation: factor A will affect all observations equally regardless of the effects of factor B. If an interaction is present, however, the effect of A will vary according to the various levels of factor B. In controlled environments where blocks are constituted by trays, plastic containers, and so on, the assumption of independence between blocks and treatments is likely to be true. In other situations, such as field manipulative experiments, the existence of a treatment by block interaction is conceivable. However, as demonstrated by Sheffé (1959, p. 95), the presence of such interaction would not invalidate the randomized block design, but rather add a caveat to its interpretation. Suppose that two species are compared in several different locations (blocks) and found to be statistically different despite the presence of a species by block interaction. Our conclusion is that, averaged over all blocks, species A did better than species B, despite the fact that in some blocks the reverse could be observed.

A potential drawback to the use of a randomized block design with one observation per cell is the difficulty of dealing with missing data. Mortality or loss of any experimental unit will lead to an imbalance in the design through a missing cell. The main difficulty in an analysis with missing cells is that no information is available for the combinations that are not observed (Shaw and Mitchell-Olds 1993). To get around that problem, we can estimate the missing value using marginal means (Mead and Curnow 1983), the mean value of a row or column in a two-way data table.

In the presence of replication, it becomes possible to estimate the interaction between block and treatment. In this case, the appropriate statistical model is

$$X_{ijk} = \mu + \tau_i + \beta_j + \tau\beta_{ij} + \varepsilon_{ijk} \tag{4.3}$$

where $\tau\beta_{ij}$ is the interactive effect of the ith level of treatment τ and the jth block. Other terms are as defined in equation 4.1. The SAS commands for this design are given in appendix 4.3. Dutilleul (1993) showed that in a randomized block design, if blocks are considered random, the treatment and block interaction is the proper error term for the treatment effect. This latter type of randomized blocked design has a distinct advantage over the classical design. If there is moderate mortality or other loss of experimental units, the imbalance does not affect the higher levels of the analysis. A simple way to analyze such data is to compute the mean response for each cell. This rebalances the data and does not inflate the degrees of freedom. The means are then used as observations and the analysis proceeds according to equation 4.1.

It is important to realize that, although blocking can be a powerful tool to reduce the size of the error, a cost is also associated with blocking. In the randomized block design, for example, partitioning of the variance into three terms (treatment, block, and error) rather than two (treatment and error) reduces the error degrees of freedom. If blocking is inappropriate and does not account for environmental heterogeneity, a design with blocks may be less efficient than a completely randomized design. The choice of the most appropriate design and the efficiency of the blocks in accounting for environmental heterogeneity depend on a good understanding of the heterogeneity of the environment. Blocking will be efficient if the variability between the experimental units in a block is smaller than the variability among blocks. Appropriate blocking will group the experimental units growing in the poorest environment into a block, those growing in the best into a second block, and so on. Each level of the treatment factor is then allocated to both good and poor growth conditions, resulting in a very accurate estimate of the true difference between the levels of the treatment factor (Mead and Curnow 1983). Another pitfall in using blocks is that the scale of environmental variation might not match the experimental constraints, such as cart or tray size. Figure 4.1 illustrates optimal and nonoptimal block placement in a randomized block design. An obvious mistake in selecting block placement is to position blocks across different environmental conditions. This discussion on the use of blocking refers to greenhouse and growth chamber experiments, yet the designs illustrated in this chapter are appropriate and useful for field experiments (see Dutilleul 1993). In the field, exercise care when blocking. In a blocked design, the treatment should be applied within natural habitat patches so that they experience the same environmental conditions, yet treatment levels should be sufficiently distant from one another to be independent.

4.3.2 Variation Between Growth Compartments: The Split-Plot Design

Often, the nature of different treatment factors necessitates that they be applied at different scales. For example, temperature or atmospheric CO_2 concentration has to be applied to a whole growth chamber, whereas species or soil nutrients can be manipulated at the pot level within a single chamber. In controlled environments, the former factors are referred to as between-chamber factors since comparison of more than one level requires the use of more than one growth compartment. In other words, the experimental unit becomes the growth chamber rather than the individual pot. If a treatment factor is applied to a whole growth chamber, replication between chambers is essential: the appropriate error term to test for between-chamber factors is the difference between chambers subjected to the same treatment level (Winer et al. 1991; Underwood 1997). Therefore, if a between-chamber factor is represented by two levels, the minimum number of

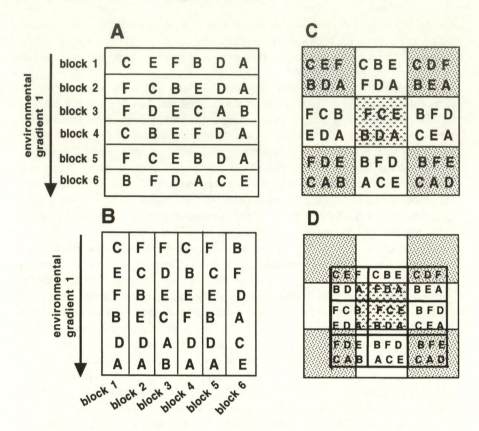

Figure 4.1 Examples of a layout for a randomized block design to compare six levels of a treatment factor (e.g., genotypes of sunflowers): (A) appropriate and (B) erroneous layout in the presence of one environmental gradient; (C) appropriate and (D) erroneous layout when the environment is patchy.

chambers is four (Potvin and Tardif 1988). Furthermore, even if 100 plants were grown in each of the four growth chambers, the analysis of the between-chamber factor would involve 3 degrees of freedom, not 399. This is because the variation between experimental units within a chamber provides no information on the between-chamber treatment factor. Physical and financial constraints usually limit the number of growth chambers available, hence the degrees of freedom for testing between-chamber factors are often small. If replication is either impossible, as in whole lake (e.g., Schindler 1974) or watershed (e.g., Likens et al. 1977) studies, or exceedingly costly, as in free air carbon dioxide experiments, "experiments involving unreplicated treatment may be also the only or the best option" (Hurlbert 1984). Subsequently, I examine the cost, in terms of Type I errors, of misusing statistical tools to provide a false sense of rigor to the data.

Often, ecological experiments involve both between-chamber and within-chamber factors, for example, temperature responses of different genotypes where temperature is a between-chamber factor and genotype is a within-chamber factor. Such experiments can be appropriately analyzed with a split-plot design. In a split-plot experiment, a single level of one treatment factor is applied to a large plot while all the levels of a second treatment factor are allocated to subplots within that main plot. Two different randomization procedures are used to allocate the treatments. The levels of the main plot treatment factor are first allocated at random, each to a different growth chamber. Growth chambers are then divided into subplots, and, in a second randomization procedure, each level of the within-chamber factor is assigned to a given subplot.

As an example, I use a theoretical split-plot experiment to compare two atmospheric CO_2 concentrations and six nutrient concentrations on plant biomass. The layout of this split-plot design is illustrated in figure 4.2 and is representative of many growth-chamber studies. The unit of comparison for CO_2 concentration is the entire growth chamber, whereas nutrient levels are compared within each chamber. The general linear model to describe the data obtained by such a split-plot design is

$$X_{k(ij)} = \mu + \alpha_i + \varepsilon_{k(i)} + \nu_j + \alpha\nu_{ij} + \varepsilon'_{k(ij)} \tag{4.4}$$

where α_i is the effect of the ith level of CO_2 concentration (the between-chamber factor), $\varepsilon_{k(i)}$ is the main plot error term and designates the effect of chamber k within level α_i, ν_j is the effect of the jth nutrient concentration (the within-chamber factor), $\alpha\nu_{ij}$ is the effect of the interaction between the ith CO_2 concentration and the jth nutrient level, and $\varepsilon'_{k(ij)}$ is the error term associated with the subplot. Other parameters are defined as in equation 4.1. Artificial data were generated and analyzed according to this model (table 4.2); the SAS commands are given in appendix 4.4.

The variance in a split-plot ANOVA is partitioned into two parts, and the analysis distinguishes two error terms: the whole plot and the subplot errors. The error term used to test the effect of CO_2 is calculated in terms of the variation between growth chambers. On the other hand, the effect of nutrient levels is independent of the variation among growth chambers. Consequently, the error term to test the nutrient effect is determined by the variation between the subplots within a growth chamber (Winer et al. 1991). The interaction term, $\alpha\nu_{ij}$, is independent of the variation due to the whole plot effect. The classic split-plot design illustrated in figure 4.2 assumes that only one observation per cell is present. As in the case of the randomized block, split-plot ANOVAs can be extended to allow replication of the experimental units in the subplots. The main advantage of the split-plot design is the relatively high statistical power for subplot factors. If main-plot replicates are few, it might be difficult to detect statistically significant main-plot differences. However, the degrees of freedom available to test for subplot interaction effects could be sufficient for a powerful statistical test.

Results from a split-plot analysis of soybean growth (Lee and Rawlings 1982) enable us to quantify the most frequent and, I would argue, damaging error that occurs in the statistical analysis of growth chamber experiments. When the split-plot nature of an experiment that involves both within- and between-chamber factors is ignored, the analysis erroneously proceeds using the subplot residual as

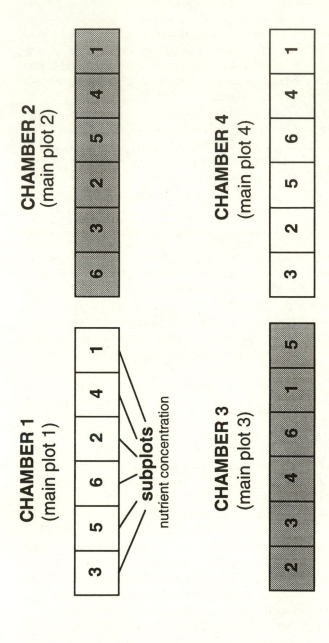

CHAMBER 1
(main plot 1)

| 3 | 5 | 6 | 2 | 4 | 1 |

subplots
nutrient concentration

CHAMBER 2
(main plot 2)

| 6 | 3 | 2 | 5 | 4 | 1 |

CHAMBER 3
(main plot 3)

| 2 | 3 | 4 | 6 | 1 | 5 |

CHAMBER 4
(main plot 4)

| 3 | 2 | 5 | 6 | 4 | 1 |

Figure 4.2 Example of layout for a split-plot design involving six nutrient levels as a subplot factor and two CO_2 concentrations as a main plot factor. A replication (block) consists of two growth chambers, each chamber being considered as a main plot. Nutrient levels are represented by numbers 1–6 and CO_2 concentration by presence or absence of shading. The chambers are then subdivided into six subplots to which the within-chamber factor is randomly allocated.

Table 4.2 A. Artificial biomass data. B. Split-plot ANOVA with CO_2 as the main plot factor and nutrient as the subplot factor

A

CO_2	Nutrients					
	1	2	3	4	5	6
350	13.3	19.9	24.5	33.6	38.9	41.4
350	12.4	20.5	22.1	29.4	36.9	42.3
675	15.8	22.4	26.9	38.0	44.8	49.2
675	13.9	21.0	27.5	35.9	46.8	49.0

B

Source of variation	SS	df	MS	F	P
Main plot					
$\quad CO_2$	130.67	1	130.67	42.959	.0225
Main plot error	6.08	2	1.93	.193	.196
Subplot					
$\quad$ Nutrient	3050.44	5	610.09	386.213	.001
$\quad$ Nutrient $\times CO_2$	35.47	5	7.09	4.491	.021
Subplot error	15.80	10	1.58		

the error term for all of the main effects. A comparison of the magnitude of the whole-plot and the subplot error MSs indicates that the use of the subplot error underestimates the real error. In the soybean experiment, the discrepancy between the errors was, at times, as large as 400-fold (table 4.3). The opposite trend can be found when the comparison involves the error degrees of freedom. In the example, the error degrees of freedom associated with the whole plot is 4 compared to 3960 for the subplot (table 4.3). Therefore, when the subplot error term is wrongly used to test the chamber and trial main effect, the analysis will proceed with artificially inflated degrees of freedom and a reduced error MS. Similar observations arise from examination of table 4.2. Needless to say, such erroneous use of the subplot in lieu of the whole-plot error term can have a drastic effect on the conclusions of the ANOVA, often leading to Type I errors.

Underwood (1997) recently provided a discussion of split-plot designs; he emphasized the difficulties that arise when the treatment factors are not independent of the plot effects and recommended that split-plot design should be abandoned. As we saw for randomized block designs (section 4.3.1), the presence of an interaction between plots and treatments would indeed complicate the interpretation and would only allow examination of the treatment effect averaged over the plots. I suggest, however, that such interaction is a lesser problem than the widespread misuse of the error term in designs where treatments are applied at different spatial scales. As demonstrated very clearly by Lee and Rawlings (1982), growth chamber experiments often dictate the use of split-plot designs.

Table 4.3 (A) Mean squares in combined split-plot analyses of variance on soybean, with a comparison of two main effects, growth chambers and trials. (B) Proportional difference in size of the two error terms[a]

Source of variation	df	Plant height	Leaf area	Petiole length	Fresh weight	Dry weight
A						
Main plot						
Chamber	3	50.05	197.72	39.30	320.63	48.10
Trial	3	48.54	310.83	36.01	464.54	48.10
Main plot error	4	125.72	127.67	50.07	211.63	31.43
Subplot						
Truck	23	9.72	16.90	5.04	30.22	9.14
Chamber × truck	69	7.93	14.59	4.75	24.10	4.08
Trial × truck	69	0.48	1.09	0.27	2.58	0.47
Chamber × trial × truck	292	0.55	1.13	0.26	2.38	0.43
Subplot error	3960	0.29	0.71	0.15	1.62	0.24
B						
Proportional difference		433.5	179.8	333.8	130.6	131.0

[a]From Lee and Rawlings (1982).

4.4 Conclusion

This chapter relies on classical experimental designs to account for environmental heterogeneity in ecological experiments. As an essential initial step for the pursuit of manipulative experiments, I recommend that both field ecologists and growth chamber users carry out a field survey or uniformity trial to document environmental variability. The techniques presented here are basic and can be incorporated into various experimental designs or used in conjunction with a wide variety of analytic techniques. For example, blocked designs are referred to in several other chapters (e.g., chapters 5, 6, 8, 15, and 16). This is because blocking, that is, the grouping of experimental units subjected to similar environmental conditions, is a powerful way to control for unwanted sources of variation. The potential cost of blocking, a reduction in the degrees of freedom of the error term, should be kept in mind. In situations where the grouping of the experimental units does not correspond to the underlying pattern of environmental heterogeneity, the completely randomized design may be more powerful because of its greater degrees of freedom.

I have considered environmental heterogeneity as a function of space alone. Variability among years is well known to ecologists. An early article on variability in controlled environments suggested that growth conditions may vary as much through time as in space (Potvin and Tardif 1988). In both field and controlled environments, it is possible to consider repetition of an experiment in time as true replication, which might provide some relief in the design of experiments.

The comparison of two levels of a treatment factor may properly be assessed by using two growth chambers or two fields in two successive trials. Whether in space or in time, replication is essential to provide an error term.

One persistent error in growth chamber experiments is the tendency to rotate plants between chambers to compensate for the absence of growth chamber replication. Such an approach loses sight of the fact that testing for a between-chamber factor can come only from the comparison of different chambers and that between-plants within-chamber variation is not an appropriate error term for between-chamber factors. If replication is impossible, it is better to avoid making statistical inferences with the data than to compute statistical analysis using the wrong error term.

Acknowledgments I dedicate this chapter to Drs. J. Antonovics and H. Wilbur. During my Ph.D. studies, their constant request for good statistical analyses sparked my interest in experimental design. The ideas developed in this chapter largely came from discussions with Dr. S. Tardif. Dr. N. Gotelli provided useful comments in reading the second version of this chapter. Dr. S. Scheiner provided the computer code for the examples.

Appendix 4.1 SAS program code for the two-way ANOVA of table 4.1.

```
PROC GLM;                        /use procedure GLM for unbalanced ANOVAs/
   CLASS A B;                              /indicates categorical variables/
   MODEL X = A B A*B/SS3;          /model statement corresponding to Table 4.1,
                                use type III sums of squares. Warning: results are
                                   only approximate if the design is unbalanced./
```

*** If factors A and B are random effects use the following
       command to obtain the correct *F* tests ***

```
RANDOM B A*B/TEST;              /Declare sites to be a random effect and ask SAS to
                                     compute the Satterthwaite correction for an
                                                       unbalanced design./
```

Appendix 4.2 SAS program code for a randomized block design (eq. 4.1).

```
PROC GLM;
   CLASS T B;                           /T=treatment levels, B=blocks/
   MODEL X = T B/SS3;                   /all terms are tested over the error/
```

Appendix 4.3 SAS program code for a randomized block design with replication (eq. 4.3).

```
PROC GLM;
   CLASS T B;
```

```
    MODEL X = T B T*B/SS3;                /now an interaction term is present/
    RANDOM B T*B/TEST;                    /the block and block-treatment
                                          interaction terms are random effects/
```

Appendix 4.4 SAS program code for a split-plot design (eq. 4.4).

```
PROC GLM;
    CLASS A E N;                          /A = CO₂ level, E = chamber (main plot error)
                                                          N = nutrient level/

    MODEL X = A E(A) N A*N/SS3;
    RANDOM E(A)/TEST;                     /CO₂ level will be tested over the chamber effect.
                                          Nutrient and the CO₂-nutrient interaction will be
                                          tested over the error term./
```

5

ANOVA and ANCOVA

Field Competition Experiments

DEBORAH E. GOLDBERG

SAMUEL M. SCHEINER

5.1 Introduction

Competition has occupied a central place in ecological theory since Darwin's time, and experiments on competition have been conducted on many different organisms from many different environments (see reviews in Jackson 1981; Connell 1983; Schoener 1983; Hairston 1989; Gurevitch et al. 1992). There are many different kinds of competition experiments. The main focus of this chapter is how to choose the appropriate designs and statistical analyses for addressing particular kinds of questions about competition. Such choices depend on many aspects of the questions and systems under study. Because the basic statistical methods appropriate for most of the designs we present, analysis of variance (ANOVA) and analysis of covariance (ANCOVA), are thoroughly covered in standard textbooks of experimental design and analysis, we provide somewhat less emphasis on the nuts and bolts of the statistical analyses than do other chapters in this volume. For an introduction to the basics of ANOVA, see chapter 4. Although we focus on competition, many of the points raised are equally applicable to experiments on other types of species interactions, such as predator–prey relationships or mutualisms.

5.2 Ecological Questions about Competition

The simplest question we can ask about competition is whether it occurs in the field. To answer this question, it is necessary to have experimental treatments in which the absolute abundance of potential competitors is manipulated and to test whether organisms in the treatments with a lower abundance of potential competitors perform better. The difference in performance between such abundance treatments is the magnitude of competition (or magnitude of facilitation if performance is better with higher abundance). Finding out whether competition occurs is an important preliminary step in any field investigation of competition, but, by itself, is rather uninteresting. Most of the important questions about competition involve comparisons of the magnitude of competition and therefore involve more complex experimental designs and analyses than a simple comparison between two or more abundance treatments (Goldberg and Barton 1992).

One group of questions requires comparing the magnitude of competition among environments (sites or times). For example, field observations might suggest the hypothesis that the distribution of a species is determined by the sum of competition from all other species at the same trophic level. A field experiment to test this hypothesis must compare the magnitude of competitive effects on that focal species in sites where it is abundant with sites where it is absent or rare (e.g., Hairston 1980; Gurevitch 1986; McGraw and Chapin 1989). Similarly, to resolve the current controversy among plant ecologists over whether the importance of competition increases with increasing productivity or stays constant, it is necessary to compare the magnitude of competition on populations or communities among sites that differ in productivity (e.g., Wilson and Tilman 1995; Twolan-Strutt and Keddy 1996). Statistical analysis of an experiment to compare magnitudes of competition among sites is described in section 5.4.1.

A second group of questions requires comparing the magnitude of competition among taxa, such as comparing competitive ability. For example, classical competition theory predicts that, for coexisting species, intraspecific competition is greater than interspecific competition. Mechanistic models of competition make predictions about particular traits that are related to competitive ability (e.g., Grime 1977; Schoener 1986; Tilman 1988; Werner and Anholt 1993) and these require comparisons between taxa that differ in these traits. Similarly, quantifying the magnitude of selection on different traits as a result of the competitive environment requires comparing the magnitude of competition among different phenotypes or genotypes. Statistical analysis of experiments to test hypotheses about rankings of competitive abilities and relationship of competitive abilities to traits of organisms is described in section 5.4.2.

In addition, some questions require comparisons of both environments and taxa. For example, many of the mechanistic models that generate predictions about traits related to competitive ability also make predictions about the ways in which those traits change between environments or about tradeoffs between competitive ability and response to other processes such as predation or disturbance (see previous references). These predictions can be tested by comparing the magnitude of competition between different taxa in different environments (e.g., different resource availabilities, natural enemy densities, disturbance rates). These are necessarily highly complex designs with many independent factors, and the main points of biological interest will often be statistical interactions between factors (e.g., taxon × competition × environment interactions to test whether competitive hierarchies change between environments). However, highly multifactorial designs will often also yield uninterpretable interaction terms, and they require enormous sample sizes. The addition of such complexity should be carefully considered in light of the ecological question of interest.

The various questions and associated experimental designs described previously are discussed in terms of testing predictions derived from models of competitive interactions, because this is the typical goal of most field competition experiments. However, an equally or perhaps even more important use is to parameterize particular models of competitive interactions, allowing exploration of the long-term dynamic consequences of interactions, as well as improving the formulation of models (Freckleton and Watkinson 1997). A full discussion relating empirical measures of competition intensity to theoretical models of competition is beyond the scope of this chapter; Laska and Wootton (1998) discuss many of the important issues (see also Freckleton and Watkinson 1997, 1999).

It is important to note that the basic competition experiment described previously cannot by itself address the mechanisms underlying competitive interactions. Negative interactions can occur through direct interactions (interference competition), through a variety of shared, limiting resources (exploitation competition), through shared natural enemies (apparent competition), and by means of other complex routes. Understanding these mechanisms is essential for developing generalizations about the role of competition in explaining evolutionary and ecological patterns (Schoener 1986; Tilman 1987). However, a broad diversity of approaches (field and laboratory, observational and experimental, manipulation of processes other than competition) is needed to examine specific mechanisms of interaction, making it difficult to provide a general discussion of experimental designs and statistical analyses. Therefore, the focus of this chapter will be on only one part of what is needed for a full understanding of competition: measurement and comparison of the magnitude of competition in the field.

To summarize, we emphasize that almost all of the important ecological questions about competition entail comparisons beyond simply demonstrating that competition occurs between a particular pair of species at a particular time and place. Therefore, the first, and perhaps most important, recommendation of this chapter is that the first step in conducting a competition experiment be careful consideration of the goal of the experiment to identify appropriate comparisons. This may sound trivial and obvious, but the literature is replete with perfectly designed and analyzed experiments that do little to explain an observed pattern, to test assumptions or predictions of theory, to parameterize a model, or to provide a sound basis for management decisions in a particular system.

5.3 Experimental Design

5.3.1 Terminology

Before diving into the details of different types of experimental designs and the questions for which each is appropriate, we present some basic terminology. A *focal taxon* is one whose response to competition we are measuring. An *associate taxon* is one whose effect on the focal taxon we are measuring; that is, it is the taxon whose abundance is being manipulated. (Note that, in some experimental designs, the same species can be both a focal and an associate species.) A *background taxon* is one that is present in all experimental treatments but is not an explicitly designated focal or associate group. Background taxa can include other potential competitors, resources, natural enemies, or mutualists. "Taxon" will often be species, but could also represent genotypes or groups of species (see section 5.3.6).

A *response parameter* is the aspect of performance of the focal species that is measured. *Individual-level* responses include aspects of behavior, morphology, and physiology, as well as components of the fitness of individuals (e.g., growth rate, survival probability, or reproductive output). *Population-level* responses include population size or growth rate, where population size can be measured as density, biomass, cover, or other measure of abundance. *Community-level* responses include parameters such as taxonomic or functional group composition, degree of dominance, or diversity.

Competitive abilities can be compared either among focal species (*competitive response*) or among associate species (*competitive effect*) (Goldberg and Werner 1983). The distinction is important because different traits can determine the ability to suppress other organisms (competitive effect) and to tolerate or avoid suppression (competitive response) (Goldberg and Landa 1991). Competitive effects can be measured as the *natural abundances* of the associate or on the basis of a *per-unit amount*. Common measures of abundance are density, biomass, and cover (for sessile organisms), but other measures may also be appropriate (e.g., total root length or leaf surface area for plants). The comparison of results using different measures of associate abundance may be informative in itself. For example, the result that species have different per-capita effects but similar per-gram effects would indicate that the major trait-influencing per-capita competitive effect is biomass per individual.

5.3.2 Basic Experimental Designs

Competition experiments fall into three general categories defined by the way in which density is controlled: substitutive designs, additive designs, and response surface designs (figure 5.1; Silvertown 1987; Gibson et al. 1999). For all categories, it is essential that densities be experimentally manipulated. If natural variation in abundance is used to test competition, environmental differences correlated with the natural density gradient of the associate may also directly affect the focal individuals, thus confounding competitive effects with other environmental factors, whether biotic or abiotic.

In substitutive experiments (replacement series), total density is kept constant and frequency of each species is varied (figure 5.1A). Numerous critiques of substitutive experiments and constraints on its use have been published over the last decade or so, and its use is not generally recommended for experiments in natural communities (Gibson et al. 1999 and references therein). A substitutive

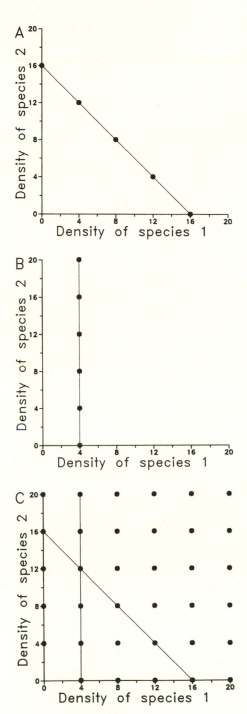

Figure 5.1 Three categories of design for competition experiments where densities of two species (or genotypes or groups of species) are manipulated: (A) substitutive experiments, (B) additive experiments, and (C) response surface experiments. Each point on the graphs represents a single experimental treatment. In part (C), the subsections of the phase plane that constitute the entire experiment in parts (A) and (B) are shown as solid lines.

experiment tests only for the relative intensity of intra- and interspecific competition. So, even if all of the rather restrictive assumptions of the design are met, it is never an appropriate design to test for the occurrence of competition or for comparisons of the absolute magnitude of competition. For questions that require only the relative magnitude of competition, however, such as whether niche separation occurs, substitutive designs can be useful if the assumptions are met (Connolly 1986).

In additive experiments, density of the focal species is kept constant and density of the associate species is varied experimentally (figure 5.1B). Although usually described with only two species but many density treatments, this definition actually fits many field experiments of competition. For example, "removal experiments" typically compare the response of a focal species between a treatment in which the associate is present at its natural abundance and a treatment in which the associate is completely removed. If the focal species is at a constant density, this is a form of additive design with only two associate densities: presence and absence. Keeping the focal density constant in all treatments is important because this keeps the number of competing conspecifics constant. However, especially when also comparing different physical environments, keep in mind that the intensity of intraspecific interactions may vary even with constant initial numbers of conspecifics as a result of variation in biomass. For this reason, Miller (1996) recommends using a density of only a single focal individual to eliminate intraspecific competition entirely in additive designs. For some situations, keeping focal density constant may mean either removing or introducing some individuals of the focal species into experimental plots. If focal individuals are added, it is critical that all focals in all treatments be manipulated the same way, that is, they must all be introduced or all be naturally occurring individuals.

Section 5.3.5 describes several important considerations in choosing associate density treatments. The major limitation on additive designs is that they confound density and frequency effects. As density of the associate increases, it also increases in frequency; in other words, it represents a greater proportion of the total mixture of the two species because the focal species is kept at a constant density. If the magnitude of competition is frequency- as well as density-dependent, this can be a serious problem.

The third design listed, the response surface experiment (or addition series, Silvertown 1987), gets around this problem by manipulating densities of both focal and associate species (figure 5.1C). This type of experiment can provide the data necessary to develop realistic population dynamic models (Law and Watkinson 1987; Freckleton and Watkinson 1997). Although this is the ideal two-species design, the large number of density combinations required makes such an experiment impractical for many field situations, especially when several species combinations must be investigated. Even in the laboratory, relatively few experiments have explored large combinations of densities (see Gibson et al. 1999 for a review). Therefore, we will not address analysis of these experiments further; the interested reader should refer to Ayala et al. (1973) and Law and Watkinson (1987).

5.3.3 Level of Response Variables and Problems of Temporal and Spatial Scale

Most field experiments on competition use individual-level response parameters such as foraging behavior or individual growth rate because of constraints on both the spatial and temporal scales of experimental manipulations. Only for small, short-lived organisms is it feasible to measure growth rates of entire populations. This constraint means that conclusions about the population consequences of competition for distribution or abundance can only be inferred from individual-level measures rather than directly demonstrated. Because interaction coefficients can differ considerably with life history stage and the demographic parameter measured (e.g., DeSteven 1991a,b; Howard and Goldberg 2000), this raises the important question of which life history stage(s) of focal individuals is most appropriate to use. Ideally, experiments should be conducted with all ages or stages, in which case models of age- or size-structured population dynamics could be used to estimate population growth rates under various experimental conditions (e.g., Gurevitch 1986; McPeek and Peckarsky 1998). In practice, this is not always feasible and the subset of stages to be studied should be chosen based on their likely importance in population regulation. Sensitivity analysis of matrix population models of demographic data can be useful in determining the best ages or stages to study (Caswell 1989).

Relatively few field experiments have measured community-level responses to ask how species composition or diversity are affected by competition (Goldberg and Barton 1992). For analysis of community-level responses, possible approaches include the use of (1) relative population abundance as the response variable in univariate analyses like those described in section 5.4, (2) diversity indices that summarize community properties in such univariate analyses, and (3) absolute abundances of all species in multivariate analyses of variance (chapter 6). Goldberg and colleagues (Goldberg 1994; Goldberg et al. 1995; Goldberg and Estabrook 1998) have described several other experimental and analytical approaches for studying community-level consequences of competition.

5.3.4 Absolute Versus Relative Response Variables: When to Standardize?

Almost any response variable can be expressed as an absolute value or standardized to the value of that variable in the absence of associates. Which of these is used in an analysis can have a major impact on the interpretation of comparisons of the magnitude of competition. Figure 5.2 shows an example in which the growth responses of two focal species to a density gradient of the same associate species are compared. Using absolute values of focal response, we find that species 2 has lower growth than species 1 in the absence of any associates (lower intercept) and also has a less steep slope. Thus, species 2 would be considered a better response competitor (less absolute decrease in growth for each associate individual added; figure 5.2A). However, when expressed as a percentage of

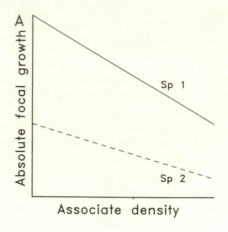

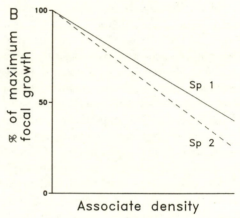

Figure 5.2 A hypothetical example of results of an additive experiment with two different types of response variables for two focal species: (A) absolute growth and (B) percentage of growth in the absence of the associate. Using absolute response, we find that species 2 is the better response competitor, whereas species 1 is the better response competitor when percentage of maximum growth is used.

growth in the absence of the associate (intercept for both is 100%), species 1 shows less of a reduction in growth and so would be considered the better response competitor (figure 5.2B).

Most plant ecologists have used a percentage-based index to standardize data from competition experiments. The most common index is $RCI = (P_r - P_c)/P_r$ where P_r is performance in a removal treatment with no associates and P_c is performance in a control treatment with associates present. However, this index is not symmetrical with respect to competitive and facilitative outcomes (Markham and Chanway 1996), and it has poor statistical properties (Hedges et al. 1999; see also Freckleton and Watkinson 1997). Therefore, Hedges et al. (1999) have recommended the use of the closely related log response ratio, $\ln RR = \ln(P_c/P_r)$, instead (either $\log_e$ or $\log_{10}$ can be used). This index has been used extensively for experimental studies of predator–prey interactions (see Osenberg et al. 1999 for further discussion of this and other indices).

The potential conflict in interpretation between standardized and unstandardized data is most obvious for comparisons of competitive response of different focal species or of the same focal species in different environments (e.g., high and low productivity) because intercepts will usually be different with unstandardized data. However, it may also appear in comparisons of competitive effect among associate species. Normally, we would expect a single focal species in a single site to have the same intercept (performance in the absence of any associates) in regressions with different associate species, so that standardized and unstandardized data would yield the same result. However, when the zero-density treatment is a result of experimental removal of the associates, two factors may lead to differences in intercepts among associate species. First, there may be residual effects of the former presence of the associate that could differ between associate taxa. This result is more likely to occur with sessile organisms that could have cumulative effects on a single location. For example, plants may have different residual effects on nutrient availability because of differences in litter quality and subsequent decomposition rates or soil organic matter accumulation. In addition, roots can seldom be removed, so the presence of dead and decaying roots could also have significant residual effects. Second, associates can be removed only where they were initially present. If different associate species were initially present or abundant in different microenvironments and these microenvironmental conditions influence the focal species directly, focals may differ in zero-density treatments for different associates, that is, their intercepts may differ. Both of these potential problems point out the importance of having separate zero-density treatments for each of the associate species used in a removal experiment.

There has been remarkably little discussion in the literature about the conditions in which response in the presence of associates should be standardized to response in the absence of associates; the following recommendations should be viewed as preliminary. For questions concerning the consequences of competition for distribution and abundance, relative values are probably most useful. For example, suppose that competition reduces individual growth of a focal species by 10 g in an unproductive site and by 100 g in a productive site, suggesting that competition is more intense in the more productive site. However, if the growth in the absence of competition were 10 g and 200 g in the unproductive and productive sites, respectively, the relative responses would be 0% and 50%, suggesting competition is more intense in the unproductive site and, in fact, intense enough there to exclude the focal species. (See Campbell and Grime [1992] for an example of contrasting results for standardized and unstandardized data in a competition experiment.)

On the other hand, if the mechanisms of competition are under investigation, absolute values may be more useful. For example, it may be of interest to relate reductions in resource availability because of consumption by an associate species to changes in resource consumption and growth in the focals. Using standardized data would make it difficult to understand the processes underlying these relationships. Absolute responses are also necessary when competitive effects and responses are being quantified to use as input in models of population interactions because most population dynamic models use absolute, not standardized, parameters. For example, in the Lotka–Volterra competition equations, equilibrium density in the absence of competition (K) is an explicit parameter, and both K and the per-capita competition coefficients (α_{ij}) have an effect on the dynamics and equilibrium outcome of the interaction.

5.3.5 Manipulation of Competition Intensity: How Many Density Treatments?

The range of density variation in an additive experiment could be simply presence versus absence of the associate or several densities could be imposed. Presence/absence comparisons are appropriate where the total magnitude of competitive effect under natural conditions is all that is required. However, unless competitive effects are completely linear, presence/absence comparisons (or comparisons of any two densities) do not allow accurate calculation of per-capita effects (figure 5.3). Where the density dependence of per capita competitive effects has been tested, it is rarely constant (e.g., Harper 1977; Schoener 1986; Pacala and Silander 1990), and so interpolating from presence/absence comparisons to per capita effects across a range of densities is probably not generally safe. Comparing competitive effects between associate species and, especially, relating these competitive effects to traits requires per capita effects so that effects of total density and per capita effects (or, more generally, total abundance and per-unit amount) are not confounded.

If multiple densities are chosen, many density treatments with low or no replication will usually be better than few, well-replicated densities so that any curvilinearity can be detected and quantified. The appropriate analysis will then be a comparison of slopes from regressions of the response variable on density rather than contrasts between means of different treatments (see section 5.4.2 and chapter 10). Because both the strongest curvilinearities and the greatest variation among replicates tend to be at the lowest densities, we recommend concentrating on density treatments at the low end. Densities above natural are also useful because (1) naturally occurring densities usually fluctuate considerably and (2) competition might occur only at densities above those normally determined by other factors such as predators. Regardless of the range of densities used in an

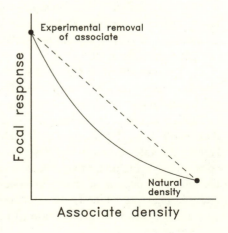

Figure 5.3 An example of inaccurate estimation of per capita competitive effects from an experiment with two treatments: the associate present at its natural density and the associate completely removed. The true per capita effect is shown as a solid line. A linear extrapolation from the two treatments (dashed line) underestimates the true per capita effect at low associate density but overestimates the true effect at high associate density.

experiment, it is always important in publications to identify the natural density range so readers can assess the applicability of results to natural conditions.

5.3.6 Pairwise Versus Total Competition: Which Associate Groups to Use?

The associate group whose abundance is being manipulated can be a single species or a group of species considered together. Which is more appropriate depends both on the biology of the species under study and on the goal of the experiment. The tendency has been for animal ecologists to measure competition between pairs of species chosen a priori as likely strong competitors, whereas plant ecologists have more often quantified total competition on a focal species from the entire plant community. Because plants all require the same few resources, all plants are potential competitors and there is often little a priori basis for choosing subsets. Therefore, the most common type of field competition experiment in plant communities is a presence/absence additive design where the associate "species" is the entire plant community except for the focal individuals (Goldberg and Barton 1992). For many animal species, on the other hand, more restricted guilds of potentially interacting species that share resources can often be defined (e.g., seedeaters or insect eaters). The danger here is that, although potential competitors are often chosen by taxonomic relatedness (e.g., congeners), distantly related taxa may just as likely share resources and compete for them (e.g., seed-eating rodents, ants, and birds in deserts; Brown et al. 1986).

When the question of interest concerns comparisons of competitive ability among species and relationships of competitive ability to traits, pairwise experiments with single species as associates and focals are usually most appropriate. The number of species also becomes important because tests of these relationships essentially use each associate or focal species as a "replicate" in a secondary statistical analysis. Use of only a single species with each value of a trait is unreplicated for tests of hypothesis about species characteristics related to competitive ability. For example, if only two associate species are compared and the one with larger leaves has a stronger negative effect, the conclusion that leaf size confers strong competitive effect is unwarranted because of lack of replication within each leaf size class. The taxa may also differ in traits such as root length or nutrient uptake rate. Because leaf size is confounded with these other variables, assigning a causal effect to leaf size alone is not justified. We recommend either replicating species within trait classes or using species with a range of values of a continuous trait.

5.3.7 Direct and Indirect Effects: What to Do with Background Species?

Response of a focal species to manipulation of abundance of an associate species represents the sum of the direct effects of the associate plus all the indirect effects and higher order interactions involving the associate and all the background species, the *net effect*. If no background species are present, only direct effects are measured (but see subsequent discussion). *Abundance-mediated indirect effects* are defined as changes in the effect of an associate on a focal species when background species change the abundance of the associate (Abrams et al. 1996). *Trait-mediated indirect effects* are defined as changes in the effect of an associate on a focal species when a background species changes the per-capita effect of the associate (Abrams et al. 1996).

If the primary question concerns the effect of the associate species on the distribution and abundance of the focal species, net effect may in fact be what is desired. For example, suppose removal of a dominant has little effect on a focal species because other, unmeasured, species respond to its removal more quickly than does the focal species and thus end up suppressing the focal species just as much as did the designated associate. The conclusion that competition from that associate species is unimportant in controlling the abundance of that focal species is reasonable, even though the associate may have strong direct effects on the focal species in a pairwise experiment with no background species. However, if the primary question is which traits of an associate species allow it to become dominant, measuring its net competitive effect will be misleading because traits of the background species will also influence the results.

Therefore, our general recommendation is that questions about the relationship of traits to competitive ability or determination of competitive hierarchies are usually best addressed in pairwise experiments that minimize the potential for abundance or trait-mediated indirect effects. In contrast, questions about the consequences of competition for abundance and distribution are usually best addressed by incorporating any nondirect effects into the response variable, that is, by leaving background species in all treatments. We point out, however, that strong nondirect effects restrict generalization across sites or times even more than is usual in ecological field experiments because the biotic environment and therefore the complex of nondirect effects differ as well as the abiotic environment and history of the sites.

In reality, almost any field experiment of any design has background species and therefore we can never measure only direct effects. It is rarely possible or even desirable to eliminate completely all the biota except for the designated focal and associate species. First, exploitation competition itself is an indirect interaction, mediated through shared food resources. When the resource is a living organism, even if the food is not an explicitly designated associate or focal species, it obviously must be present in the experiment. Similarly, negative interactions may be mediated by shared natural enemies (apparent competition; Holt 1977; Connell 1990) or mutualists, and removing background species would preclude detecting an important interaction between two competitors (but appropriate if quantification of only exploitation competition is desired). Monitoring the dynamics of these intermediary nonmanipulated species, whether resources, natural enemies, or mutualists, can be an important step toward understanding the mechanisms of interactions. Second, microorganisms are part of the biota and may be important links in chains of indirect effects, but they are very difficult to remove completely. In terms of designing and interpreting field experiments, it is critical to recognize which background species are present and to incorporate this knowledge when interpreting results.

5.4 Statistical Analysis and Interpretation of Results

5.4.1 Comparison of Habitats: Effects of Competition on Distribution and Abundance

One of the most common questions about competition is whether it influences the distribution and abundance of a focal species. The recommended approach to address this question is to repeat a presence/absence additive design with background species in differing habitats, those where the focal species is present or abundant and those where it is absent or rare. This is really a population-level question, and ideally population-level response parameters should be used (section 5.3.3). A minimum of two replicate sites in each habitat category (focal abundant versus rare) should be used so that the relevance of habitat differences to focal species distribution can be inferred from the results. With only a single site in each habitat category (admittedly the most common design actually used), the only valid inference we can make is whether sites differ in the magnitude of competition. That is, site is confounded with habitat.

The associate species can be a single species chosen a priori, a subset of species, or the entire community (see section 5.3.6). Since the question concerns the net consequences of the presence of some associate species or group of species, it is appropriate to leave all other species present as background species in all treatments, in other words, to quantify the sum of all direct and indirect (both abundance- and trait-mediated) effects.

The actual experiment would involve establishing an experimental area in each site within which the focal species would be added and the associate species removed as appropriate. It is important to add individuals of the focal species back into its native habitat as well as the habitat from which it is absent to account for transplanting effects. Such a design also ensures that the density of the focal species is the same across all treatments (section 5.3.2).

For an experiment with a single focal species, a single associate species or group, and several sites of each habitat type, the appropriate ANOVA model is

$$X_{ijkl} = \mu + \tau_i + \theta_j + \tau\theta_{ij} + \Sigma(\theta)_{jk} + \tau\Sigma(\theta)_{ijk} + \varepsilon_{ijkl} \tag{5.1}$$

where μ is the overall mean, τ_i is the deviation due to the ith associate treatment (present at natural abundance or removed), θ_j is the deviation due to the jth habitat (focal species naturally abundant or rare), $\tau\theta_{ij}$ is the deviation due to the interaction of the ith treatment with the jth habitat, $\Sigma(\theta)_{jk}$ is the deviation due to the kth site nested within the jth habitat, $\tau\Sigma(\theta)_{ijk}$ is the deviation due to the interaction of the ith treatment with the kth site nested in the jth habitat, and ε_{ijkl} is the deviation due to the lth replicate within each site–treatment combination. Treatment and habitat are both fixed factors and site is a random factor, so the overall model is

Table 5.1 ANOVA for a completely balanced cross-nested design

Effect	df	E(MS)	F-ratio
Treatment	$t-1$	$\sigma_e^2 + n\sigma_{\tau\Sigma(\theta)}^2 + hsn\sigma_\tau^2$	$MS_T/MS_{TS(H)}$
Habitat	$h-1$	$\sigma_e^2 + tn\sigma_{\Sigma(\theta)}^2 + stn\sigma_\tau^2$	$MS_H/MS_{S(H)}$
Treatment × habitat	$(t-1)(h-1)$	$\sigma_e^2 + n\sigma_{\tau\Sigma(\theta)}^2 + sn\sigma_{\tau\theta}^2$	$MS_H/MS_{S(H)}$
Site (habitat)	$h(s-1)$	$\sigma_e^2 + tn\sigma_{\Sigma(\theta)}^2$	$MS_{S(H)}/MS_e$
Treatment × site (habitat)	$(t-1)h(s-1)$	$\sigma_e^2 + n\sigma_{\tau\Sigma(\theta)}^2$	$MS_{TS(H)}/MS_e$
Residual (error)	$ths(n-1)$	σ_e^2	

a mixed cross-nested design. The degrees of freedom, estimated mean squares E(MS), and F-tests for this model are shown in table 5.1; SAS commands are in appendix 5.1A. See Searle (1971) for a detailed explanation of the underlying statistical theory. More complex designs, which include blocking factors within sites, can also be used (chapter 4).

The important result from this ANOVA is whether the treatment × habitat interaction is significant, that is, whether the magnitude of competition differs between habitats where the focal species is abundant and where it is absent or rare. The expected pattern of results is shown in figure 5.4, with a larger magnitude of competition where the focal species is absent or rare than where it is abundant. Notice that, in this analysis, the treatment × habitat interaction is tested over the treatment × site interaction, and the power of the test is a function of the number of sites. Thus, the most powerful design in this case maximizes the number of sites within habitat types (s in table 5.1) and minimizes the number of replicates within sites (n in table 5.1). Obviously, such a strategy creates the potential for logistical difficulties if the sites are at any distance apart.

When the number of replicates is not the same for each level of all treatment factors, the experimental design is termed "unbalanced" (Searle 1987). Unbalanced

Figure 5.4 Pattern of predicted results from a presence/absence additive design if competition influences the distribution and abundance of the focal species. The key result is that the magnitude of competition (depression due to presence of competitors) is much greater where the focal species is normally absent than where it is normally present (i.e., habitat × competition interaction). In this case, the focal species also performs better in the habitat where it is abundant than where it is absent, even in the absence of associates. Competition is present in both types of habitats, so both habitat and competition would probably be significant main effects.

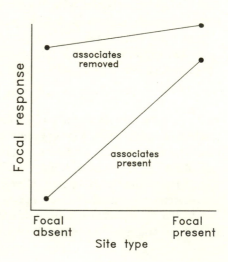

designs are very common in ecological experiments because of insufficient material (e.g., it may be more difficult to get replicates from one site, species, or genotype than others) or loss of replicates through mortality. Unbalanced ANOVA designs are more difficult to analyze, and, unlike balanced designs, there is no single correct way to analyze them. Readers may wish to consult Milliken and Johnson (1984), Fry (1992), Shaw and Mitchell-Olds (1993), and Newman et al. (1997). SAS 6.0 and above will construct an approximate F-test for unbalanced designs using the Satterthwaite approximation for the degrees of freedom (Satterthwaite 1946; Hocking 1985; see appendix 5.1A). If SAS procedure GLM is used for the analysis, Type III sums of squares (SS) must be requested; the F-tests provided by the other types of SSs (I, II, and IV) are incorrect. Because there are several approaches to this problem and no general agreement, the individual researcher must evaluate whether the approach taken by SAS is appropriate for the particular design employed (see chapter 4).

An alternative approach to the analysis of unbalanced designs that contain both fixed and random effects—mixed models—is maximum likelihood (McLean et al. 1991), which is implemented in SAS as procedure MIXED (appendix 5.1B). One potential advantage of this method is that it will likely prove to be more robust to unbalanced designs than the standard least-squares approach, although this is not yet proven. A disadvantage is that computations for complex designs can take a long time and sometimes fail to converge on a solution. Be warned that the documentation of the technique is not completely clear, especially for those without extensive statistical training; consultation with a statistician is highly recommended.

5.4.2 Comparison of Species: The Relationship of Traits to Competitive Ability

Comparing competitive abilities among species (or other groups) and determining the relationship of competitive abilities to traits will usually require per capita measurements. The recommended approach to compare species is an additive design with multiple densities of the associate, using either a single focal species and several associate species that differ in value of some trait (comparison of competitive effect) or a single associate species and several focal species (comparison of competitive response; figure 5.2A). A significant relationship (nonzero slope) between density of the associate and response of the focal species indicates competition (negative slope) or facilitation (positive slope).

The analysis is stepwise (see Zar 1996, figure 17.1). The first step is to test for differences in density effects among focal or associate species using a test for homogeneity of slopes. The ANCOVA model is

$$X_{ijk} = \mu + \Delta_i + \alpha_j + \Delta\alpha_{ij} + \varepsilon_{ijk} \tag{5.2}$$

where μ is the mean intercept (the value of the response parameter at 0 density), Δ_i is the deviation due to the ith density treatment (the average slope of the line that predicts the effect of associate species density on the response parameter), α_j is the deviation from the mean intercept due to the jth associate or focal species (differences in intercepts among associate or focal species), and $\Delta\alpha_{ij}$ is the deviation due to the interaction of species identity and density (differences in slopes among species; figure 5.2A). In addition, Δ_i is a continuous variable, and α_j is a categorical factor.

In this first analysis, we are concerned only with the interaction term. We ignore the main effects of density and species identity at this juncture because such main effects are meaningful only if the interaction effect is declared not statistically significantly different from 0. The SAS commands for performing this analysis are given in appendix 5.2. If differences among slopes are found, then a post hoc multiple comparison test can be performed (Zar 1996, section 17.6) to discover which particular species differ from each other. Alternatively, one can test a priori hypotheses concerning differences between particular species or sets of species, by conducting a series of specified comparisons (contrast tests) rather than a single overall ANCOVA.

A potential problem in this analysis arises if only a single zero-density treatment (no competitors) is used for all associates, because it is impossible to assign a treatment level (identity of associate) to replicates at this density. In this case, it is reasonable to randomly allocate all of the zero-density replicates among the associate species. (For example, if there are three associate species and 30 zero-density replicates, then 10 replicates would be randomly assigned to each species treatment.) This problem should arise only when all associate individuals are added (e.g., to cages or in greenhouse experiments), because in field removal experiments, there will usually be distinct zero-density treatments for each associate species (see section 5.3.4).

If slopes are homogeneous, we can then proceed to (1) measure the average slope (Δ_i) and (2) test for differences in intercepts (α_j) among species using the following model:

$$X_{ijk} = \mu + \Delta_i + \alpha_j + \varepsilon_{ijk} \qquad (5.3)$$

Note that the model is identical to that in equation (5.2) except that the interaction term has been deleted. The SAS commands are given in appendix 5.3. Significant density effects (a nonzero slope) indicate that competition or facilitation is occurring. The average slope measures the general per capita effect of all associates on the focal species.

In comparisons of focal species, significant species effects (α_j) indicate simply that the focal species differ consistently across all densities of the associate species (e.g., in maximum growth rates or sizes). In comparisons of associate species, significant species effects are more problematic because the focal species remains the same and thus should not differ in response to the associates when no associates are present. Biological interpretations of significant associate main effects were given in section 5.3.4. An alternative interpretation of either significant focal or associate effects is that it could be an artifact of using linear regressions on nonlinear data. Figure 5.5 shows an example where two focal species have the same slopes but different intercepts in linear regression analyses. However, inspection of the graphs shows that the intercepts are actually very similar but that species B is strongly nonlinear at low associate densities. If transformations do

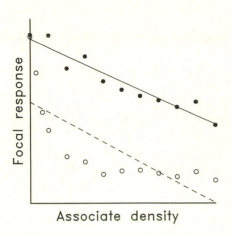

Figure 5.5 An example of misleading results of comparison of slopes (individual competition coefficients) using linear regression models. Using linear regressions, the two species appear to have different intercepts, but similar slopes; that is, they have similar competitive effects if the curves represent two associate species. However, inspection of the data reveals the intercepts are actually very similar, and species B (open circles) has a strongly nonlinear effect on the focal species at low densities. Inspection of the residuals of the linear regression for species B would show a nonrandom distribution with associate density.

not succeed in linearizing the data, a nonlinear regression analysis is warranted (chapter 10).

Relating per capita effects or responses to values of a particular trait can be conducted as a second-level statistical analysis by correlating slopes with trait values (e.g., Goldberg and Landa 1991). With fewer species or with few discrete values of the trait, slopes can be compared with a priori contrasts among groups of the associate or focal species (Day and Quinn 1989) or in the context of the ANCOVA where species are nested within species types in a way exactly analogous to sites nested within habitats (see equation 5.1).

5.4.3 The Boundary Constraint Effect

The performance of an organism is likely to be affected by many factors (e.g., genotype, weather, predators) other than the density of competitors. Usually in field experiments such additional factors are either poorly controlled or not controlled at all, and the effects of competitors may be obscured by the large amount of variation caused by these other factors. This general problem may arise in any type of analysis of field experiments; in regression analyses of competition (including ANCOVAs), it may be manifested as an "envelope effect" (Firbank and Watkinson 1987; Guo et al. 1998). Competition may act to keep the response of the focal species below some maximum, while within this competitively determined envelope other factors act to further depress the focal species. The result is that a plot of the response variable against density will be a scatter of points forming a triangle in the lower left-hand portion of the graph (figure 5.6); such bivariate distributions have also been referred to as "triangular" or "polygonal" (Scharf et al. 1998), or "factor ceiling" distributions (Thomson et al. 1996) or as reflecting "boundary constraints" (Guo et al. 1998).

Two general approaches have been used to analyze such envelope effects. The more traditional approach is to treat this effect as a problem that causes consider-

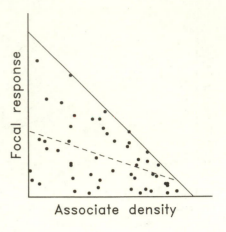

Figure 5.6 Example of the "envelope effect" where the variance in focal species response is much greater at low associate densities than at high associate densities. The solid line at the upper edge of the envelope represents the constraint on maximum focal response imposed by competition. The dashed line represents the competitive effect estimated by a standard least-squares linear regression.

able difficulty in quantifying density effects and to search for solutions that allow the competitive effect to be isolated from other factors. This approach treats relationships such as those in figure 5.6 as problematic for two reasons. First, the effect itself will be obscured and the slope of the line will be decreased by the larger amount of variation caused by other factors at low density. Second, the important statistical assumption of homoscedasticity—variation around the regression line is the same everywhere—is being violated. Clearly, the variation decreases with increasing density. In this case, there may very well be no transformation of the data capable of making the variance homogeneous. Under the traditional approach, the solution to these problems will require somehow finding the "true" competitive effect despite the noise and the violation of standard statistical assumptions. For example, one purely statistical solution was developed by Pacala and Silander (1990). They note that the residuals (the variation around the regression line) of such a distribution are distributed as a gamma function, which has the properties of being unimodal, left-skewed, and bounded at zero (see chapter 13 for a description of the gamma function). In their experiment, competitive effects followed a hyperbolic function. Thus, rather than the more familiar least-squares method, they used a maximum-likelihood estimator and iterative Newton method, as described in chapter 10. A second solution aimed at finding the "true" competitive effect is to monitor other potentially interacting factors and use them as covariates in a multiple regression. The possible drawbacks of this approach are (1) the logistical difficulty of measuring many additional variables, (2) the potential failure to measure the key factor, and (3) the fact that multiple regression brings with it an additional host of statistical assumptions and difficulties.

An alternative approach to analyzing data with distributions such as those in figure 5.6 is to recognize that it implies a different sort of role for competition: imposing a boundary constraint on maximum potential performance rather than a deterministic effect that can be applied to every individual in the population. In the last several years, this notion of boundary constraints has been applied to a number of problems in ecology besides species interactions, and a number of different statistical approaches have been used to quantitatively describe the boundary. Thomson et al. (1996) suggested using partitioned regressions, whereby standard regression techniques are used to partition a data set into those with positive and negative residuals. A subsequent regression uses only data points with positive residuals from the first regression, with as many cycles of such regressions as possible to approach the upper boundary of the data set before sample size declines to too small levels. Blackburn et al. (1992) suggested dividing a data set into classes along the abcissa and conducting regressions using only the highest value within each class. Scharf et al. (1998, p. 451) have promoted the use of quantile regression, which "estimates quantiles of the dependent variable ranging from 0 to 100, conditional on the values of the independent variable." Scharf et al. (1998) compared least-square regressions using maximum values within classes to quantile regressions (which use entire data sets) and found results of quantile regressions to be much more consistent and less contingent on arbitrary decisions on analysis procedures. Finally, Garvey et al. (1998) have suggested adopting an approach from astronomy based on two-dimensional Kolmogorov–Smirnov tests. This approach asks whether the distributions of two variables could have arisen independently as opposed to the usual regression analysis, which asks (in part) how much variance can be explained in one variable by the values of another.

Boundary constraint analysis is clearly a rapidly expanding area in statistical ecology; the reader should check for recent literature before beginning any analyses. We expect the development of new techniques and rigorous evaluation of existing techniques to be an ongoing process.

5.5 Related Techniques

5.5.1 Isotonic Regression

An alternative and intermediate approach to the ANOVA and ANCOVA models presented in previous sections is isotonic regression, which can be thought of as a form of ANOVA to be used when there are ordered expectations (Barlow et al. 1972). Gaines and Rice (1990) present a cogent description of this technique and offer a microcomputer program for implementing it (see also Rice and Gaines 1994a,b and references therein). The advantage of isotonic regression for detecting density-dependent responses is twofold. First, there are no assumptions about the shape of the density response except that it is monotonic. For this reason, isotonic regression can be more powerful than standard least-squares regression analysis when the shape of the response is irregular or a step function. Second, fewer density treatments are needed than in ANCOVA because the precise shape of the density effect is not of interest. Instead, replication efforts can be placed elsewhere in more associate or focal species, more sites, or more replicates per site. However, several limitations to isotonic regression restrict its usefulness to relatively simple questions about competition. The most general is that, because no particular shape of the density response is assumed, per capita effects cannot be calculated. Therefore, isotonic regression is not appropriate if comparisons of species effects are necessary independent of their total abundance.

5.5.2 Spatial Heterogeneity

This chapter has not dealt with the analysis of spatial heterogeneity except when spatial variation is an explicit part of the ecological question of interest (e.g., variation in magnitude of competition between sites or habitats). Chapter 4 discusses ANOVAs with block designs that explicitly account for variation due to spatial heterogeneity. Chapter 15 discusses an alternative approach to analysis of spatial heterogeneity using variograms. Both of these methods require replication within a treatment, which is impossible using the multiple, unreplicated density design recommended in this chapter when comparisons of per capita competitive abilities are required. There is currently no satisfactory general method that can both account explicitly for spatial heterogeneity and maximize investigation of potential nonlinearities. Therefore, the individual investigator must decide which is more important in the study and sacrifice number of blocks (replicates per density) or number of distinct densities accordingly. If an unreplicated regression design is chosen and the important environmental variables are known, these can be measured in individual experimental units and used directly as covariates in an ANCOVA. A third possible approach for incorporating spatial heterogeneity in designs with unreplicated densities is the permutation method described in chapter 16, although this method, to our knowledge, is untried for this application.

5.5.3 Beyond Simple Dependent Variables

The analysis of competitive effects potentially involves examining the response of many different types of response variables. The previous discussion was in the context of examining a single variable that most likely meets the standard normality assumptions of parametric statistics. Time-based response variables such as survival time or time to emigration, although still single measures, often do not meet standard normality assumptions and require different analytic methods (chapter 13).

Often the response variable is more complex than a single measure of focal performance. If individual-level measures are being used, for plants one might measure vegetative growth, seed number, and seed size. If community-level measures are being used, one might measure the densities of several focal species. If the response variables (traits or species) are likely to be correlated with each other or to interact with each other, then a multivariate analysis of variance is appropriate (chapter 6). If data are collected over time, for example, as growth curves of individuals or populations, then a repeated-measures analysis should be used (chapter 8). For long-term experiments, time series analysis may be most appropriate (chapter 9).

For all of these techniques, the basic design criteria outlined in this chapter are still appropriate, although different statistical techniques will be most powerful for different numbers of treatment levels versus replicates within treatments. In all cases, careful thought should be given to the type of analysis to be used at the time of experimental design, rather than waiting until the experiment is complete.

5.6 Some Final Comments on Experimental Design

We have argued that answering questions besides "Does competition occur?" is critical to developing and testing general models of competition and that answering these questions will require more complex experimental designs than have often been used in field competition experiments. Nevertheless, there is a real danger that experimental designs will become so complex that insufficient resources are available to detect any but enormous effects. Decisions on allocation of resources will always depend on the details of any particular study, but a few preliminary steps will aid in the process. We strongly recommend conducting pilot experiments in the field to estimate variances and thus the replication necessary to detect effects of a given magnitude (chapter 2); results of such power tests may quickly demonstrate that experiments for some questions are simply logistically impossible in that system. If several factors are experimentally manipulated, replication should be focused on the factor(s) that will be used as the error variance(s) (denominator MS) for the most important test(s). As the example in table 5.1 showed, this is not always at the level of replication of the smallest experimental units or the most detailed treatment combination (e.g., field plots). It is critical to work out the estimated mean squares for any design before beginning an experiment so that replication will be done at the proper level. Brownlee (1965) and Searle (1971) present examples for many common designs; consultation with a statistician at this stage is also highly recommended. The important point is simply that there will always be tradeoffs between the complexity of questions that can be asked from an experiment and the statistical power available for testing particular effects. Which of these is emphasized and which sacrificed is the most important decision underlying any experimental design.

Acknowledgments We thank Brad Anholt, Jessica Gurevitch, and Earl Werner for valuable discussions of many of the issues raised in this chapter and Yoram Ayal, Drew Barton, Norma Fowler, Jessica Gurevitch, Ariel Novoplansky, David Ward, and Earl Werner for their helpful comments on previous drafts. The National Science Foundation provided support for the research on which this chapter was based.

Appendix 5.1 SAS Program Code for Analysis of the Cross-nested ANOVA Model in Equation 5.1

A. Analysis Using Least Squares

```
PROC GLM;
    CLASS T H S;           /T = associate treatment, H = habitat, S = sites within habitat/
    MODEL X = T H T*H S(H)   /use type III sums of squares. Warning: results are
      T*S(H)/SS3;              only approximate if the design is unbalanced./
```

*** If the design is unbalanced use the following command to
   obtain the correct *F* tests for the main effects ***

```
RANDOM S(H) T*S(H)/TEST;        /Declare sites to be a random effect and ask SAS to
                                compute the Satterthwaite correction for an
                                unbalanced design./
```

*** If the design is balanced use the following commands to
   obtain the correct *F* tests for the main effects ***

```
TEST H = T  E = T*S(H);   /test associate treatment over the treatment-site interaction/
TEST H = H  E = S(H);                                        /test habitat over site/
TEST H = T*H  E = T*S(H);                    /test treatment-habitat interaction
                                             over the treatment-site interaction/
```

B. Alternative Analysis Using Maximum Likelihood

```
PROC MIXED;                              /mixed model analysis using reduced
                                         maximum likelihood (the default)/


CLASS T H S;
MODEL X = T H T*H;        /only fixed effects are placed in the model statement/
RANDOM S(H) T*S(H);                 /random effects are specified separately/
```

Appendix 5.2 SAS Program Code for Analysis of the ANCOVA Model in Equation 5.2

```
PROC GLM;
   CLASS A;                           /A = associate or focal species treatment/
   MODEL X = D A D*A/SS3;             /D = density, use type III sums of squares/
```

Appendix 5.3 Analysis of the ANCOVA Model in Equation 5.3

```
PROC GLM;
   CLASS A;                           /A = associate or focal species treatment/
   MODEL X = D A/SS3;                 /use type III sums of squares/
```

6

MANOVA

Multiple Response Variables and Multispecies Interactions

SAMUEL M. SCHEINER

6.1 Ecological Issues

Ecological questions very often ask how multiple components might respond to some change in the environment or might differ among groups. Ecologists are often most interested in the interactions among those components and how the interactions might change as the environment changes. Studies in population biology commonly address issues of constraints or trade-offs among traits, for example, seed size versus seed number, sprint speed versus endurance, or development time versus size at maturity. Studies in community ecology often address questions regarding how a number of coexisting species respond to the removal of a keystone predator or to a change in the environment. In statistical parlance, these multiple components are referred to as multiple response variables.

In this chapter, I present situations in which the independent variables (experimental manipulations) consist of categorical variables, for example, competitors present or absent, or three levels of nutrient availability. When only one response variable has been measured, such data are examined by analysis of variance (ANOVA) (chapters 4 and 5). When more than one response variable has been measured, the most appropriate method of analysis is usually multivariate analysis of variance (MANOVA), in which all dependent variables are included in a single analysis. I am happy to report that the use of multivariate analyses by ecologists has increased in recent years. For the first edition of this book, I did a quick survey of volume 72 (1991) of *Ecology* and found that, of 62 articles that measured more than one response variable for which a multivariate test would have been appropriate, only 9 actually used MANOVA. In contrast, in volume

79 (1998), of 62 papers, 40 used MANOVA, repeated-measures analysis (chapter 8), or another multivariate technique. The other studies all used multiple univariate analyses, with each response variable being examined separately, although many dealt explicitly with interactions among the response variables.

6.2 Statistical Issues

Multivariate analysis is preferred over multiple univariate analysis for two reasons. First, the ecological questions are often multivariate, involving interactions among response variables. Differences that exist among groups may not be a feature of any one response variable alone, but, rather, of the entire suite of variables. Second, performing multiple univariate tests can inflate the α-value—the probability of a Type I error—leading us to conclude that two groups or treatments are different with respect to one or more of the dependent variables, even though the differences are simply due to chance. I address each of these issues in turn.

6.2.1 Multivariate Responses

We must be concerned about correlations among response variables in two cases that are opposite sides of the same coin. The more obvious case is when a research question explicitly deals with interactions among response variables. Such interactions are indicated by either (1) shifts in correlations among variables from one treatment to another or (2) correlations across treatments that differ from correlations within treatments. Correlations among response variables can be revealed only by some form of multivariate analysis. This need for multivariate analysis is especially true when more than two response variables are measured. Simple pairwise correlations cannot reveal the entire pattern of changing relationships among a multiplicity of variables.

A common alternative to multivariate analysis, especially in community ecology, is to analyze a composite variable such as a diversity index (Magurran 1988). Although such an index has the virtue of simplicity, it can actually hide the features of the data that we are attempting to discover. For example, two samples may have identical diversity values even though individual species may have very different abundances; this would occur if an increase in one species exactly matched a decrease in another species. Such interactions become apparent only by following the behavior of individual response variables within a multivariate analysis.

A not-so-obvious need for multivariate analysis arises when the question is whether two or more groups or treatments differ. For example, we may measure abundances of five species of small mammals on plots with and without predators. Or we may measure 10 aspects of plant morphology in a nutrient and light manipulation experiment. It is possible that tests of individual response variables will fail to reveal differences among groups that a multivariate test would reveal. One form of this effect occurs when the correlation of variables differs in its sign within and between treatments; this is termed Simpson's paradox (Simpson 1951).

For example, the number and size of seeds may be positively correlated within treatments because larger plants have more total resources to devote to reproduction. However, across different nutrient levels, the correlation may be negative as plants switch from a strategy of many, small seeds at high resource levels to few, large seeds at low resource levels.

Conversely, a multivariate test might reveal that what appears to be a number of independent responses is, actually, a single correlated response. For example, if the removal of a predator led to (univariate) significant increases in two prey species, we might be tempted to conclude that the predator affected two independent species. However, if the prey species abundances were correlated due to other interactions (e.g., mutualism), then there is only a single (albeit complex) response.

6.2.2 The Problem of Multiple Tests

The most important statistical issue involves the use of multiple statistical tests to address a single hypothesis. For example, consider a comparison of two groups that do not actually differ. If one uses a typical nominal Type I error rate of $\alpha = 0.05$ and 10 response variables are measured, then the probability of declaring the two groups different for at least one variable is 40%. Two approaches, plus one hybrid approach, can be used to solve this problem.

The simplest solution is to conduct the standard univariate ANOVAs but use a Bonferroni correction in setting the α-level, the probability level at which an effect is deemed to be statistically significant. For example, if five tests are to be done, then the correct $\alpha' = \alpha/5 = 0.01$. This procedure is actually somewhat conservative; a less conservative procedure, a sequential Bonferroni correction, is now recommended (Rice 1990).

The alternative approach is to use MANOVA. Only a single analysis is performed and thus no correction of α is needed. The hybrid approach is to first perform the MANOVA, proceeding to the ANOVA only if the MANOVA yields a significant result. This approach is referred to as a protected ANOVA. The underlying logic is the same as that in which an ANOVA is performed before doing multiple comparison tests of means (section 1.2.4). The one difficulty is that, unlike multiple comparison procedures like Tukey's test and the Student–Newman–Keuls test, there is no generally accepted way to correct for multiple tests even within the "protected" framework. That is, some spurious significant differences are still likely to be declared even when the overall rate of such errors has been lowered. See Harris (1985, section 4.2) and Stevens (1992, section 5.5) for discussions of these issues.

Considerable debate exists as to the correctness of each of these approaches. As is often the case in statistics, the best approach depends on the questions being asked. Those who argue for a strictly multivariate approach emphasize the interrelatedness of the dependent variables. Univariate analyses are not able to capture and dissect those relationships except in a very superficial way. On the other hand, there are instances in ecology when the behaviors of individual variables are of interest in their own right. For example, theory may make predictions regarding changes in specific traits across treatments. Or predictions as to the probability of local extinction of species require examination of the abundances of each species in a study. Or further theory development may require a detailed examination of individual variable behavior as a preliminary to further refinement of predictions. Huberty and Morris (1989) have a cogent discussion of the relevant considerations regarding the appropriateness of multivariate tests versus multiple univariate tests.

This discussion of univariate versus multivariate approaches is couched in terms of determining the effects of individual response variables on the outcome of the MANOVA. For example, of the many traits that were measured, which were actually responsible for the differences among populations? Such information is available directly from MANOVA; univariate ANOVA is not necessary. In addition, the carriers of the information, the eigenvectors of the MANOVA, also contain information on patterns of variable correlations. Thus, properly interpreted, a MANOVA can provide all necessary information.

6.3 Statistical Solution

6.3.1 A Simple Example

To elucidate the basic principles and procedures for MANOVA, I present data from a study of the herbaceous perennial *Coreopsis lanceolata* L. (Asteraceae). These data have been modified to better illustrate several points. We will pretend that the data are derived from an experiment consisting of plants grown in individual pots in a greenhouse at three different nutrient levels, low, medium, and high, with 67, 48, and 80 plants, respectively. Treatments were randomized on the bench with no further blocking. The plants were allowed to flower and set seed. From each plant, one flower head was collected (for simplicity of analysis), and the number of seeds and mean seed mass were determined. The results are presented in figure 6.1A. The data were analyzed using the following model:

$$(X_{1ij}\ X_{2ij})^{T} = \mu + \tau_i + \varepsilon_{ij} \qquad (6.1)$$

where $(X_{1ij}\ X_{2ij})$ is a vector of measurements of traits 1 (mean seed mass) and 2 (number of seeds) of the jth replicate in the ith treatment, T indicates vector transpose, μ is the grand mean of all measured individuals, τ_i is the deviation from the grand mean due to the ith treatment, and ε_{ij} is the deviation of the jth individual from the mean of the ith treatment. These last deviations (ε_{ij}) are assumed to be independently and randomly distributed with mean 0 and variance σ_e^2. These are the same assumptions as for ANOVA, except that the error variance is now assumed to be multivariate normal (see also section 6.4.3). In this instance, I am interested in whether differences in nutrient levels affect patterns of reproductive allocation.

The first step is to look for overall differences among the treatments. The basics of MANOVA are just like ANOVA except that, rather than comparing the means of groups, we compare their centroids. A centroid is a multivariate mean, the center of a multidimensional distribution. Thus, the programming steps are

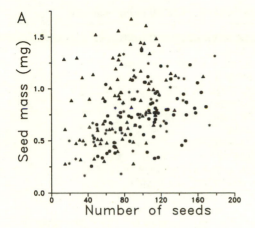

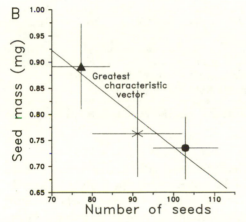

Figure 6.1 (A) Number of seeds and mean seed mass for each individual in the three nutrient treatments: high (●), medium (*), and low (▲). (B) Bivariate means and 95% confidence intervals for each treatment (note the change in scale from A). The line is the first eigenvector, the greatest characteristic vector.

very simple, although the actual computations are more complex: (1) choose SAS (SAS Institute Inc. 1989a,b) procedures ANOVA (for balanced data) or GLM (for unbalanced data or covariate analyses), (2) construct the usual MODEL statement, (3) include all relevant response variables on the left-hand side of the equation, and (4) add the MANOVA statement (appendix 6.1A). (See chapters 4 and 5 for details on constructing model statements.) As with ANOVA, the test for significant differences is based on examining the variation among groups to see whether it is larger than would be expected by chance given the variation within groups, using the standard F-statistic. The actual formula is based on the ratio of the among-group sums-of-squares/cross-product (SS&CP) matrix (**H**) divided by the pooled within-group (error) matrix (**E**) (table 6.1). The diagonals of these matrices are the among-group and error sums of squares of the univariate ANOVAs. SAS takes care of the actual calculations; for details see Harris (1985, section 4.2).

SAS (and other computer packages) supplies four statistics for the test of significant differences among the groups: Wilk's lambda, Pillai's trace, Hotelling–Lawley trace, and Roy's greatest root. The latter is often referred to as the greatest

Table 6.1 SAS MANOVA output from the program listed in the appendix[a]

GENERAL LINEAR MODELS PROCEDURE
MULTIVARIATE ANALYSIS OF VARIANCE
H = TYPE III SS&CP MATRIX FOR TREATMENT

	NUM	MASS
NUM	24250.912821	−150.8244582
MASS	−150.8244582	1.0084091912

E = ERROR SS&CP MATRIX

	NUM	MASS
NUM	218033.46667	829.14072
MASS	829.14072	18.269723565

MANOVA TEST CRITERIA AND F APPROXIMATIONS FOR
THE HYPOTHESIS OF NO OVERALL TREATMENT EFFECT
H = TYPE III SS&CP MATRIX FOR TREATMENT   E = ERROR SS&CP MATRIX
S = 2   M = −0.5   N = 94.5

STATISTIC	VALUE	F	NUM DF	DEN DF	PR > F
WILKS' LAMBDA	0.782757	12.442	4	382	0.0001
PILLAI'S TRACE	0.217648	11.723	4	384	0.0001
HOTELLING-LAWLEY TRACE	0.277018	13.158	4	380	0.0001
ROY'S GREATEST ROOT	0.275136	26.413	2	192	0.0001

NOTE: F STATISTIC FOR ROY'S GREATEST ROOT IS AN UPPER BOUND.
NOTE: F STATISTIC FOR WILKS' LAMBDA IS EXACT.

[a]Only those portions of the output directly relevant to the multivariate analysis are shown. **H** is the among-group matrix, and **E** is the within-group matrix.

characteristic root or the first discriminant function. All four measures are based on the eigenvectors and eigenvalues of the matrix derived from dividing **H** by **E**. Eigenvectors are linear combinations of all dependent variables [see Harris (1985) or Morrison (1990) for further discussion]. In the analysis of differences among groups, the first eigenvector is arrayed along the axis of maximal among-group variation (one can think of it as a least-squares regression of the group centroids) (figure 6.1B). It is the linear combination of variables that results in the greatest amount of among-group to within-group variation and, thus, the greatest *F*-value. Subsequent eigenvectors are each orthogonal (at right angles and uncorrelated) to the previous vector, with each arrayed along the axis of remaining maximal variation.

The number of eigenvectors will be equal to either the number of response variables (v) or one less than the number of groups (k − 1), whichever is smaller. [SAS actually gives you as many eigenvectors as dependent variables, but *if* v > (k − 1) then the last v − (k − 1) eigenvectors will explain 0% of the variation. For factorial or nested MANOVAs, the number of eigenvectors is based on the numerator degrees of freedom for the test of each factor rather than k − 1.]

The eigenvalues (λ) indicate the amount of variation explained by each eigenvector [$r^2 = \lambda/(1 + \lambda)$]. The first three MANOVA statistics, Wilk's lambda, Pillai's trace, and Hotelling–Lawley trace, are based on either the sum or the product of all eigenvalues. Roy's greatest root is the first eigenvalue. As can be seen from the output in table 6.1, the statistical conclusions about whether the groups differ are identical for all four measures, as will usually be the case. [Although SAS indicates that the F-statistic for Roy's greatest root is an upper bound, SYSTAT (Wilkinson 1988) uses a different calculation, which provides an exact probability (theta) value for this parameter.] When they differ, preference should be given to either Pillai's trace or Roy's greatest root. The former has been shown to be the most robust to violations of assumptions, whereas the latter has the greatest power. In addition, Roy's greatest root leads directly and naturally to post hoc procedures. See Harris (1985, section 4.5) for additional discussion.

6.3.2 Post hoc Procedures

Once a difference among groups has been established (table 6.1), two questions are immediately raised: (1) Which of the several groups actually differ from each other? and (2) Which traits and what relationships among them are responsible for the differences?

The choice of which statistical procedure to use to establish which groups differ depends on whether the sets of comparisons were decided on prior to analyzing the data. (In statistical terminology, such comparisons are called contrasts.) These comparisons can be made either between pairs of groups or between sets of groups. The procedure will also depend on what combination of response variables is used. The biggest difference in either case is based on whether the decisions on the form of the comparisons were made a priori or are a posteriori (after the fact). If they are a posteriori, then you must adjust the α-level used to determine significance and accept a decrease in power.

Typically, pairwise contrasts would be tested based on the linear combination of variables obtained from the overall analysis. This procedure will result in $k(k-1)$ different pairwise comparisons. A valid statistical test of these comparisons requires that the α-level be adjusted to correct for the large number of comparisons. The simplest correction to use is the Bonferroni correction mentioned previously. However, this procedure tends to be conservative, especially if the number of groups, and subsequently the number of potential comparisons, is large. For the univariate case, tests such as Scheffé's, Tukey's, and Student–Newman–Keuls have been developed to specifically correct for these multiple comparisons. The closest equivalent in the multivariate case is presented by Harris (1985, section 4.2). In his procedure, the critical value for the F-statistic is determined from the formula

$$F = df_{eff} \ (\theta/1 - \theta) \tag{6.2}$$

where df_{eff} is the numerator degrees of freedom from the MANOVA and θ is the critical value for the greatest characteristic root. A table of critical values is available in Harris (1985).

This procedure can also be used with other combinations of dependent variables that might be suggested by examination of the data. For example, an apparent trade-off between two traits or species (X_1 and X_2) could be tested by creating a new variable, the difference ($X_1 - X_2$) in values for each replicate, and then performing a univariate ANOVA. This is a valid a posteriori test as long as the critical F-value is properly adjusted. Such a procedure greatly strengthens the potential interpretive power of MANOVA and is especially appropriate if it is used to generate hypotheses for future experiments.

For the *Coreopsis* example, I performed two types of follow-up tests, univariate and multivariate. Such follow-up tests should be done only if the MANOVA indicates the existence of some significant differences. The univariate tests are not actually necessary, as I indicate, but are included here for comparative purposes (table 6.2A). The pairwise contrasts of the individual response variables, using a Bonferroni corrected $\alpha' = 0.05/3 = 0.017$, show that the high and low treatments differ from each other for both traits and that the medium treatment differs from neither other treatment for neither variable. (The same results are obtained using a Tukey's test.) These differences can also be seen by examining the overlap among confidence intervals (figure 6.1B). Thus, the differences among groups are ambiguous. More powerful are the multivariate follow-up tests (table 6.2B). The multivariate comparisons show that the high and medium treatments do not differ from each other and they both differ from the low treatment. The difference in results for the comparison of the low and medium treatments shows how univariate tests can be misleading. These conclusions would be the same using either a Bonferroni correction ($\alpha' = 0.017$) or the critical value from Harris ($F = 8.5052$). Note that the formula from Harris applies only to the greatest characteristic root, although when there are only two groups, the F- and P-values will be exactly the same for all four test statistics.

The comparisons were made using the CONTRAST option in the SAS procedure GLM (appendix). The alternative would be to break the data into three sets containing each pair of treatments and redo the MANOVA for each pair. (A two-sample MANOVA is also called a Hotelling's T^2-test, just as a two-sample ANOVA is the equivalent of a t-test.) The advantage of using the CONTRAST statement is the

Table 6.2 Pairwise contrasts of the three treatments (high = H, Medium = M, low = L) based on individual traits and the combination of both traits using the CONTRAST statements in the appendix

A. Individual traits

		Number of seeds			Seed mass		
		CONTRAST			CONTRAST		
CONTRAST	DF	SS	F VALUE	PR > F	SS	F VALUE	PR > F
H VS M	1	3915.907	3.45	0.0648	0.02139	0.22	0.6359
H VS L	1	24083.173	21.21	0.0001	0.88943	9.35	0.0026
M VS L	1	5768.533	5.08	0.0253	0.49557	5.21	0.0236

B. Multivariate analysis showing Roy's greatest root

CONTRAST	VALUE	F	NUM DF	DEN DF	PR > F
H vs M	0.02773	2.65	2	191	0.0734
H vs L	0.26597	25.40	2	191	0.0001
M vs L	0.09166	8.75	2	191	0.0002

increased power of the test. Note that the denominator degrees of freedom in the pairwise contrasts is 191, just 1 less than that for the full analysis. When contrasts are carried out, the pooled error SS&CP matrix is used, just as the pooled error variance is used in a univariate post hoc Tukey's test. Because this error matrix is pooled across all groups, the accuracy of the estimation of its elements is greater, resulting in a more powerful test. The appendix lists the contrast statements necessary to make these comparisons. If other types of comparisons are desired (such as the comparison of high and medium versus low), those contrast statements can also be added or substituted. See Neter and Wasserman (1974), Milliken and Johnson (1984), Hand and Taylor (1987), or Snedecor and Cochran (1989) for a discussion of how to construct contrast statements (see also chapter 15).

In some instances, all of the contrasts will be specified before the experiment is carried out. For example, we might plan to compare each of several treatments to a single control treatment. In that case, the critical F-value is determined as

$$F = \frac{v \cdot df_{err}}{df_{err} - v + 1} F \qquad (6.3)$$

where F is an F-statistic with degrees of freedom v and $df_{err} - v + 1$, v is the number of response variables, and df_{err} is the error degrees of freedom from the MANOVA.

Determining which combinations of response variables are responsible for the differences among groups can be done by examining the greatest characteristic vector and a related parameter, the standardized canonical variate, which are shown in table 6.3. Examining figure 6.1B, we see that plants grown at low nutrient levels are ripening fewer and heavier seeds than those grown at higher nutrient levels. In this case, there are two eigenvectors; the first explains 22% of the total variation (the "squared canonical correlation" reported in table 6.3B) or virtually all of the explainable variation (the "percent" reported in table 6.3A). That is not surprising as the three centroids nearly lie on a straight line (figure 6.1B).

The canonical analysis has the potential for providing a more complete picture of the statistical significance of the individual eigenvectors and is obtained using the CANONICAL option of the MANOVA statement (second MANOVA statement in the appendix.) SAS reports a test, given as the likelihood ratio, for the significance of the sum of the ith to nth eigenvectors. That is, if there were five eigenvectors, SAS would sequentially do the following: test all five together, then delete the first and test the remaining four, delete the second and test the remaining three, and so forth. The first test is the equivalent of Wilk's lambda (compare tables 6.1 and 6.3B). However, Harris (1985, section 4.5.3) cautions against the reliability of such a procedure. Generally, only the greatest characteristic root is of interest, the significance of which is tested separately as Roy's greatest root.

The numbers in the two right-hand columns in table 6.3A under the headings NUM and MASS are the coefficients of the eigenvectors. However, they are not very useful because the magnitudes of the coefficients are dependent on the scale of the data. A better comparison uses standardized coefficients, obtained by multi-

Table 6.3 The eigenvector (default) and canonical analyses of the MANOVA option from SAS procedure GLM showing the parameters and associated statistics of the characteristic vectors and canonical variates

A. Default eigenvector analysis from the first MANOVA statement in the appendix

CHARACTERISTIC ROOTS AND VECTORS OF: E INVERSE * H, WHERE
H = TYPE III SS&CP MATRIX FOR TREATMENT = ERROR SS&CP MATRIX

CHARACTERISTIC ROOT	PERCENT	CHARACTERISTIC VECTOR V'EV = 1	
		NUM	MASS
0.27513581	99.32	−0.00211225	0.19920128
0.00188219	0.68	0.00103995	0.16269926

B. Alternative canonical analysis from the second MANOVA statement in the appendix

CANONICAL ANALYSIS
H = TYPE III SS&CP MATRIX FOR TREATMENT   E = ERROR SS&CP MATRIX

	CANONICAL CORRELATION	ADJUSTED CANONICAL CORRELATION	APPROX STANDARD ERROR	SQUARED CANONICAL CORRELATION
1	0.464510	0.456687	0.056304	0.215770
2	0.043343	.	0.071661	0.001879

TEST OF H0: THE CANONICAL CORRELATIONS IN THE
CURRENT ROW AND ALL THAT FOLLOW ARE ZERO

	LIKELIHOOD RATIO	APPROX F	NUM DF	DEN DF	PR > F
1	0.78275689	12.4419	4	382	0.0001
2	0.99812135	0.3614	1	192	0.5484

STANDARDIZED CANONICAL COEFFICIENTS

	CAN1	CAN2
NUM	−1.0343	0.5092
MASS	0.8701	0.7107

BETWEEN CANONICAL STRUCTURE

	CAN1	CAN2
NUM	−0.9949	0.1008
MASS	0.9862	0.1657

plying each of the original coefficients by the standard deviation of each response variable. (This standardization is the equivalent of going from partial regression coefficients to standardized regression coefficients in multiple regression.) Unfortunately, SAS does not calculate the standardized coefficients for you; for these data, the standardized coefficients are −0.075 and 0.063 for number of seeds and seed mass, respectively.

Most informative are the standardized canonical coefficients of the canonical variates (table 6.3B). (In the SAS output, the canonical variates are arranged vertically, whereas the eigenvectors are arranged horizontally.) Canonical variates are simply eigenvectors scaled to unit variance by multiplying the eigenvector coefficients by the square root of the error degrees of freedom. Again, the most useful form is obtained by standardizing the coefficients by multiplying each by the standard deviation of the variable; in this instance, SAS does the standardization for you. For the first variate, the standardized coefficients for NUM and

MASS are −1.03 and 0.87, respectively (see table 6.3B under the heading CAN1). The opposite signs indicate that the two variables are negatively correlated across groups; that is, the pattern of resource allocation is changing across treatments. This contrasts with the positive correlation within groups (mean $r = 0.42$), which is reflected in the similar signs for the coefficients of the second variate. The greater magnitude of the coefficient for NUM indicates that NUM explains more of the variation among groups than does MASS. Interpreting eigenvector or canonical coefficients must be done cautiously because the magnitudes, and even the signs, can change, depending on what other response variables are included in the analysis.

SAS also provides information on the canonical structure of the data under three headings, TOTAL, BETWEEN, and WITHIN. The between canonical structure is of greatest interest and is shown in table 6.3B. These values are the correlations between the individual response variables and the canonical variates; they are sometimes referred to as canonical loadings. In this case, because the three centroids lie nearly on a single line, the magnitudes of the correlations for the first canonical variate are close to ±1. Whereas the canonical coefficients indicate the unique contribution of a given variable to the differences among groups (i.e., they are analogous to partial regression coefficients), the canonical correlations indicate how much of the variation each variable shares with the canonical variate, ignoring correlations among response variables. Such correlations can also be calculated for the eigenvector coefficients. The canonical correlations are touted as being more stable to sampling effects than the canonical coefficients; however, I feel that the canonical coefficients are more informative. See Bray and Maxwell (1985, pp. 42–45) or Stevens (1986, pp. 412–415) for a discussion of the use of canonical correlations versus canonical coefficients.

6.3.3 A Complex Example

The above example was simple and straightforward to interpret. As the number of response variables increases, interpreting differences among groups becomes more complex, and simple graphic examination becomes very difficult. In addition, the right-hand side of the equation, the independent variables, can also become more complex. Any model that can be used in ANOVA or ANCOVA (nested, factorial, and so on; see chapters 4 and 5 for examples) can also be used in MANOVA. In this example, I present the results from a $2 \times 2 \times 2$ factorial experiment. The data are part of a study on trophic interactions among inhabitants of pitcher-plant leaves in which changes were examined in the prey species' community structure as predator species changed (Cochran-Stafira and von Ende 1998). In this experiment, the presence of three protozoan predators (*Cyclidium, Colpoda*, and *Bodo*) was manipulated in laboratory test tube cultures. The response variables were the densities of four bacterial prey species (designated A, B, C, and D) after 72 hours of growth. There were three replicates per treatment; the design was completely balanced.

The ANOVAs (not shown) indicate that species A had a significant response to all treatment effects and interactions, species B did not respond significantly to any treatments, species C had a significant response to the main effects of *Colpoda* and *Bodo* and the three-way interaction, and species D had significant responses only to the main effects of *Colpoda* and *Cyclidium*. But this description does not capture the extent of the interactions among the species. Examination of the results of the MANOVA (table 6.4) indicates that all the main effects were significant, the *Colpoda–Cyclidium* interaction was significant, and the three-way interaction was significant. Thus, predator-species identity affected prey-species community structure. (Note that in determining the proper numerator and denominator matrices for the tests, the same rules as for ANOVA apply; in the current case, all factors were fixed so the **E** matrix was used for all tests.)

The canonical coefficients show that significant differences were primarily due to changes in the abundance of species A (figure 6.2). Additionally, the abundance of species D tended to be positively correlated with that of species A, whereas species C tended to be negatively correlated with both of them. The correlation of A and D was reversed, however, for the *Colpoda–Bodo* and *Cyclidium–Bodo* interactions. Finally, the full extent of changes in species C and D was more clearly revealed in the MANOVA as indicated, for example, by the relatively large canonical coefficients for the *Colpoda–Cyclidium* interaction, which was not significant for the ANOVAs of either species. This last is an example of how two variables that are correlated can be important when combined even though they are not individually significant.

Table 6.4 Multivariate analysis of the effects of protozoan species on bacterial densities[a]

A. Roy's greatest root

SOURCE	VALUE	F	NUM DF	DEN DF	PR > F
COLPODA	75.9066	246.7	4	13	0.0001
CYCLID	8.4839	27.6	4	13	0.0001
COLPODA*CYCLID	9.3905	30.5	4	13	0.0001
BODO	6.5291	21.2	4	13	0.0001
COLPODA*BODO	0.9522	3.1	4	13	0.0539
CYCLID*BODO	0.8017	2.6	4	13	0.0848
COLPODA*CYCLID*BODO	2.5207	8.2	4	13	0.0016

B. Standardized canonical coefficients

	A	B	C	D
COLPODA	7.2868	0.2261	−1.8530	2.4058
CYCLID	7.2648	0.5033	−2.5893	2.3589
COLPODA*CYCLID	7.5991	0.3505	−2.3772	2.1048
BODO	4.5085	−0.3285	0.9958	−0.2532
COLPODA*BODO	5.9543	−0.0961	−0.0045	−0.5063
CYCLID*BODO	4.5032	−0.2423	0.9595	−0.9633
COLPODA*CYCLID*BODO	6.9132	0.2715	−2.6776	1.3263

[a]The significance tests for Roy's greatest root and the standardized canonical coefficients for each main effect and interaction are shown. Because each main effect has only two levels (treatments), only one eigenvector exists for each effect.

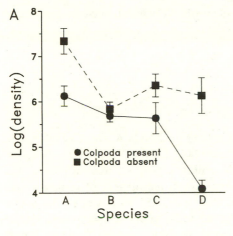

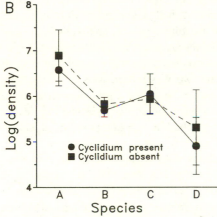

Figure 6.2 (A) Bacterial species densities (mean ± 95% CI) with and without *Colpoda*. (B) Bacterial species densities (mean ± 95% CI) with and without *Cyclidium*.

Several different types of ecological interactions could be responsible for these patterns. The bacterial species could be directly competing for resources. The bacterial species could be affecting each other—positively or negatively—indirectly through the production of secondary metabolites. The protozoans could also be producing secondary metabolites, creating indirect correlations; for example, as *Colpoda* eats species A and increases in density, the protozoan's waste products might benefit species C. At this stage, any ecological interpretations of the results must be considered tentative hypotheses to be tested by further experiments. The MANOVA indicates which potential interactions are worth pursuing, information not extractable from the ANOVAs.

In this case, because there are only two treatments (presence versus absence) for each main effect in the MANOVA, post hoc tests of the main effect differences are not necessary. However, the three-way effect is significant, and we might wish to determine which of the eight treatments differ from each other. No single best procedure exists. Beyond a simple examination of treatment means, Harris (1985) describes two alternatives: (1) using a method of contrasts (section

4.7) or (2) recoding the data as a one-way MANOVA followed by a series of appropriate contrasts (section 4.8).

6.4 Other Statistical Issues

6.4.1 Experimental Design and Power

The same basic design considerations that hold for ANOVA and ANCOVA, and discussed in detail in chapters 4 and 5, also hold for MANOVA. The one additional consideration is that of power, the ability to detect a true difference. The power of MANOVA declines with an increase in the number of response variables. Stevens (1992, sections 4.9, 4.10, 4.12, and 5.15) has a detailed discussion of power considerations in MANOVA, including tables for suggested sample sizes for different effect sizes; more extensive tables can be found in Lauter (1978). For example, the *Coreopsis* experiment had a 90% probability of detecting a small to moderate difference in the treatment means. (No actual formulas are available to determine power; rather, values are listed in the appropriate table.) On the other hand, the power for the protozoa–bacteria experiment was low, so that conclusions about the lack of significant two-way interactions involving *Bodo* must be viewed with caution. In designing an experiment, a priori power analyses are strongly recommended so that you will have the ability to detect the desired effect. If a conclusion of no differences is reached, a post hoc power analysis can assess the robustness of that conclusion.

Because the power of MANOVA declines with an increasing number of response variables, the distribution of effort among response variables versus replicates must be carefully considered. Very often experimenters will simply measure all possible response variables, examine them in a single huge analysis, and hope that something appears. In addition to problems of power, such an approach will often lead to results that are hard to interpret. A better approach is to start with clearly delineated hypotheses involving specified sets of response variables. Because not all of the response variables would be included in any single analysis, the power of the analysis would not be overly diminished. The interpretation of the results would be simplified because they would be compared with prespecified conjectures.

6.4.2 Missing Data and Unbalanced Designs

Missing data can take two forms in MANOVA: either the value of one or more response variables is missing for a subject (replicate) or an entire subject is missing. MANOVA can be performed only when all subjects have been measured for all response variables. If some measurements are missing, the entire subject is deleted from the analysis. Because we all are loathe to throw away data, one solution is to estimate the missing values using some form of multiple regression analysis of the other response variables. One assumes that the missing values do not deviate in any unusual way from the general distribution of the other values.

A graphical analysis (chapter 3) of the other values for that subject is recommended to determine whether that subject was either unusual or an outlier. Other information should also be considered (e.g., the data are missing because the animal was sick and died). Such procedures are generally robust as long as the missing data are sparse (<5% of all data, missing values distributed among all response variables, and generally no more than one missing value per subject).

When an entire subject is missing, estimating missing values is more problematic because very restrictive assumptions must be made about the entire covariance structure of the data; in particular, we must assume that it is uniform (as opposed to merely similar) across treatments or groups. However, estimating missing values is not necessary because SAS procedure GLM (and other computer packages) will handle unbalanced designs. In such a case, one must use Type III sums of squares as they are designed to account for unbalanced designs. Generally, unequal sample sizes are not a big problem for one-way MANOVA, but they might bias the results for factorial or nested designs. Unfortunately, little work has been done on this problem for MANOVA, thus statements about the robustness of ANOVA to unbalanced designs (Shaw and Mitchell-Olds 1993) should be extrapolated to the multivariate case with caution. As a rule, as long as designs are not too unbalanced (the smallest cell not less than 50% the size of the largest cell), results are reliable. As always, conclusions based on probability values close to the critical value ($0.01 < P < 0.10$) should always be tempered.

A relatively new approach, maximum likelihood (McLean et al. 1991; SAS procedure MIXED), has the advantage that having complete data for all subjects is not necessary, so missing values need not be estimated. However, assumptions about homogeneity of subjects (missing values are not unusual) still hold, so examination of the data before analysis and consideration of the reasons for the missing data are still important. A disadvantage is that computations for experiments with either many response variables or complex designs can take a very long time for MANOVA and sometimes will fail to converge on a solution. Be warned that the documentation of the technique is not completely clear, especially for those without extensive statistical training; consultation with a statistician is highly recommended.

6.4.3 Assumptions

The assumptions of ANOVA also hold for MANOVA: subjects are independent, all random effects (particularly the within-group or within-cell error effects) are normally distributed, and the variances of those error effects are equal among groups or cells (homoscedasticity). In addition, for MANOVA it is assumed that the error effects are multivariate normal and that the covariances are equal among groups. These latter two assumptions are just multivariate extensions of the last two assumptions from ANOVA. Unfortunately, it is virtually impossible to actually test for multivariate normality because of the very large sample sizes necessary. The best we can do is to test each of the dependent variables separately for univariate normality and then (reasonably) assume that multivariate normality holds. There are tests for equality of variance–covariance matrices, but statisticians are cautious about their use because (1) the tests are sensitive to departures from multivariate normality and (2) MANOVA is generally robust to deviations such that a significant departure from equality will be found before it will bias the conclusions. In particular, Pillai's trace has been found to be very robust.

My general advice is that if visual inspection of the data indicates overall agreement with the assumptions, go ahead and do the MANOVA. The obvious case where the assumptions are being violated is when the sign of the correlation between two dependent variables differs among groups. In that case, the most appropriate analysis strategies are to (1) specifically address questions regarding differences in correlations and (2) compare means using univariate analyses or multivariate analyses that do not use that particular pair of dependent variables simultaneously. The other situation in which the assumption of normality will likely be violated is in community-level studies that involve species that are not being directly manipulated (e.g., the presence of predators is manipulated while prey species freely move on and off plots). In that instance, several of the species are likely to have low or very variable abundances, resulting in a data matrix with many zeros. In such a case, no transformation will succeed in normalizing the distribution. One potential solution to the general problem of violation of assumptions is the use of randomization tests (chapters 7, 14, and 16), which are distribution-free. Such randomization tests for the case of a one-way MANOVA are presented by Mielke et al. (1981) and Zimmerman et al. (1985).

6.5 Related Issues and Techniques

I have discussed briefly the very complex subject of multivariate analysis. A number of books explain the basics and provide an entry into the more esoteric aspects of this subject. Nontechnical introductions can be found in Barker and Barker (1984), Bray and Maxwell (1985), and Manly (1986b). More detailed treatments are given in Harris (1985), Stevens (1992), and Morrison (1990). A complete mathematical treatment can be found in Johnson and Wichern (1988). In chapter 8, von Ende discusses another aspect of this topic, repeated-measures analysis, in which the same subject is measured more than once. He explains a procedure called profile analysis, which can also be used with MANOVA to dissect effects of particular variables (see also Simms and Burdick 1988). More complex situations in which the same subject is measured more than once for more than one trait, called doubly multivariate designs, can also be envisaged (see example 9 of SAS procedure GLM). In some instances, the response variables themselves may be further related in some cause-and-effect schema; in that case, the problem should be broken into a series of subanalyses as part of a path analysis (chapter 12). Multiple regression analysis involves independent variables consisting of several continuous effects. If there are also multiple response variables, the result is referred to as a canonical analysis. Such an analysis is actually the most general form of parametric statistical analysis, and all other procedures can be thought of as special cases of this general approach. See Gittens (1985) for a useful description of canonical analysis in an ecological context.

The basic message I have tried to convey in this chapter is that MANOVA is no more difficult, and often more informative, than ANOVA. Its implementation must be encouraged in ecology because the questions that we ask are very often multivariate ones.

Acknowledgments   I thank Liane Cochran-Stafira for kindly letting me use her data and assisting in the interpretation of its analysis. Discussions with Carl von Ende were invaluable for my understanding of MANOVA. Comments by Charles Goodnight, Jessica Gurevitch, Richard King, and Mark VanderMuelen helped improve the text.

Appendix SAS Program Code for Multivariate Analysis of Variance (MANOVA)

A. Analysis of the *Coreopsis* Experiment

```
PROC GLM;                              /use GLM for unbalance designs/
   CLASS TREAT;                        /TREAT = treatments/
   MODEL NUM MASS = TREAT/SS3;         /use type III sums of squares
                                        for unbalanced designs/

   CONTRAST 'H VS M' TREAT -1 1 0;     /contrast statements for
   CONTRAST 'H VS L' TREAT -1 0 1;      obtaining the tests of
   CONTRAST 'M VS L' TREAT 0 -1 1;      pairwise differences/
   MANOVA H = SITE/PRINTH PRINTE;     /MANOVA statement for obtaining the
                          multivariate and the coefficients analysis of the greatest
                              characteristic roots. PRINTH and PRINTE request
                              the among-group and within-group SS&CP matrices/
   MANOVA H = SITE/CANONICAL;     /alternative MANOVA statement for obtaining
                              the multivariate analysis and the canonical coefficients/
```

B. Analysis of the Protozoan–bacteria Experiment

```
PROC ANOVA;
   CLASS COLPODA CYCLID BODO;
   MODEL A B C D = COLPODA|CYCLID|BODO;     /the "|" specifies a model with
                                             all possible interactions/

   MANOVA H = _ALL_/CANONICAL;             /the "_ALL_" requests multivariate
                                             tests for all model effects/
```

7

ANCOVA

Nonparametric and Randomization Approaches

PETER S. PETRAITIS

STEVEN J. BEAUPRE

ARTHUR E. DUNHAM

7.1 Ecological Issues

Ecological data often fail to meet required parametric assumptions. When this occurs, randomization approaches can provide a good alternative to more familiar parametric methods such as analysis of covariance (ANCOVA) and regression analysis. Randomization methods are quite easy to use, and because standard parametric ANCOVA is well known to ecologists, we use it to motivate our discussions of the benefits and problems associated with using nonparametric and randomization methods. We do so by examining randomization and nonparametric approaches to the analysis of sex-specific and site-specific variation in body size among populations of rattlesnakes, with age as a confounding factor.

Variation among populations in body size is common in many animals (e.g., invertebrates: Paine 1976; Lynch 1977; Sebens 1982; Holomuzki 1989; amphibians: Nevo 1973; Berven 1982; Bruce and Hairston 1990; squamate reptiles: Tinkle 1972; Dunham 1982; Schwaner 1985; Dunham et al. 1989; mammals: Boyce 1978; Melton 1982; Ralls and Harvey 1985) and has been of great interest to evolutionary ecologists because body size covaries with many reproductive traits, such as age at maturity, number and size of offspring, and amount of maternal investment in offspring (Stearns 1992; Roff 1980, 1992). Explanations for variation in body size include differences in resource seasonality, quality, and availability (e.g., Case 1978; Palmer 1984; Schwaner and Sarre 1988), size-specific predation (Paine 1976), population density (Sigurjonsdottir 1984), character displacement (Huey and Pianka 1974; Huey et al. 1974) and clinal variation in developmental rate (Roff 1980). Geographical variation in body size, however, may often result from the interplay of size-specific growth rates and population age structure. King (1989), for example, suggested that population differences in age structure were an important aspect of the variation in body size of the water snake *Nerodia sipedon insularm*. Therefore, making sense out of temporal and geographic patterns in body size and ultimately growth rate requires knowledge of, and correction for, the ages of animals so that animals of similar age may be compared.

Traditionally, patterns of growth and sexual size dimorphism in reptiles have been analyzed by using nonlinear growth modeling techniques (Andrews 1982; Stamps 1995). Accurate fit of nonlinear models requires large samples with observations well distributed over the total size range, a requirement that is frequently not met in field studies (chapter 10). Furthermore, because a separate model is fit to each line, the form of the best-fit model (e.g., von Bertalanffy versus logistic by length, or others) may vary, complicating comparisons. Likewise, as fitted parameters are compared among several groups, the probability of Type I error increases, in a manner analogous to multiple pairwise *t*-tests.

For small to moderate data sets with multiple groups for comparison, ANCOVA using age as the covariate seems the best alternative approach for comparing body size among groups. Yet field data on body size and age are often messy. Analyses frequently require adjustment for one or more covariates, and residuals of fitted models rarely meet the assumption that they are independently, identically, and normally distributed (Sokal and Rohlf 1995; Zar 1996).

Conventional nonparametric statistics based on ranks or other types of randomization tests can provide good alternatives to parametric analyses. The parametric analysis assumes errors are normally distributed, an assumption that is relaxed in nonparametric tests based on ranks and other randomization tests. On the other hand, parametric procedures, randomization procedures, and nonparametric tests based on ranks all require the errors to be independently and identically distributed. Randomization methods and conventional nonparametric tests are sensitive to heterogeneity of variances, and it is a common misconception that the problem of heterogeneity of variances can be solved by using nonparametric tests (Hayes 1996).

Typical nonparametric tests use ranks of the original data; the null hypothesis specifies that ranks are randomly assigned across treatment levels. For small samples, the exact probability of the observed ranking can be calculated because all possible rankings can be enumerated. Thus, a conventional nonparametric test is a randomization test of the ranking of the original observations. For large sample sizes, calculation of significance levels for most commonly used nonparametric tests are estimated using a χ^2-distribution. The χ^2-distribution of the test statistic is based on the assumption that ranked data for each treatment level are samples drawn from distributions that differ only in location (e.g., mean or median). The underlying distributions are assumed to have the same shape (i.e., all other moments of the distributions—the variance, skewness, and so on—are identical). These assumptions about nonparametric tests are often not appreciated, and ecologists often assume that such tests are distribution-free.

Other types of randomization tests are based on a reshuffling of the original data (chapter 14). These tests also require assumptions about the population distribution. There is often confusion about which procedures constitute randomization tests and which constitute permutation tests. Kempthorne and Doerfler (1969) use the term *permutation* for tests based on all possible orderings of the data. Randomization tests typically use only a randomly chosen subset of all possible permutations. Conventional nonparametric tests, in the strict sense, are permutation tests.

In the next section, we discuss the advantages and disadvantages of parametric, nonparametric, and randomization approaches to problems normally analyzed using ANCOVA. To illustrate these issues, we use data on sexual and geographical variation in body size in the mottled rock rattlesnake (*Crotalus lepidus*). There are very few data on sexual dimorphism in snakes. Beaupre (1995) undertook a study of sexual dimorphism in the mottled rattlesnake at two sites in Texas. After adjusting for age, he found that females were significantly smaller than males at both sites, and snakes from the low-elevation site were significantly smaller than those from the high-elevation site (Beaupre 1995). He also detected a significant site by sex interaction. Beaupre used nonparametric methods because significant departures from normality were detected.

7.2 Statistical Issues

7.2.1 The Data

The data set we used consists of mark-recapture observations of the age and size of male and female *Crotalus lepidus* from two populations at different elevations in Big Bend National Park, Texas, collected over a 6-year period. Our data set is not identical to the set used by Beaupre (1995). We included 87 observations of males and females of which 33 were recaptures. Beaupre (1995) had 99 observations with 31 recaptures. A more detailed description of the data may be found in Beaupre (1995). Relative age of each captured snake was estimated from rattle morphology (i.e., number of rattle segments adjusted for shedding frequency; see methods in Beaupre 1995), and snout-vent length (SVL) was used as an estimate of body size. There are four variables: site (Boquillas and Grapevine Hills), sex, relative age, and size (SVL). Site and sex were treated as fixed effects, with age as the covariate. Site was considered a fixed effect because of our interest in microclimatic effects that result from elevational differences between these two particular localities (Dunham et al. 1989).

7.2.2 Conventional ANCOVA

Tests for significant main effects (sex and site, in this case) and their interaction can be analyzed as a two-factor ANCOVA, with body size as the dependent variable and age as the covariate. Before the ANCOVA is run, tests of the assumption of homogeneity of slopes should be carried out. These are tests of the similarity of linear dependence of body size on age among the treatment levels. If the homogeneity of slopes criterion is met, the ANCOVA procedure is valid.

ANCOVA is known to be robust to minor violations of model assumptions, especially in tests of significance for fixed effects. Under most situations, ANCOVA is the favored parametric approach. However, serious violations of the assumptions are often encountered with data on field-captured animals.

First, the dependent variable, body size, may not meet the assumptions of parametric statistics. Distributions of body size in populations of reptiles are often highly skewed, and the variances of distributions of male and female body size can differ considerably (e.g., Beaupre et al. 1998). Thus, it is highly unlikely that the errors are normally distributed as required for parametric analysis. Second, the covariate, age of each snake, is not known precisely, yet the use of ANCOVA, as with other model I regression methods, assumes that the error in measurement of the covariate is small. The estimation of age of field-caught animals, even under the best of circumstances, is problematic. In most cases, ecologists use a proxy for age and assume that the proxy has a linear, or at least monotonic, relationship with age. Model I regression may be used even if the independent variable is measured with error as long as the error distribution for the independent variable (or covariate, in the case of ANCOVA) is much narrower than the error distribution of the dependent variable (LaBarbara 1989). It may often be the case, however, that the age of field-caught animals is estimated with as much or more uncertainty than body size and, as a result, we would expect age to have a larger measurement error than body size. Third, factorial designs involving field-caught animals are rarely balanced. It is usually impossible to capture the same number of males and females at every site. Unbalanced ANOVAs and ANCOVAs are very sensitive to heterogeneity of variances, which may be an issue when comparing males and females.

Experimental ecologists usually try to correct for these difficulties, which involve assumptions about the parametric model itself. The most common approach is to transform the dependent variable to make the error variances homogeneous and to use Type III sums of squares for unbalanced designs. Most hope that the error distribution of the covariate is narrow enough. There are some cases where the covariate itself is transformed in a mistaken attempt to reduce its variance; however, such transformations of the covariate should only be applied when relationships require linearization.

Although it corrects one problem, transforming the dependent variable can create another. For example, body size may be transformed to reduce the heterogeneity among the error variances of the different treatment levels, but the transformation may make the error distributions nonnormal. Transformations also change the relationships among the dependent and independent variables. Log transformation of body size may reduce the heterogeneity of the variance and normalize the error distributions, but the transformation changes the model from one with additive effects to one with multiplicative effects. This can be a serious problem, particularly in experiments where ecologists used the tests of interactions within ANOVA and ANCOVA to make inferences about "nonadditive" ecological effects, such as higher order interactions (Wootton 1994).

The difficulties of unbalanced designs can be easily solved by throwing out data. Balancing the design reduces the effects of heterogeneous variances. Most ecologists, however, loathe discarding hard-won information. A potential pitfall of discarding data is that the reduced data set may lead to a significant loss of power. The advantages of creating a balanced design by discarding data rarely outweigh the disadvantages.

7.2.3 Nonparametric Approaches

Two general approaches to nonparametric analysis of covariance have been developed. The first, termed *matched pairs* involves the restriction of the data set to pairs matched for the value of the covariate and the generation of a transformed data set based on differences (Quade 1982). The matching approach involves some arbitrariness in determining which data values constitute a matched set, and it apparently has not been generalized beyond a one-way analysis.

The second approach, formalized by Shirley (1981), is a nonparametric ANCOVA based on ranking the dependent variable. Shirley's approach is based on the work of Bennett (1968), who developed nonparametric tests of general linear hypotheses for ranked data. For two-factor ANOVA, the most familiar example of Bennett's test is the Scheirer–Ray–Hare test (Scheirer et al. 1976), which is an extension of the Kruskal–Wallis test. The assumption of normality is relaxed in conventional nonparametric tests based on ranked data. It is not widely recognized that nonparametric tests, such as the Kruskal–Wallis test, may not detect true differences in location (i.e., differences among average rankings) if the groups being compared differ in scale (i.e., in variance) or in shape (Lehmann 1975). In extreme cases, the covariate may also need to be ranked (Shirley 1981).

Nonparametric ANCOVA is performed like any ANCOVA except that the ranks of the observations are used as the dependent variable. As usual, ties are assigned the average of the spanning ranks. As with the standard ANCOVA (chapter 5), two models are used: (1) the full model with the interaction between the covariate and treatment effects as a test of slope heterogeneity (called the homogeneity-of-slopes model in SAS; equation 5.2) and (2) the model without these interactions for tests of the adjusted averages (called the analysis-of-covariance model in SAS; equation 5.3).

The test statistic for a fixed effects model is the sum of squares for the appropriate main effect or interaction divided by the total mean square (i.e., the total sums of squares, SS, divided by the total number of degrees of freedom). The test value is compared to a critical value from a χ^2-distribution at the desired α-level and degrees of freedom of the effect under consideration. Using a χ^2-distribution to determine the significance level gives us an approximation based on the assumption that the central limit theorem applies to the ranked data (Lehmann 1975). This can be safely assumed only if the sample size is large and there are few tied ranks. Procedures for post hoc comparison of adjusted mean ranks are described by Shirley (1981, 1987).

The test statistics are χ^2-distributed, not F-distributed, because the parametric variance is known for ranked data (Mood and Graybill 1963; Lehmann 1975; Sokal and Rohlf 1995). The parametric variance equals $N(N + 1)/12$ where N is the total number of observations in the experiment. If there are no ties, then the total SS from the ANCOVA of the ranked data divided by the total degrees of freedom equals the parametric variance (e.g., see Sokal and Rohlf's 1995 discussion of the Scheirer–Ray–Hare test). The parametric variance must be corrected if there are ties; the corrected parametric variance is $[N(N+1)/12] - C$, where $C = (t_i^3 - t_i)/12(N-1)$, summing over i from 1 to S, S is the number of sets of tied values, and t_i is the number of tied values in the ith set. The total SS/total df = $[N(N+1)/12] - C$. Note that C is different from the correction D found in Sokal and Rohlf (1995, box 13.6), but it is easily shown that $DN(N+1)/12 = [N(N+1)/12] - C$. The two equations give the same result.

7.2.4 Randomization Approaches

Randomization tests are carried out by randomizing the observations and recalculating the appropriate test statistic many times, generating a distribution of possible outcomes. If all possible outcomes are enumerated, a randomization test is a permutation test. Hypothesis tests are conducted by direct estimation of the probability of the observed ordering based on the distribution of randomly derived outcomes (Manly 1997). The outcomes of a parametric test and a randomization test are asymptotically equivalent if the data meet the assumptions of the parametric model. See chapters 14 and 16 for other examples of randomization tests.

The test statistic used in a randomization test need not be a conventional one, such as a t-or F-statistic (Manly 1997). For example, with ANOVA and ANCOVA, treatment mean squares or sums of squares may be as suitable as F-statistics. In one-way ANOVAs, the distribution of F-statistics and sums of squares based on randomizations differ by a constant factor. This is not true for more complex designs. Edgington (1995) favors using sums of squares, but Manly (1997) favors using the F-ratio because his simulations suggested that randomizations based on sums of squares tended to have lower power. We will show subsequently that sums of squares and the F-ratio will often give different answers because they test different hypotheses. It is not simply a matter of differences in power.

A more difficult question than the choice of the test statistic is how to randomize the observations. There are two different approaches for factorial designs, depending on the null hypothesis (Manly 1997). On one hand, suppose our null hypothesis about differences due to sex and site is based on the assumption that the observation of a particular snake body size for any sex × site combination is drawn from a single population. To the extent that this assumption is true, we expect that any observation could be drawn from any sex × site combination and, thus, we should randomize the observations across all cells. This approach is advocated by Manly (1997) because of its ease of calculation, and his simulations suggest that results are similar to the second approach. On the other hand, we could assume that sex and site should be tested independently. Thus, we would test for the difference between male and female body size while holding constant the effect of site. To do this, we randomize the observations between sexes but within each site separately, and observations from the two sites are not mixed. This approach, known as restricted randomization, is advocated by Edgington (1995).

Two approaches can be used to randomize the residuals rather than the original data. Ter Braak (1992) suggests calculating the residuals for each observation from the overall model and then randomizing the residuals. In our example, the residual body mass would be adjusted for the effects of sex, site, sex × site interaction, and age, the covariate. Either complete or restricted randomizations could be used. A hybrid approach was suggested by Still and White (1981) in which the main effects are tested using an overall randomization and the interaction effects are tested using ter Braak's method by randomizing the residuals from the overall main effect model.

7.3 Statistical Solutions: Comparison of Several Approaches

7.3.1 Parametric ANCOVA

Recall that the data are snout-vent length (SVL) of male and female rattlesnakes from two sites in Texas. The covariate is age, which is estimated from the number of rattle segments. The null hypotheses are as follows:

1. no differences in adjusted body size (i.e., SVL adjusted for rattle count) due to sex or site, and
2. no sex × site interaction. Site and sex are considered to be fixed effects.

The complete data set consisted of 87 observations, of which 33 were recaptures, so, to avoid nonindependence, each animal was used only once in the analysis. A single observation for each individual was randomly drawn from the complete data set. This gave an unbalanced design with 54 rattlesnakes (figure 7.1).

Preliminary analysis revealed significant slope heterogeneity between sites (table 7.1). The ANCOVA showed significant effects of sex, site, and age. There was no detectable interaction between sex and site. These results must be viewed with caution for several reasons. First, the heterogeneity of slopes violates the assumptions of ANCOVA and suggests that the interpretation of differences in body size between the two sites—Boquillas and Grapevine Hills—depends on the age of the snake. Second, the covariate, which is the number of rattle buttons as a proxy for age, is likely to have substantial measurement error. Third, the residuals show some signs of heterogeneous variances and nonnormality. Plots of residuals exhibit a systematic increase in the residual with increasing SVL (figure 7.2A) suggesting some heterogeneity of variances. Tests of the residuals for heterogeneity of variances were not significant (Levene's test, $P = 0.201$; Bartlett's test, $P = 0.087$), but lack of significance may be due to the low power of the tests. More important, the distribution of residuals with respect to values of the covariate is hump-shaped (figure 7.2B), suggesting that the residuals fail to meet the assumption of normality. Log_{10}-transformation of SVL did not improve the residual plots.

These observations suggest that the data do not meet the assumptions of parametric ANCOVA. Clearly, another approach to analyzing these data is necessary.

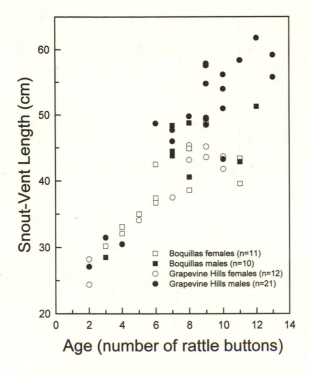

Figure 7.1 Relationship between SVL and age for groups of rattlesnakes defined by site and sex. Numbers in parentheses are sample sizes.

7.3.2 Nonparametric ANCOVA

We repeated Beaupre's (1995) analysis and applied Shirley's technique for nonparametric ANCOVA on the ranked SVL. First, we note that there are three sets of tied ranks, with two observations in each tie. The correction for the parametric variance equals 0.028 (table 7.2). Obviously, the correction is very small and will have little effect on the outcome of the analysis. The importance of tied values increases with small sample size and the number of tied values.

Table 7.1 Results of tests of homogeneity of slopes and ANCOVA

Test	df	SS	MS	F	P
Slope homogeneity					
Age × sex	1	13.07	13.07	0.77	0.384
Age × site	1	109.79	109.79	6.48	0.014
Age × sex × site	1	<0.01	<0.01	<0.01	0.989
Error	46	779.46	16.94		
ANCOVA					
Sex	1	218.13	218.13	11.53	0.001
Site	1	87.61	87.61	4.63	0.036
Sex × site	1	48.98	48.98	2.59	0.114
Age	1	2101.08	2101.08	111.06	<0.001
Error	49	927.01	18.19		

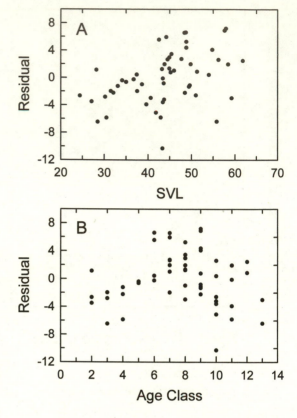

Figure 7.2 Plots of residuals
from the parametric
ANCOVA. The top panel
(A) suggests systematic
changes in the variance of
model residuals with in-
creases in SVL. The bottom
panel (B) shows distinct nonnormality of residuals over
the values of the covariate
(age class).

Table 7.2 Results of the nonparametric ANCOVA on ranked SVL[a]

Tests	df	SS	MS[b]	χ^2	P[c]
Slope homogeneity					
Age × sex	1	72.79	72.79	0.294	0.588
Age × site	1	293.40	293.41	1.186	0.276
Age × site × sex	1	0.60	0.60	0.002	0.961
Total	53	13116.00	247.47		
ANCOVA					
Sex	1	1070.21	1070.21	4.325	0.038
Site	1	525.62	525.62	2.124	0.145
Site ×T\sex	1	95.86	95.86	0.387	0.534
Age	1	4852.34	4852.34	19.608	<0.001
Total	53	13116.00	247.47		

[a]Test values = SS of an effect divided by the total MS. For example, for age × sex test of heterogeneity, 72.79/
247.47 = 0.294.
[b]Total MS equals the uncorrected parametric variance minus the correction. The uncorrected parametric variance = $N(N + 1)/12 = 54(55)/12 = 247.500$. There were three two-way ties, so $C = 18/12(54 - 1) = 0.028$. Thus
$247.5 - 0.028 = 247.475$, which is the total MS.
[c]Probability levels are calculated using the PROBCHI function in SAS (e.g., $P = 1 - PROCHI(0.294,1)$ give
$P = 0.587$.

The nonparametric analysis detected no heterogeneity of slopes among groups (table 7.2). The ANCOVA revealed significant effects of age and sex. There was no detectable effect of site and no interaction of site × sex (table 7.2). It is noteworthy that Beaupre's (1995) original analysis using Shirley's method (Beaupre 1995) detected significant effects of age, site, sex, and site × sex interaction. The differences in results are most likely due to the use of two different subsamples of data; we did not use the same set of observations used by Beaupre (1995). The differences are unsettling and raise the question of whether the subsets used by us or by Beaupre (1995) are representative of the population as a whole. We show later in this chapter how randomization procedures can be utilized to address this question.

7.3.3 Randomization Tests

Randomization tests are quite easy to carry out in SAS. It is possible to write a seamless listing of SAS code so that randomization tests can be carried out in a single run (see also chapter 14), but for clarity of presentation, we will break the analysis into four distinct steps. First, a conventional analysis must be run to obtain the observed values for the test statistic. Table 7.1 gives the SS and F-ratios for the conventional ANCOVA for SVL of rattlesnakes with age (the number of rattle segments) as the covariate and the tests for homogeneity of slopes. Second, the data must be randomized a large number of times and stored in a form that can be used by SAS. Third, the appropriate SAS procedure (in this case, procedure GLM) is run many times using the randomized data sets. Finally, the large number of iterations must be summarized. See the appendix and the Website [http://www.oup-usa.org/sc/0195131878/] for more details about writing the SAS code.

The second step, randomization of the data, can be done in a variety of ways. We chose to assign observations randomly to the four site × sex cells with the restriction that the design remained unbalanced in the same fashion as the original data set. A thousand or more sets of randomized data can be created very quickly with a number of programming languages. Sets of randomized data can also be created in SAS, but this tends to be slower than doing the randomization using BASIC or some other programing language and then importing the data sets into SAS. On the Website, we demonstrate one way to do a randomization within SAS that is easy to code and comprehend. The biggest problem with conducting the randomizations entirely within SAS is the sorting of large data files. Procedure SORT maintains two complete files during a sort, and it is very easy to run out of memory if the data file is large and if the analysis is done on a personal computer. See the Website for more details.

The randomizations were done differently for each of the tests of homogeneity of slopes, the test of the covariate, and tests of the effects of sex and site. The tests of sex, site, and site × sex interaction are concerned with the effect of treatment levels on SVL when adjusted for age. Here we preserved the observed pairing of age and SVL for each snake, but we randomized snakes across treatments. We did this because we assumed that each snake with its own unique combination of SVL and age was the unit of interest. The tests of sex, site, and site × sex interaction thus hold the effects of age constant. In contrast, for tests for the covariate and homogeneity of slopes, we randomized age with respect to SVL but did not randomize SVL with respect to the treatment classes. This restricted randomization tests the hypothesis that there is no effect of age on SVL while holding the effects of sex and site constant.

Tests for homogeneity of slopes were not significant across sex and across site × sex, but they were significant across sites (table 7.3). The probabilities from the F-ratio randomizations tend to match the probabilities from the parametric analyses, whereas those from the SS randomization are always greater. Thus, slopes between sites are homogeneous based on the SS randomization but are heterogeneous based on the F-ratio randomization.

The parametric and randomization ANCOVAs gave similar results for the effect of sex (always significant) and the site × sex interaction (never significant) but not for the effect of site (table 7.3). The effect of site was significant at $P < 0.05$ for the F-ratio randomization but was not significant for the SS randomization. Figure 7.3 shows the distribution of the F-ratio and the error mean square for the test of the effect of site. Although the mean square from the parametric analysis (18.19) is in the lower tail of the distribution generated by the randomization, this is not the cause of the discrepancy between the F-ratio and SS randomizations. F-ratio randomizations tend to give probability levels closer to those from the parametric analysis. As shown in figure 7.4, probability levels based on the SS randomizations were larger.

Age was significant in all analyses. Neither the F-ratio nor the SS randomization gave a single case greater than or equal to the observed test statistic, but this is not surprising since the probability from the parametric analysis was <0.0001. It is possible that with more iterations (>10,000), F-ratio and SS randomizations would differ, but there is no clear advantage of doing more iterations since the probability levels with 1000 iterations are so small.

Table 7.3 Probability levels from parametric, nonparametric, and randomization analyses[a]

			Randomization	
	Parametric	Nonparametric	Based on F	Based on SS
Slope homogeneity				
Age × sex	0.384	0.588	0.338	0.607
Age × site	0.014	0.276	0.006	0.112
Age × sex × site	0.989	0.961	0.989	0.992
ANCOVA				
Sex	0.001	0.038	0.001	0.002
Site	0.036	0.145	0.038	0.063
Sex × site	0.114	0.534	0.129	0.192
Age	<0.001	<0.001	0.001	0.001

[a]Probabilities for randomization tests are based on 1000 iterations plus the original observed values (see table 7.1).

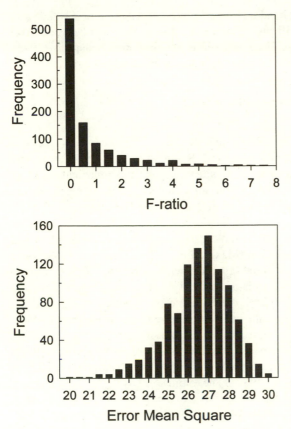

Figure 7.3 Distributions of F-ratios for the site effect and error mean squares from 1000 randomizations. The observed value for the error mean square in the parametric analysis (18.19) is well below all 1000 randomizations.

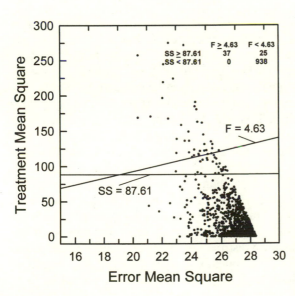

Figure 7.4 Treatment mean squares for the site effect versus the error mean squares from the 1000 randomizations. Lines for SS = 87.61 and F = 4.63 delineate $P = 0.036$, which is the significance level of the test of a site effect in the parametric ANCOVA (see table 7.1).

With respect to the data at hand and based on the distribution of F-values, we conclude that there were significant differences between sites and between sexes in SVL when adjusted for age. Our conclusions are similar to Beaupre's (1995), although we did not detect a significant site × sex interaction as he did.

7.3.4 Causes for the Differences Between F and SS Randomizations

The randomizations based on the F-ratio and the SS differed because in factorial designs they test slightly different hypotheses. Randomization with SS as the test statistic tests the effect of one factor (e.g., site), ignoring the contributions of all other effects (e.g., sex and site × sex interaction). In contrast, using the F-ratio as the test statistic tests the effect of site while partitioning the effects of all other factors.

The difference between the two is most easily expressed in terms of partial and unpartial measures of association (Maxwell et al. 1981; Petraitis 1998). As a test statistic, SS is related to η^2 (eta squared), which Friedman (1968) recommended as a measure of the strength of association (i.e., the proportion of explained variation; see Petraitis 1998). The estimate of η^2 equals $SS_{Treatment}/SS_{Total}$, and η^2 reflects the strength of a treatment effect relative to the total variance (Maxwell et al. 1981). Since SS_{Total} is constant, randomization of SS for any treatment is, in effect, identical to a test for η^2. In contrast, the F-ratio is related to a partial measure of association, $\eta^2_{Partial}$, which is a measure of a treatment effect relative to the error variance (Maxwell et al. 1981). The estimate of $\eta^2_{Partial}$ equals $SS_{Treatment}/(SS_{Treatment} + SS_{Error})$. In terms of the F-ratio in a two-way ANOVA, $\eta^2_{Partial}$ equals $(df_{Treatment})F/[(df_{Treatment})F + df_{Error}]$, where $df_{Treatment}$ and df_{Error} are the degrees of freedom for the treatment effect and the error, respectively. Figure 7.4 shows how the treatment and error sums of square covary to produce the differences between the F-ratio and SS randomizations.

The choice of SS versus F-ratio depends on exactly what you wish to test. If you wish to test the effect of a treatment regardless of the other factors in the design, then use the SS. If you wish to hold the effects of other factors constant, then use the F-ratio. In the tests for homogeneity of slopes, for example, the SS tests are equivalent to tests of simple regression coefficients, and the F-ratio tests are equivalent to tests of partial regression coefficients (table 7.3). The outcomes for the SS and F-ratio randomizations will differ only in complex designs; the two test statistics give identical results for one-way ANOVAs.

One additional concern with the use of the F-ratio as a test statistic is the estimate of the error mean square. Estimates of the error mean squares from randomizations tend to be much larger than the error mean squares from the original parametric analysis. In our randomizations, the distribution of the error mean squares did not include the error mean square estimate from the parametric analysis (figure 7.3). As a result, the F-ratio is small when the estimate of the error mean square is large. The broad distribution for the error mean square arises from the fact that randomizations in complex designs alter not only the SS of the treatment under consideration (e.g., SS_{Site}) but also the other SS values (e.g.,

SS_{Sex} and $SS_{Site \times Sex}$). Recall that the effect of site, $F = (49)SS_{Site}/SS_{Error} = (49)SS_{Site}/(SS_{Total} - SS_{Site} - SS_{Sex} - SS_{Site \times Sex})$. Randomization breaks up the correct assignment of observations and thus tends to reduce all of the treatment SS values (i.e., SS_{Site}, SS_{Sex}, and $SS_{Site \times Sex}$). Since the total sum of squares remains constant, the estimate of the error mean square increases with a decrease in the other sums of squares. Restricted randomization has been suggested as a solution to this problem (Manly 1997), but it does not completely solve it. For example, under restricted randomization for the effect of site, SS_{Site} would vary because of the randomization and SS_{Sex} would remain constant because of the restriction. The interaction sums of squares, however, would still vary because the interaction of site and sex would be altered with each randomization, and so the estimate of the error mean square would remain large and variable.

It is not clear how the distribution of the error mean square would vary in complex designs and for data in which the errors are not normally distributed. Also, given that randomizations based on the F-ratio and SS are testing slightly different hypotheses, we suggest that the error mean square estimates should always be examined; if these estimates are variable, plots of treatment versus error mean squares should be made (e.g., figure 7.3). Choose the appropriate statistic with care, basing your decision not only on the data, but also on the hypothesis under consideration.

7.4 Related Issues

7.4.1 Is There Something Unusual About the Chosen Subset of Data?

The results of our nonparametric analysis differ from the results of Beaupre (1995), who used a nearly identical data set but drew a different subset of observations for his analysis. Is it possible that we or Beaupre chose an unusual set of observations drawn from the complete set? One advantage of randomization methods is that this possibility can be checked by randomly selecting different subsets of observations.

We did the randomization in two ways. In the first case, 1000 sets of 54 different observations were selected from the whole data set of 88. Each observation was placed in the sex × site cell to which it rightfully belonged (e.g., an observation for a male from Grapevine Hills was placed in the male–Grapevine Hills cell). Thus we had 1000 sets of data in which a conventional ANCOVA could be carried out, and we asked whether our original selection of data (figure 7.1) was unusual when compared to these 1000 sets. In the second case, 1000 sets of 54 different observations were again selected, but the 54 observations were assigned to treatment cells at random. This assignment is equivalent to a single randomization for each random selection. With these 1000 randomizations, we are asking whether the distribution of the error mean squares seen in the original randomization (figures 7.3 and 7.4) is unusual.

Both randomizations suggest that the selection of the original subset of data and the patterns seen in the mean squares were not unusual. When observations

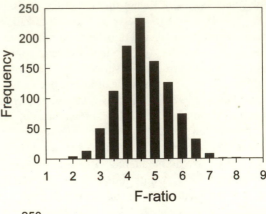

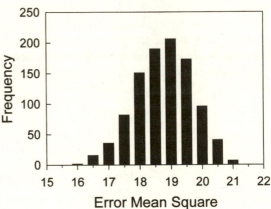

Figure 7.5 Distributions of *F*-ratios for the site effect and for the error mean squares from 1000 randomizations in which observations were selected from the complete data set and assigned to the correct cell. The observed values for the *F*-ratio and error mean square from the parametric analysis ($F = 4.63$ and error MS = 18.19) are very close to the averages of the distributions.

were assigned to the correct treatment cell, the averages for the *F*-ratio and the error mean square were within 5% of the values observed in the original ANCOVA (compare figure 7.5 with results in table 7.1). Moreover, the plot of the treatment mean squares against the error mean squares gives an ellipsoid cloud of points with the original subset of data near the center of the cloud (figure 7.6). All in all, the original subset of the data that we used appears to be a good representation of possible subsets. When observations were assigned to treatment cells randomly, the patterns were remarkably similar to the patterns seen in the randomization of the original subset of data (compare figures 7.4 and 7.7). It appears that the variation of the treatment mean squares and the error mean squares is a reflection of structure found in the entire data set and not simply an artifact of our subsampling. These two comparisons demonstrate the effectiveness of randomization methods for data exploration and their advantages over a single analysis.

7.4.2 Randomization Methods and Statistical Inference

Randomization methods force a researcher to consider how a sample is drawn from a population and to what extent the sampled population matches the popula-

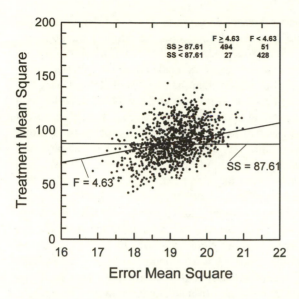

Figure 7.6 Treatment mean squares for the site effect versus the error mean squares from the 1000 randomizations in which observations were randomly selected from the complete data set but assigned to the correct cell. The number of observations per cell matched the number of observations per cell in the original unbalanced design.

tion that is under consideration. For example, our choice of randomizing the observations across all cells was based on the null hypothesis that all snakes were from the same population. Different null hypotheses require different types of randomization. Although we must be careful how we randomize the data, there is a real advantage to using randomization methods because the test can be tailored to very specific hypotheses if necessary.

In the last section, we showed how randomizations can be used to ask whether the data at hand are representative of the population under consideration. There

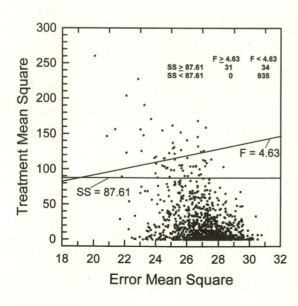

Figure 7.7 Treatment mean squares for the site effect versus the error mean squares from the 1000 randomizations in which observations were doubly randomized. Observations were first randomly selected from the complete data set and then randomly reassigned to any cell. As in figure 7.6, the original unbalanced design was preserved.

is always a concern about whether the sample is representative of the target population. Often, researchers do not realize that the population sampled should match the population under consideration (Mood and Graybill 1963). If a random sample is drawn, then valid probability statements can be made about the sampled population. However, Mood and Graybill (1993) note that strict probability statements about a target population cannot be made unless the target population is the sampled population. The fact that we had additional observations allowed us to ask whether the subsample we used was in some way unusual compared to the complete data set. Clearly, our subsample was not unusual, so we have some confidence that the inferences based on our sample of rattlesnakes (i.e., snakes are sexually dimorphic and differ between the sites) can be extended to all rattlesnakes at Grapevine Hills and Boquillas.

Finally, we should note that some authors take an even more radical view of statistical inferences. Edgington (1995) asserts that randomization methods can be used to make statistical inferences about a collection of data without assuming that the data are a random sample from a known population (also see Lehmann 1975, pp. 63–65, for a very cogent discussion of the problem). However, any such inferences are specific to that collection of data alone. Inferences that extend beyond the original data set must, per force, be logical arguments based on the plausibility that other possible collections of data will be similar. Edgington (1995) argues quite strongly that most data are not random samples from known populations and thus randomization methods are far more appropriate than parametric tests. More often that not, ecological data are not truly a random sample from a population that matches the target population, and thus, Edgington's criticisms must be considered. In many situations, truly random samples are impossible, or unlikely, and ecologists are hard pressed to make inferences beyond the data at hand (Dunham and Beaupre 1998). Regardless of Edgington's views, the value of randomization approaches is that they force us to deal with these questions in an explicit manner. In conclusion, even the simple choice of how the data should be randomized may not be that simple. The answer depends on our assumptions about how observations should be randomized across different treatment levels.

Acknowledgments   We would like to thank Sam Scheiner and Jessica Gurevitch for their help and encouragement during the development of this chapter. We also benefited from the comments of Marti Anderson and thank her for her insights.

Appendix

Once the data file has been made, simply run PROC GLM and save the output to a separate file. There are several tricks to doing this. The first is to use a BY statement to run the 1000 randomized iterations within a single PROC GLM statement. The second is to suppress the output. The NOPRINT option in the PROC GLM statement suppresses the description of the analyses, and the NOUNI

Table Example of output from OUTSTAT option in PROC GLM: the first 15 lines of a run of 1000 iterations of PROC GLM[a]

OBS	C	_NAME_	_SOURCE_	_TYPE_	DF	SS	F	PROB
1	1	SVL	ERROR	ERROR	49	1338.58	0	0
2	1	SVL	SEX	SS3	1	20.56	0.753	0.38983
3	1	SVL	SITE	SS3	1	15.62	0.572	0.45316
4	1	SVL	SEX*SITE	SS3	1	24.71	0.904	0.34628
5	1	SVL	AGE	SS3	1	2546.22	93.207	0
6	2	SVL	ERROR	ERROR	49	1333.06	0	0
7	2	SVL	SEX	SS3	1	0	0	0.9918
8	2	SVL	SITE	SS3	1	0.05	0.002	0.96655
9	2	SVL	SEX*SITE	SS3	1	55.34	2.034	0.16015
10	2	SVL	AGE	SS3	1	2880.39	105.876	0
11	3	SVL	ERROR	ERROR	49	1139.44	0	0
12	3	SVL	SEX	SS3	1	133.76	5.752	0.02032
13	3	SVL	SITE	SS3	1	71.08	3.057	0.08666
14	3	SVL	SEX*SITE	SS3	1	45.53	1.958	0.16805
15	3	SVL	AGE	SS3	1	2776.46	119.398	0

[a]Each iteration cycle is identified in the column labeled "C". Variable C was created with the file of randomized data. Refer to the Website for instructions on how to create a file of randomization data suitable for PROC GLM.

option in the MODEL statement suppresses the output of the analyses to the OUTPUT window. If these options are not specified, the output window will overflow if done on a personal computer. Third and most important, the OUTSTAT option in the PROC GLM statement creates an output file that can be read as a data file (appendix table).

The final step involves summarizing the data in the file created by the OUTSTAT option. In our example, the file contains 5000 lines of observations (i.e., the number of sources of variation times the number of iterations), so it must be simplified in some fashion. To do this, the observed F-ratios and SS values from the original data ordering are used as cut-off points in a series of IF . . . THEN OUTPUT . . . statements. Values for F and SS from the randomizations that are greater than the observed values are saved in temporary files. The files are sorted by the sources of variation, and then the MEANS procedure is used to list the number of iterations in which the randomized data gave values of F (or SS) greater than or equal to the observed statistic for the original data. Since 1000 iterations were run, the probabilities are estimated as the number of iterations with values equal to or greater than the observed outcome plus the observed outcome divided by 1001. (For example, for the effect of site, the probability that the observed F of 4.63 could be drawn by chance alone $= (37 + 1)/1001 = 0.038$; see figure 7.4.

8

Repeated-measures Analysis

Growth and Other Time-dependent Measures

CARL N. VON ENDE

8.1 Repeated-measures Questions

There are many situations in which ecologists make repeated measurements on the same individual, on the same experimental unit, or at the same sampling site. Most commonly, some characteristic or factor is measured at several different times. For example, we may be interested in how body size, plant size, survivorship, clutch size, population size, nutrient level, or pollutant level change over time for different populations, locations, or experimental treatments. A second type of study involving repeated measurements exposes each individual organism to different levels of some treatment and measures its response at each level. For example, in plant ecophysiological studies, the same plants often are exposed to a series of different CO_2 levels, and their photosynthetic rate is measured at each level (Potvin et al. 1990b). In both kinds of studies, there is an explicit interest in the pattern or shape of the response over time or over the levels of an experimental treatment.

The same questions could be addressed without repeated measurements on individuals. Assume we want to study the effect of diet on growth in squirrels, in particular, to examine whether the pattern of growth in squirrels fed acorns is different from that of squirrels fed hickory nuts over a 3-month period. Assume the squirrels were fed independently and were weighed on the starting date and at 2-week intervals for a total of six measurements. With six dates and two feeding regimes, there would be 12 diet $\times$ date combinations. The experiment could be conducted in either of two ways. We could assign, for example, three squirrels to each diet $\times$ date treatment combination. All squirrels would be started at the same time and fed regularly, but a different subset of squirrels would be weighed on each date for each diet. Thus, each squirrel would be weighed only once, and a total of 36 squirrels would be required for the experiment. Alternatively, six squirrels could be assigned to each diet and all squirrels measured on each date, so that each was measured six times. Only 12 squirrels would be required for this second design, a repeated-measures design.

The first experiment could be analyzed as a two-factor ANOVA (diet, date) because the squirrels within a diet treatment level at any one date would be independent of those at other dates, but obviously the experiment would require many more animals than the repeated-measures design. It would be inappropriate, however, to analyze the second design as a two-factor ANOVA because the same animals would be measured repeatedly during the experiment, and the weights of each squirrel on the different dates would not be independent of one another. (Independence of replicates is a basic assumption of ANOVA.) Rather, a *repeated-measures analysis*, which takes into account the correlation among dates because animals are reweighed, should be used for the analysis of the second experiment. Historically, although many ecologists have collected data in a repeated-measures design, they frequently have analyzed them incorrectly with an "ordinary" ANOVA, "plugged" the data into a repeated-measures analysis in a statistical package without considering the underlying assumptions of some repeated-measures analyses, or thrown out intermediate data and analyzed only the final measurements with ANOVA. As the level of ecologists' statistical knowledge has increased, the first two alternatives have become unacceptable and the last undesirable. Repeated-measures designs have been used in psychology and agriculture for some time (Snedecor and Cochran 1989; Winer et al. 1991), but they have received the explicit attention of ecologists only relatively recently (Gurevitch and Chester 1986; Potvin et al. 1990b). In this chapter, I discuss various parametric methods of analyzing repeated-measures data.

8.2 Statistical Issues

Repeated-measures designs can be analyzed by parametric methods using either univariate or multivariate approaches. The univariate analyses are ANOVA designs (randomized block, split-plot) that involve blocking (chapter 4). MANOVA (chapter 6) is used for the multivariate analyses. Although the univariate approach is computationally simpler and generally considered to be more powerful than MANOVA, it also has more restrictive assumptions. However, Mead (1988) discusses the philosophical problem of treating time as a "split-unit" factor in which repeated measurements are taken on the same, not different, experimental units. With the development of sophisticated statistical software over the last decade, the use of MANOVA is no longer limited by computational complexity. Both approaches can handle factorial designs. I briefly review the univariate and the multivariate approaches, highlight important aspects, and then discuss several examples in more detail.

8.2.1 Univariate Repeated-measures Analyses

The basic univariate designs for repeated measures are randomized complete block and split-plot designs. In the simplest repeated-measures designs, we are interested in whether different treatment levels applied to the same individuals have a significant effect (the plant photosynthesis example mentioned previously). These can be analyzed as a randomized block ANOVA (chapter 4). Each individual is considered a different block within which the treatment is applied. This is analogous to an agricultural experiment in which different fertilizers are applied to adjacent areas within each of several blocks. In repeated-measures jargon, the treatment is referred to as the *repeated factor* or by psychologists as the *within-subject factor*. The purpose of blocking is to make the analysis more sensitive by removing variance among blocks, or among subjects, from the error term. Conditions are assumed to be more homogeneous within a block than between blocks, so it is within a block, or to the same subject, that the treatment levels are applied. When time is the within-subject factor, there are observations at different times for each subject.

The split-plot design (chapter 4) is the natural extension of the randomized block design. A typical agricultural example would be a situation in which we are interested in the effect of irrigation on a fertilization treatment. Imagine several fields or plots, half of which were irrigated and half of which were not. Each whole plot would be divided into several subplots, and within each subplot one of the fertilizer levels described in the randomized block example would be applied. In repeated-measures jargon, the irrigation/no irrigation treatment is called the *between-subjects* (whole-plot) factor, and again, the fertilizer treatment the within-subject (split-plot) factor. The previous squirrel example in which there are repeated observations on the same squirrels fed different diets can be treated as a split-plot design. Diet is the between-subjects factor, individual squirrels are the whole-plots, or subjects, within each food type, and date is the within-subject factor. Again, the ecological question is whether the squirrels on the two different diets have the same patterns of weight change over time.

Although randomized block and split-plot designs differ from ordinary ANOVA in that they assume some correlation among treatment levels within a block, many ecologists are unaware that these designs make certain assumptions about the variances and covariances of the levels of the within-subject factor and the relationship among these variances and covariances. Specifically, these designs assume what is called *circularity* among the levels of the within-subject factor. We can construct a square variance–covariance matrix for the within-subject factor in these designs. The variances for each level of the within-subject factor will fall along the diagonal of the matrix, and the covariances between all the different levels will occupy the remainder of the matrix. A circular variance–covariance matrix has the property that the variance of the *difference* between any two levels of the within-subject factor equals the same constant value (Winer et al. 1991). This means that if we take two levels of the within-subject factor and subtract the scores for one level from the scores for another level, the resulting scores must have the same variance for every pair of levels (Maxwell and

Delaney 1990). The assumption of circularity is less restrictive than the assumption of *compound symmetry* in which the variances are all assumed to be equal to one another and the covariances are also assumed to be equal to one another. For many years, it was thought compound symmetry was required for the validity of the *F*-statistic in repeated-measures ANOVA; however, Huynh and Feldt (1970) showed that circularity was sufficient. All matrices that have compound symmetry are circular, but not all circular matrices have compound symmetry.

Another matrix characteristic, *sphericity*, is used in assessing the circularity of a variance–covariance matrix. An orthogonal transformation creates transformed variables that are independent (orthogonal) of each other. If these transformed variates are then scaled such that the sum of squares of the coefficients for each variable is 1, this constitutes a set of *orthonormal* variables (Stevens 1996). A circular variance–covariance matrix that is transformed to its normalized orthogonal form is *spherical* (Winer et al. 1991). The transformation can be accomplished by first constructing a matrix with the coefficients of orthogonal contrasts as the rows and then normalizing the coefficients (see Winer et al. 1991, p. 244, for an example). This matrix is then used to transform the variance–covariance matrix to an orthonormalized variance–covariance matrix. The circularity of a variance–covariance matrix can be assessed by testing the sphericity of an orthonormalized form of the matrix. If the transformed variance–covariance matrix is spherical, the original variance–covariance matrix is circular. In a spherical variance–covariance matrix, the variances of the transformed variables are equal and their covariances are 0.

There are tests for sphericity, the most popular of which is Mauchly's (Crowder and Hand 1990). The test statistic *W* can range from 0 to 1 (equation in Winer et al., 1991). It uses the orthonormalized form of the variance–covariance matrix. The closer *W* is to zero, the greater the departure from sphericity (Winer et al. 1991). Huynh and Mandeville (1979) showed that *W* is sensitive to violations of normality and that these tendencies are amplified with increased sample sizes. O'Brien and Kaiser (1985), Stevens (1996), and Winer et al. (1991) recommend not using *W*.

In the case of the plant experiment, if the sequence of the CO_2 levels presented to a plant were randomized and there were no carry-over effects on photosynthetic rate from one CO_2 concentration to the next, the assumption of circularity probably could be met. However, when time is the within-subject factor, usually data collected on adjacent sampling dates are more highly correlated than are data from separated sampling dates, and the circularity condition is not met. There are numerous instances in the ecological literature in which ecologists have conducted a repeated-measures ANOVA analysis with time as the within-subject factor without mentioning the assumption of circularity (or sphericity). The consequence of failing to meet the assumption is that *F*-statistics for the within-subject factors (and their interactions) are inflated, so we are more likely to conclude that effects are statistically significant when they are not. When this assumption cannot be met, we have two alternatives: (1) adjust the degrees of freedom of the *F*-test so that it is more conservative or (2) use the multivariate approach to analyze the data. These are discussed subsequently.

An additional assumption is involved when using the split-plot design because of between-subjects factors. The variance–covariance matrices of the *differences* among the levels of the within-subject factor should be homogeneous (equal) for all levels of the between-subjects factor. That is, if we had three levels of a between-subjects factor and calculated the variance–covariance matrix of *differences* for the within-subject factor for each of those levels, those matrices should be identical because the test of the significance of the within-subject factor (and its interaction with the between-subjects factor) is based on the pooled variance–covariance matrix. The matrices should be equal to be pooled. This pooled variance–covariance matrix of differences must meet the criterion of circularity. The significance tests of the within-subject factor, and its interaction with the between-subjects factor, are sensitive to the circularity assumption as described previously for the simple within-subject design. They are robust to the second assumption of homogeneity of variance–covariance matrices as long as sample sizes are equal. However, as sample sizes become less equal, the test becomes less robust (Maxwell and Delaney 1990).

When within-subject or between-subjects effects are found to be significant, follow-up tests often are desirable. O'Brien and Kaiser (1985), Maxwell and Delaney (1990), and Stevens (1996) discuss post hoc tests for repeated-measures designs. It is important to realize that planned (a priori) contrasts involving the specific error term for each contrast, rather than an overall error term, are not subject to the circularity assumption. See O'Brien and Kaiser (1985) for an example.

Although I have described only the simple experimental designs with a single within-subject and between-subjects factor, repeated-measures designs can be expanded to include additional within- and between-subjects factors. See Maxwell and Delaney (1990), Winer et al. (1991), and Stevens (1996), and section 8.4 for examples.

8.2.2 Multivariate Repeated-measures Analyses

Repeated-measures data also can be analyzed by using MANOVA (chapter 6). In this approach, the response variable for each level of the within-subject factor is treated as a different dependent variable. In terms of the previous plant example, the photosynthetic rate at each of the CO_2 concentrations would be considered a different response variable. So, if there were five CO_2 levels, the response of each plant would be characterized in terms of five photosynthetic rates, one for each of the CO_2 concentrations. MANOVA is designed to simultaneously analyze the response of several correlated dependent variables. As is usually the case with repeated-measures data, MANOVA does not require the dependent variables to be equally correlated as repeated-measures ANOVA does. It assumes what is called an "unstructured" variance–covariance matrix, which means there is no particular pattern required of the matrix.

To analyze the squirrel example as a MANOVA, the dependent variables would be the weights on each of the six dates and the treatment would be diet. The analysis tests whether the mean response vectors for the two diets are different (chapter 6). Although the problem of circularity inherent in repeated-measures ANOVA is avoided by doing MANOVA, MANOVA also has its limitations. In the repeated-measures design for the squirrel example, I suggested the analysis could be done with six squirrels (n) for each diet. However, the number of dependent variables (k) that can be analyzed in a MANOVA depends on the total number of samples (subjects, N) and the number of between-subjects treatment levels (groups, M) (Potvin et al. 1990b). The constraint is that $N - M > k$. Because the squirrel example would have weights on six dates, six squirrels, and two treatment levels, this condition would be satisfied ($12 - 2 = 10 > k = 6$). Although $N = 12$ would satisfy the condition, the test would have a low power. It would be best to increase the sample size even more or to decrease the number of dates on which the squirrels were weighed. Power increases as the ratio $n{:}k$ increases (Potvin et al. 1990b). This limitation on the number of levels of within-subject factors and associated problems of low power are inherent in using MANOVA (chapter 6). Stevens (1996) has a detailed discussion of power in MANOVA and mentions the need to use a larger α if low power is suspected. Maxwell and Delaney (1990) discuss the power of MANOVA in repeated-measures designs and provide tables of estimated sample sizes for different levels of power for one within-subject design. They recommend doing MANOVA only when $N - M > k + 9$ (Maxwell and Delaney 1990, p. 676). These restrictions highlight the need to consider the amount of replication and the number of levels of within-subject factors in the initial stages of experimental design rather than after an experiment has been completed. Potvin et al. (1990b) recommend using the minimum number of levels of the within-subject factor required to characterize a response adequately and to increase the amount of replication to yield a more powerful analysis.

The MANOVA analysis combines two sources of differences in the within-subject data when comparing the mean response curves of the levels of the between-subjects treatment: it takes into account differences in the *shapes* of the response curves, as well as differences in the *levels* of the response curves (Harris 1985). Consider the squirrel example, but for only three dates (figure 8.1). The growth curves could be parallel (figure 8.1A), indicating the squirrels had similar patterns of growth on the two diets (shape), or they could diverge (figure 8.1B), indicating different patterns of growth. If they were parallel and one response curve was consistently higher than the other (figure 8.1A), one group of squirrels would be consistently gaining more weight as time progressed (levels). If the curves were parallel and nearly overlapping there would be no effect of diet. Finally, if the curves were parallel, a significant slope (i.e., slope > 0) would indicate there was a change in weight over time (figure 8.1A), whereas horizontal response curves would indicate no effect of time on weight.

Profile analysis, a methodology that enables us to test the significance of these three aspects of the multivariate response, is the approach most commonly used to analyze repeated-measures data with MANOVA (Harris 1985; O'Brien and Kaiser 1985). It uses both univariate (ANOVA) and multivariate tests (MANOVA). The three aspects are examined as tests of specific hypotheses. A comparison of the shapes of response curves is a test of the *parallelism* hypothesis; a comparison of the levels of the curves is a test of the *levels* hypothesis. Determining whether

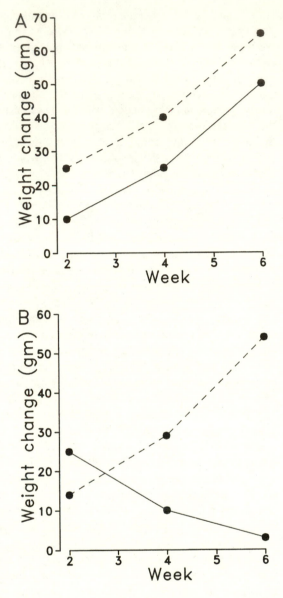

Figure 8.1 Profile analysis: hypothetical weight change over 6 weeks for squirrels fed hickory nuts (dashed lines) and acorns (solid lines). (A) Curves show parallelism and levels effects. (B) Curves show absence of parallelism, that is, diet × time interaction.

the response curves have an average slope different from zero is a test of the *flatness* hypothesis. As implied previously, the parallelism hypothesis should be tested first because the levels hypothesis becomes moot if the curves are not parallel. This is similar to examining interactions before main effects in the analysis of factorial designs (Keppel 1991). The plots in figure 8.1 are analogous to interaction plots for analysis of factorial designs (Keppel 1991). O'Brien and Kaiser (1985) give an extensive primer on using profile analysis in repeated-measures analyses. Their approach will be followed in the examples that follow.

Profile analysis also can be applied to multivariate analyses other than repeated-measures (Morrison 1990).

Often an ecologist is interested in measuring more than one response variable for a subject or experimental unit over time. For example, plant height and leaf area could respond in some correlated manner that would be of interest in studies of plant competition. In community ecology studies, an experimental unit (pen, bag, mesocosm) often has a array of species that is subjected to some treatment. Analysis by MANOVA is appropriate because of the potential correlation of the responses of the species co-occurring in the pens. These kinds of examples in which there is one within-subject factor, but for which more than one response variable is measured for each subject (experimental unit), are referred to as *doubly multivariate* designs and can have any number of between-subjects factors, as in an ordinary MANOVA. A doubly multivariate example is given in the SAS manual (SAS Institute Inc. 1989b, chapter 24, example 9) and in the Command Reference section of the SPSS Advanced Statistics manual (Norusis 1990).

8.3 Example: One Between-subjects and One Within-subject Factor

The first example is a split-plot design (table 8.1). Assume that a plant species was grown under two nutrient regimes (low, high) in pots in a greenhouse, and the number of leaves per plant was monitored for five weeks. The example will be analyzed first by repeated-measures ANOVA (8.3.1) then by MANOVA and profile analysis (8.3.2).

8.3.1 Repeated-measures ANOVA

Recall that, in a split-plot design applied to repeated-measures data, subjects are plots and levels of the within-subject factor are subplots. Each subject is assigned

Table 8.1 Growth of a hypothetical plant (number of leaves per plant) over 5 weeks at low (L) and high (H) nutrient levels

Plant	Nutrient level	Week 1	Week 2	Week 3	Week 4	Week 5
1	L	4	5	6	8	10
2	L	3	4	6	6	9
3	L	6	7	9	10	12
4	L	5	7	8	10	12
5	L	5	6	7	8	10
6	H	4	6	9	9	11
7	H	3	5	7	10	12
8	H	6	8	11	10	14
9	H	5	7	9	10	12
10	H	5	8	9	11	11

to one of the between-subjects treatment levels. In terms of table 8.1, the within-subject factor is Time, and the between-subjects factor is nutrient level. The model for the data in table 8.1 is

$$X_{ijk} = \mu + \nu_i + \psi_{k(i)} + \tau_j + \nu\tau_{ij} + \psi\tau_{jk(i)} = \varepsilon_{m(ijk)} \qquad (8.1)$$

where ν_i is the effect of the nutrient treatment on plant growth, $\psi_{k(i)}$ is the subject effect nested within the respective nutrient levels, τ_j is the time effect, $\nu\tau_{ij}$ is the nutrient $\times$ time interaction, $\psi\tau_{jk(i)}$ is the subject $\times$ time interaction, and $\varepsilon_{m(ijk)}$ is the error term. In addition, m is a dummy subscript included to indicate that the experimental error is nested within the individual observation. The subject $\times$ time interaction is included in the model statement to emphasize that the potential exists for this interaction to be present; however, because there is no replication of subject $\times$ time cells, the interaction cannot be estimated and becomes part of the error term for testing the within-subject factor and its interaction with the between-subjects factor (Winer et al. 1991). The expected means squares for this model assuming nutrient and time are fixed factors, and subjects (plants) are random, as shown in table 8.2.

When repeated-measures designs (i.e., not doubly multivariate) are analyzed in SAS (SAS Institute Inc. 1989a,b), both the repeated-measures ANOVA and profile analysis can be done in one analysis (see the appendix). The steps are (1) select SAS procedure ANOVA for balanced designs or procedure GLM for unbalanced designs, (2) include the CLASS statement to list the between-subjects factor(s), (3) write the MODEL statement in a MANOVA form, with the dependent variables describing the levels of the within-subject factor on the left side of the equation and the between-subjects treatment on the right, and (4) use the REPEATED statement to list the "label," and the number of levels, of the within-subject factor. The results of this analysis of the data in table 8.1 are presented in table 8.3.

Notice that the sources of variation are divided into between-subjects (nutrient) and within-subject effects. The latter includes the within-subject main effect (time)

Table 8.2 Expected mean squares for repeated-measures ANOVA (one between- and one within-subject factor)[a]

Source of variation	df	E(MS)
Between-subjects	$np - 1$	
Nutrient	$p - 1$	$\sigma_e^2 + q\sigma_S^2 + nq\sigma_N^2$
Subjects within groups	$p(n - 1)$	$\sigma_e^2 + q\sigma_S^2$
Within-subject	$np(q - 1)$	
Time	$q - 1$	$\sigma_e^2 + \sigma_{TS}^2 + np\sigma_T^2$
Nutrient $\times$ time	$(p - 1)(q - 1)$	$\sigma_e^2 + \sigma_{TS}^2 + n\sigma_{NT}^2$
Subjects $\times$ time within groups	$p(n - 1)(q - 1)$	$\sigma_e^2 + \sigma_{TS}^2$

[a] σ_N^2 = variance due to nutrients
σ_S^2 = variance due to subjects
σ_T^2 = variance due to time
σ_e^2 = error variance

Table 8.3 Repeated-measures ANOVA for a split-plot design[a]

A. Between-subjects

Source	df	MS	F	P > F
Nutrient	1	16.82	2.60	0.1453
Error	8	6.46		

B. Within-subject

Source	df	MS	F	P > F	Adj. P > F G–G	Adj. P > F H–F
Time	4	66.85	158.22	0.0001	0.0001	0.0001
Nutrient × time	4	1.27	3.01	0.0326	0.0666	0.0353
Error(Time)	32	0.42				
		Greenhouse–Geisser $\varepsilon = 0.5882$				
		Huynh–Feldt $\varepsilon = 0.9531$				

[a]$P > F$ is unadjusted probability, G–G and H–F are Greenhouse–Geisser and Huynh–Feldt adjusted probabilities respectively based on the respective epsilons (see text for details).

and its interaction with the between-subjects factor (nutrient × time). We usually use a repeated-measures design because we are interested in the effect of the between-subjects treatment over time, that is, the nutrient × time interaction. Hence, it should be the first treatment examined.

The probabilities for the respective F statistics $(P > F)$ in column six (table 8.3) are calculated assuming the data meet the circularity assumption. Based on these probabilities, there is a statistically significant nutrient × time interaction $(P < 0.0326)$ that indicates plants in the fertilized conditions added leaves at a faster rate than the unfertilized plants. The statistically significant time effect $(P < 0.0001)$ indicates the average number of leaves increased over time. The nutrient main effect was not statistically significant based on this F-test, which is not unusual since the interaction was statistically significant.

The F-statistics for within-subject factors (and their interactions) are inflated when the sphericity condition is not met. Several estimators (epsilons, ε) have been developed for decreasing the degrees of freedom of the F-statistic according to the severity of the violation of the sphericity assumption. The degrees of freedom for the critical F-statistic are multiplied by the ε. Box (1954) originally proposed the equation for ε for a within-subject design (no between-subjects factors) based on the "population" variance–covariance matrix. Geisser and Greenhouse (1958) extended this to the split-plot design. However, since ε is not known (because the "population" variance–covariance matrix is unknown), ε has to be estimated from the sample variance–covariance matrices. This introduces the added complication of not knowing how inaccuracies in the sampling data will affect the estimate of ε, and hence how this will influence the adjustment of the F-statistic degrees of freedom. For that reason, Greenhouse and Geisser (1959) suggested the very conservative approach of replacing ε with the smallest value it could take, $1/(k - 1)$, where k is the number of levels of the within-subject factor.

Collier et al. (1967) and others have examined the bias introduced by using the sample variance–covariance matrix for estimating ε (Crowder and Hand 1990). The estimator based on sample data is referred to as $\hat{\varepsilon}$. For $\hat{\varepsilon}$ greater than 0.75, and n (sample size) less than $2k$, $\hat{\varepsilon}$ tends to overcorrect the degrees of freedom and produce a conservative test (Crowder and Hand 1990; Winer et al. 1991). Huynh and Feldt (1976) proposed a less biased estimate of ε based on sample data ($\tilde{\varepsilon}$) (described in Winer et al. 1991), although it can be overly liberal in some cases (Type I error). SAS provides estimates of both Box's ε ($\hat{\varepsilon}$) and Huynh–Feldt's adjustment to ε ($\tilde{\varepsilon}$), as well as the respective probabilities for the adjusted F-statistics of the within-subject factors, as part of its repeated-measures ANOVA output (table 8.3). They refer to Box's ε as the Greenhouse–Geisser ε (as I will). Maxwell and Delaney (1990) call it the Geisser–Greenhouse ε. Its value ranges from 0 to 1, whereas $\tilde{\varepsilon}$ can be greater than 1. When the latter is >1, 1 is used as the value. The smaller the value of $\hat{\varepsilon}$ and $\tilde{\varepsilon}$, the greater the departure from sphericity. As $\tilde{\varepsilon} \geq \hat{\varepsilon}$, generally $\hat{\varepsilon}$ is the more conservative adjustment. They converge as sample size increases (Maxwell and Delaney 1990).

The epsilons for the analysis of the data in table 8.1 are at the bottom of table 8.3. One can see that the adjustment is greater for $\hat{\varepsilon}$ than for $\tilde{\varepsilon}$. The probabilities for the modified F-statistics are in the two right-hand columns of the within-subject treatment combinations. When the adjustments are made for the F-statistics, the nutrient × time interaction is no longer statistically significant according to the Greenhouse–Geisser adjustment, whereas it is significant for the Huynh–Feldt adjustment. The probabilities for the time effect are unchanged. There is not universal agreement among statisticians about the use of these adjustment procedures, especially compared with the alternative of profile analysis. (See section 8.8.2 for a comparison of alternative recommendations.)

8.3.2 Profile Analysis

The MANOVA approach to the analysis of repeated-measures data can consist of simply estimating the multivariate test statistic for data from all levels of the within-subject factor considered simultaneously. However, profile analysis is more informative because it analyzes the pattern of response of the within-subject factor. Profile analysis addresses hypotheses about parallelism, levels, and flatness by transforming within-subject repeated-measures data to a set of contrasts or differences, and then doing univariate (t-test, ANOVA) or multivariate (Hotelling's T^2, MANOVA) analyses on the contrasts. Potvin et al. (1990b) termed this approach MANOVAR. I will present the details of profile analysis of the nutrient experiment (table 8.1) first using data from the first 2 weeks and then from all five dates.

For comparative purposes, I will estimate the overall multivariate statistic for the first 2 weeks of data and then analyze it by profile analysis. The multivariate comparison of the number of leaves in the low and high nutrient levels for the first 2 weeks (two dependent variables) using Hotelling's T^2 gives $F = 5.69$, df =

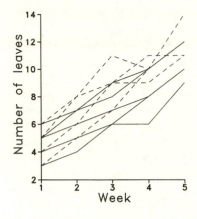

Figure 8.2 Growth of hypothetical plants grown under high-nutrient (dashed lines) and low-nutrient (solid lines) conditions for 5 weeks.

2,7, and $P = 0.034$. This shows that the number of leaves is different between low and high nutrient levels when weeks 1 and 2 are considered simultaneously.

To analyze the first 2 weeks of data by profile analysis, first the "parallelism" hypothesis will be addressed by testing whether the shapes or slopes of the response curves are the same for both groups (figure 8.2). In terms of repeated-measures ANOVA, this is a test of whether there is a significant nutrient × time interaction. Comparing the shapes of the response curves is equivalent to asking whether the mean change in number of leaves between weeks 1 and 2 is the same for both nutrient levels; in other words, how do the mean differences (d) compare for the two groups (table 8.4)?

The hypothesis tested is H_o: $(\mu_{N1} - \mu_{N2}) = (\mu_{W1} - \mu_{W2})$, where the first ($i$th) subscript refers to low and high nutrient and the second (jth) to week 1 and week 2.

Table 8.4 Difference in number of leaves per plant for the example in table 8.1

Plant	Nutrient level	Difference between adjacent weeks			
		$2-1$	$3-2$	$4-3$	$5-4$
1	L	1	1	2	2
2	L	1	2	0	3
3	L	1	2	1	2
4	L	2	1	2	2
5	L	1	1	1	2
Mean		1.2	1.4	1.2	2.2
6	H	1	3	0	2
7	H	2	2	3	2
8	H	2	3	−1	4
9	H	2	2	1	2
10	H	3	1	2	0
Mean		2.0	2.2	1.0	2.0

For weeks 1 and 2, d is 1.2 for the low-nutrient treatment level and 2.0 for the high-nutrient treatment level. A t-test shows that the two groups are significantly different ($t = 3.54$, df $= 8$, $P = 0.008$) in terms of the means changes. This indicates that there is a significant nutrient $\times$ time interaction and that the number of leaves increased faster in the high-nutrient treatment level.

Since the nutrient $\times$ time interaction was found to be significant, the tests of the levels and flatness hypotheses are less meaningful, but both will be explained for demonstration purposes. The test of the levels hypothesis is a test of the nutrient main effect. The week 1 and week 2 values are averaged for each subject [(week 1 + week 2)/2], and then the means compared between groups with a t-test. One is asking whether the levels of the response curves are different, that is, whether, when averaged over weeks 1 and 2, the high-nutrient plants had more leaves than the low-nutrient plants. In this example, there was no uniform effect of the nutrient treatment in the first 2 weeks ($t = 0.66$, df $= 8$, $P = 0.53$), which was consistent with the significant nutrient $\times$ time interaction.

Finally, for a test of the flatness hypothesis, or the time effect, the contrast variable (differences between weeks 2 and 1) is used again, but it is averaged over nutrient levels. The test is whether the grand mean of the contrast variable is different from 0, which is essentially a paired t-test (Zar 1996, chapter 9), but without regard to the nutrient treatment grouping. In terms of the original response variables, it is a test of whether there is an increase in the number of leaves between weeks 1 and 2 (slope > 0) when all plants are averaged together for each week. In our example, time was statistically significant between weeks 1 and 2 ($t = 7.97$, df $= 9$, $P < 0.0001$), so there was an overall slope different from 0. In summary, in the multivariate analysis of a simple design with one between-subjects factor (nutrient) and a within-subject factor with two levels (weeks 1, 2), we can use (1) Hotelling's T^2 to test for an overall difference between groups or (2) profile analysis (three separate univariate tests) to separately examine the parallelism, levels, and flatness hypotheses.

Next, I will analyze the data in table 8.1 for all five dates using profile analysis, because this is the more informative approach and is used by SAS (procedure GLM and procedure ANOVA). Profile analysis of the time effect (flatness) and the nutrient $\times$ time interaction (parallelism) is based on differences (contrasts) of adjacent weeks. There are four differences because there are data for 5 weeks. Because these differences can be treated as four dependent variables, MANOVA is appropriate for the analysis. SAS refers to these differences (contrasts) as a *transformation* of the original within-subject data. This is somewhat different from the kinds of transformation most ecologists are familiar with in statistical analyses, for example, log, square root, and arcsine. In the difference transformation, the values of two levels of the within-subject factor are transformed to one. For a test of the parallelism hypothesis, the set of differences used is week 2 − week 1, week 3 − week 2, week 4 − week 3, and week 5 − week 4 (table 8.3). The desired transformation is identified for the within-subject factor in the REPEATED command (appendix). PROFILE is used for this analysis because it generates the contrast variables based on differences of adjacent levels of the within-subject factor. Other transformations available in SAS are described subsequently.

Table 8.5A shows the results of the MANOVA analysis of the differences. The values of four different multivariate test statistics are shown in the second column and their equivalent F-statistics in the third. The probabilities for the F-statistics are in the last column. The probabilities ($P = 0.025$) indicate that the nutrient $\times$ time interaction is statistically significant. Therefore, the slopes of the growth curves in the high- and low-nutrient levels are different according to profile analysis, which agrees with the repeated-measures ANOVA analysis (table 8.3). (A discussion of the different multivariate test statistics is presented in chapter 6.)

The test for the flatness hypothesis (time effect) tests whether there is a significant increase in the number of leaves over time when averaged over both nutrient levels. This is a test of whether the grand mean (averaged over nutrient levels) of the set of the four time differences is zero. Again, the statistically significant F-values in the MANOVA analysis indicate that there is a increase in the number of leaves over time ($P < 0.0001$) (table 8.5B).

The test for the between-subjects effect(s) in profile analysis (e.g., nutrient levels) is the same as in repeated-measures ANOVA: both are based on a comparison of the mean within-subject response across the levels of the between-subjects factor (time) (SAS Institute Inc. 1989b; Potvin et al. 1990b).

If we are interested in identifying the particular time intervals in which the treatment effects are different, individual ANOVAs (F-tests) can be done on each of the contrasts. For example, we can test the significance of the nutrient $\times$ time interaction or the time effect for each contrast in table 8.4. The results of such an analysis are shown in table 8.6. Mean refers to a test of the flatness hypothesis of whether there was a significant time effect, and nutrient refers to the test of the parallelism hypothesis of whether there was a significant nutrient $\times$ time

Table 8.5 MANOVA of the nutrient $\times$ time interaction and the time effect of the data in table 8.1[a]

Statistic	Value	F	Num. df	Den. df	$P > F$
A. Nutrient $\times$ time					
Wilks' lambda	0.14477212	7.3843	4	5	0.0250
Pillai's trace	0.85522788	7.3843	4	5	0.0250
Hotelling–Lawley trace	5.90740741	7.3843	4	5	0.0250
Roy's greatest root	5.90740741	7.3843	4	5	0.0250
B. Time					
Wilks' lambda	0.00848656	146.0417	4	5	0.0001
Pillai's trace	0.99151344	146.0417	4	5	0.0001
Hotelling–Lawley trace	116.83333333	146.0417	4	5	0.0001
Roy's greatest root	116.83333333	146.0417	4	5	0.0001

[a]Num. df and Den. df refer to numerator and denominator degrees of freedom, respectively.

Table 8.6 ANOVAs on each of the contrasts of within-subject factor, time, for the data in table 8.1

Source	df	MS	F	P > F
Contrast variable: week 2 – week 1				
Mean	1	28.9	144.50	0.0001
Nutrient	1	2.5	12.50	0.0077
Error	8	0.2		
Contrast variable: week 3 – week 2				
Mean	1	32.4	64.80	0.0001
Nutrient	1	1.6	3.20	0.1114
Error	8	0.5		
Contrast variable: week 4 – week 3				
Mean	1	12.1	7.56	0.0251
Nutrient	1	0.1	0.06	0.8089
Error	8	1.6		
Contrast variable: week 5 – week 4				
Mean	1	44.1	40.09	0.0002
Nutrient	1	0.1	0.09	0.7707
Error	8	1.1		

interaction in the respective ANOVAs for each contrast variable. Notice that although the multivariate F-statistics (MANOVA) for the nutrient × time interaction as well as the time effect were both statistically significant when all four contrasts were analyzed together (table 8.6), the latter was only significant for week 2 – week 1 in the individual ANOVAs. This indicates that a significant difference in the change in the number of leaves due to the nutrient treatment occurred only between week 1 and week 2, which can be seen when the means for the contrasts are compared (table 8.3). The differences between the mean differences for the low- and high-nutrient levels for the other dates were not sufficiently large themselves, but they contributed to an overall difference when all were analyzed simultaneously.

Because repeated ANOVAs are done when individual contrasts are analyzed, the experimentwise error rate should be adjusted according to the number of tests run. To maintain an overall α of 0.05, a Bonferroni adjustment of $\alpha = 0.05/4 = 0.0125$ for each contrast could be used for table 8.6. If the investigator is interested in testing only a single contrast, then α would not be adjusted. Although a statistical package such as SAS automatically does the MANOVA and ANOVAs for all contrasts, we could examine the ANOVA only for the contrast of interest.

It should be emphasized that different transformations (contrasts) can be used for the within-subject factor(s), and this will not affect the outcome of the MANOVA (multivariate F) because of the invariance property of the multivariate test statistic (Morrison 1990). The particular transformation used depends on the patterns we are looking for in the within-subject effect(s). SAS gives five choices of the transformation: Profile, Contrast, Helmert, Means, and Polynomial. As described previously, I used the Profile transformation in which differences were constructed from adjacent levels of the time factor. The Contrast transformation is the default condition, one level of the within-subject factor is considered the control and all other levels are compared to it. Potvin et al. (1990b) used the Helmert transformation: it compares the level of a within-subject factor to the mean of subsequent levels. The Mean transformation compares each level of the within-subject factor to the mean of all the other levels of the within-subject factor.

The Polynomial transformation is particularly useful if we are interested in examining trends in the within-subject data to see whether they conform to a particular form. For example, does the number of leaves per plant change in a linear, quadratic, or cubic manner over time? This method relies on a set of polynomials of increasing order that are constructed as a set of independent (orthogonal) contrasts and often is referred to as trend analysis (Winer et al. 1991). Orthogonal polynomials enable us to ask whether there is a significant linear (first-order), quadratic (second-order), or cubic (third-order) trend in the data (Gurevitch and Chester 1986; Hand and Taylor 1987). If k is the number of levels of the within-subject factor, $k - 1$ polynomials can be constructed, although usually we would be interested in the lower order (linear or quadratic) trends only. In the analysis of the within-subject factor(s), MANOVA considers simultaneously all orders of the polynomials that can be obtained from the data. As with the profile transformation, we can examine particular polynomials by individual ANOVAs. In examining orthogonal polynomials, normally we proceed from higher to lower order polynomials in testing for significance, and we stop at the order of polynomial at which significance is found. Because higher order polynomials would seem inappropriate for most ecological data and examining more tests increases our chance of committing a Type I error, it would be appropriate in many cases to begin with the cubic or quadratic analysis for testing for significance.

The plant growth data from table 8.1 were analyzed using orthogonal polynomials (table 8.7). Only the individual F-tests are shown because the multivariate F-values are the same as in table 8.5. Only the quadratic and linear trends were

Table **8.7** Individual ANOVAs for first- and second-order orthogonal polynomials for the data in table 8.1

Source	df	MS	F	P > F
Contrast variable: first order				
Mean	1	265.690	432.02	0.0001
Nutrient	1	2.890	4.70	0.0620
Error	8	0.615		
Contrast variable: second order				
Mean	1	0.007	0.03	0.8651
Nutrient	1	2.064	8.89	0.0175
Error	8	0.232		

examined in this example, although SAS also gives the cubic trend. Also, α was set at 0.025 because there were two separate analyses (0.05/2). There was a significant quadratic trend for the nutrient $\times$ time interaction ($P = 0.0175$). Remember that nutrient in the SAS output indicates the nutrient $\times$ time interaction. So the low- and high-nutrient treatment levels differed significantly in their quadratic responses of number of leaves over time. Time (mean) did not have a significant quadratic response, but the linear trend for time was statistically significant ($P < 0.0001$), indicating there was an overall linear increase in the number of leaves during the experiment.

8.4 Multiple Within-subject Factors

Situations often arise in ecological experiments and data collection in which repeated observations on the same individual can be classified factorially or the same individual is observed under more than one set of conditions. In terms of the plant growth experiment (table 8.1), it may be of interest to run the experiment for a second year to see whether the high-nutrient plants also produced more leaves than the low-nutrient plants in the second year (table 8.8). This would be a design with two within-subject factors, weeks and years, and one between-subjects factor, nutrient.

The within-subject factors are treated as follows: (1) averaging over years, we can test whether there is a significant week effect; (2) averaging over weeks, we can test whether there is a significant year effect; (3) using the appropriate contrasts, we can test whether there is a significant week $\times$ year interaction, or, in other words, averaging over the nutrient treatment, did the plants grow differently between years?

Table 8.8 Growth of plants in table 8.1 for years 1 and 2[a]

Plant	Nutrient level	Year 1					Year 2				
		1	2	3	4	5	1	2	3	4	5
1	L	4	5	6	8	10	4	4	5	7	9
2	L	3	4	6	6	9	3	4	6	8	10
3	L	6	7	9	10	12	3	4	6	7	9
4	L	5	7	8	10	12	3	4	6	8	10
5	L	5	6	7	8	10	5	5	6	7	10
6	H	4	6	9	9	11	4	4	5	7	10
7	H	3	5	7	10	12	5	5	6	8	11
8	H	6	8	11	10	14	4	4	6	7	10
9	H	5	7	9	10	12	3	4	5	7	9
10	H	5	8	9	11	11	5	5	7	9	11

[a]Number of leaves per plant over 5 weeks at low- (L) and high-nutrient (H) levels for years 1 and 2. Plants add leaves for a 5- to 6-week period each spring, with an initial flush of leaves within the first week of emergence. Leaves die back during the winter.

The between-subjects factor (nutrient) is cross-classified with each of these, so we must examine the interactions of each of the within-subject main effects (week, year), and their interaction (week × year), with the nutrient treatment. In fact, it is the three-factor interaction (nutrient × year × time) that we should examine first. It should be emphasized that designs with greater than one within-subject factor can be analyzed both by repeated-measures ANOVA (Winer et al. 1991) and profile analysis (MANOVA) (O'Brien and Kaiser 1985).

Although the analysis at first may seem complicated, it really is not if we keep track of between- and within-subject factors (table 8.9). In the MODEL statement in the SAS program, each of the weeks is treated as a different dependent variable for a total of 10 when data for both years are combined (appendix, part B). Also, the number of levels of each of the within-subject factors is listed in the REPEATED command. I have included only the results of the profile analysis (MANOVA) in the results of the analysis (table 8.9) and have condensed the SAS output. Under the within-subject effect, the three-way interaction (nutrient × week × year) was statistically significant. The interpretation of that significance is that the high-nutrient plants produced more leaves than low-nutrient plants during year 1, but not during year 2, so the treatment produced different effects in different years. As in the analysis of the first year's data, the nutrient main effect was not statistically significant. The ANOVA or orthogonal polynomials analyses would result in the same conclusions. See O'Brien and Kaiser (1985) for a detailed example with two between-subjects factors and two within-subject factors.

8.5 Other Ecological Examples of Repeated Measures

I have concentrated on repeated-measures analyses in which time is the repeated (within-subject) factor. Other repeated-measures situations arise in ecological

Table 8.9 Results of a profile analysis (MANOVA) of the data in table 8.8[a]

A. Between-subjects				
Source	MS	df	F	$P > F$
Nutrient (N)	13.69	1	3.83	0.0861
Error	3.58	8		

B. Within-subject				
Source	F	Num. df	Den. df	$P > F$
Year (Y)	14.47	1	8	0.0052
Week (W)	135.14	4	5	0.0001
Year × week	14.34	4	5	0.0060
Nutrient × year	0.97	1	8	0.3530
Nutrient × week	0.52	4	5	0.7301
N × Y × W	5.66	4	5	0.0425

[a]The F-statistics for all four MANOVA test criteria were identical for all treatment combinations. Num. df and Den. df are numerator and denominator degrees of freedom, respectively.

studies in which individual organisms or experimental units are exposed to several different conditions and some response is measured. In the example used by Potvin et al. (1990b), the photosynthetic rate of the same plants exposed to different concentrations of CO_2 was compared (within-subject factor). There were two between-subjects factors: plants were from two different populations and were either chilled or kept at the ambient temperature before measurement of photosynthetic rates.

Investigators of plant–herbivorous insect interactions often are interested in the fitness of different populations of an insect species on different plant species or on different plant populations. A typical protocol is to raise siblings from the same egg masses (subjects) from different populations (between-subjects factor) on leaves of different plant species or plant populations (within-subject factor). Horton et al. (1991) give a detailed analysis of the increase in power when such data are analyzed as a repeated-measures design. The egg mass is considered the block or subject, and there are replicate egg masses from each insect population.

In some frog species, males and females are different colors, and many also change color on different colored backgrounds. King and King (1991) examined differences in color between male and female wood frogs (*Rana sylvatica*) (between-subjects) in which the same individuals were exposed to different colored backgrounds (within-subject). Because color was assessed in terms of three response variables, this was a doubly multivariate design.

8.6 Alternatives to Repeated-measures ANOVA and Profile Analysis

The power of MANOVA decreases as the number of dependent variables increases and the sample size decreases. One reason for the low power of MANOVA is the lack of restrictions on the structure of the variance–covariance matrix (Crowder and Hand 1990). Repeated-measures ANOVA may have more power in some circumstances, but also has much more restrictive conditions for the variance–covariance matrix (Khattree and Naik 1999). One alternative has been to fit curves using the general linear model and then to compare parameters of these curves by multivariate procedures, what statisticians regard as the analysis of growth curves (Morrison 1990; Khattree and Naik 1999). An obvious advantage of fitting data to a curve it that it reduces drastically the number of parameters to be compared for within-subject effects. Also, neither repeated measures ANOVA nor profile analysis can handle missing data for subjects, whereas curve fitting can.

Another alternative, which has been an active area of statistical research, is to take the structure of the variance–covariance matrices into account in the analysis (Crowder and Hand 1990; Hand and Crowder 1996). These regression-based methods often are referred to as random effects models, random coefficient models, or the general linear mixed model (Littell et al. 1996). The advantage of the approach is that it makes available a variety of covariance matrices whose structure is intermediate between the unrestrictive multivariate approach (unstructured) and the restrictive univariate approach (compound symmetry). An additional advantage of the mixed model approach is that it can accommodate subjects with missing data (assumed randomly missing), as well as repeated observations that are unequally spaced (Littell et al. 1996).

Just as the general linear model is a generalization of the basic fixed-effect ANOVA model, the general linear mixed model is a generalization of the mixed-model ANOVA. Repeated measures data are handled easily by the mixed-model approach, since subjects usually are considered a random effect and the within-subject effects a fixed factor (Winer et al. 1991). The method relies on computer-intensive likelihood estimation techniques, rather than those of analysis of variance. The development of software specifically designed for the analysis of data using the mixed model (e.g., procedure MIXED) now makes the method readily accessible to ecologists.

The mixed model analysis involves two basic steps: first, a variance–covariance matrix appropriate for the analysis is selected by comparing fit criteria for alternative covariance structures (>20 available in procedure MIXED), then an analysis of the fixed effects is evaluated using the selected covariance structure. Procedure MIXED is described with examples in SAS/STAT 6.12 Software and Changes and Enhancements (SAS Institute, Inc. 1996). Wolfinger and Chang (1995) give a detailed comparison of a repeated-measures analysis using procedure GLM and procedure MIXED for a classical data set, pointing out important differences and similarities. Littell et al. (1996) compare repeated-measure analyses using procedure GLM and procedure MIXED for a variety of examples and also describe how procedure MIXED can be used for repeated-measures data with observations that are heterogeneous within subject and between subjects. Wolfinger (1996) has a more detailed discussion of the subject with examples and procedure MIXED SAS code. Khattree and Naik (1999) present examples of the classical repeated-measures analyses using MANOVA, ANOVA, and growth curve fitting, emphasizing how subtle difference in the analyses may be required, depending on the experimental design and how data are collected. This is followed by examples of how to formulate analyses of repeated-measures data analysis using procedure MIXED. Dawson et al. (1997) present two graphical techniques to help assess the correlation structure of repeated measures data to aid in selecting the best covariance structure for mixed model analyses. When designs are unbalanced and group covariance matrices differ, based on simulations and observed power differences, Keselman et al. (1999) recommend using the Welch-James type test described by Keselman et al. (1993). The general linear mixed model analysis is an active and evolving area of theoretical and applied statistical research that has many potential applications in ecological research. Ecologists should become familiar with the methodology and should monitor developments in the applied statistical literature that may be appropriate for the analysis of their data.

Potvin et al. (1990b) also used nonlinear curve fitting for repeated-measures analysis and discussed the pros and cons of the procedure. A problem with curve fitting is how sensitive the fitted parameters are to significant changes in the data. See Cerato (1990) for an analysis of parametric tests for comparing parameters in the von Bertalanffy growth equation. Potvin et al. also used a nonparametric approach suggested by Koch et al. (1980), which avoids assumptions of normality, but generally has less power. Manly (1997) describes two examples of using bootstrapping and randomization methods for comparing growth curves. Edgington (1995) also discusses randomization tests for simple repeated-measures designs.

Muir (1986) presents a univariate test for the linear increase in a response variable with increasing levels of the within-subject factor (e.g., selection over time). The advantages of this test are that it increases in power as the number of levels of the within-subject factor increases and it does not have to meet the assumption of circularity because it is testing a specific contrast (see section 8.2.1 and O'Brien and Kaiser 1985). Because the analysis is not straightforward using SAS, a statistician should be consulted when using this technique.

Crossover experiments are considered by some to be related to repeated-measures experiments (Mead 1988; Crowder and Hand 1990, chapter 7; Khattree and Naik 1999, chapter 5). As a very simple example, a group of subjects is first exposed to condition A and then to condition B. A second group is treated in just the opposite sequence. Feinsinger and Tiebout (1991) used such a design in a bird pollination study.

Categorical data also can be analyzed as repeated measures (Agresti 1990, chapter 11). Contingency tables can be analyzed as repeated-measures data based on Mantel–Haenszel test statistics for a single sample, that is, a single dependent variable and one or more explanatory variables (Stokes et al. 1995). Weighted least squares can handle multiple dependent variables (repeated measures) for categorical explanatory variables (Stokes et al. 1995). Both procedures are available in procedure CATMOD (SAS Institute Inc. 1989a). The generalized estimating equation (GEE) approach is a recent methodology for regression analysis of repeated measurements that can be used with categorical repeated-measures data and continuous or discrete explanatory variables. GEE is available in procedure GENMOD (SAS Institute Inc. 1996). Johnston (1996) has an introduction to the procedure that preceded the availability of GEE in procedure GENMOD. Finally, time series analysis is a special kind of repeated-measures analysis in which there are many repeated measurements and may be appropriate in some cases (chapter 9).

8.7 Statistical Packages

Repeated-measures analyses can be done with the major mainframe computer statistical packages: SAS and SPSS. Crowder and Hand (1990) and Hand and Crowder (1997) summarize the methods and capabilities of the routines and give examples. Both packages automatically generate the appropriate contrasts necessary for multiple within-subject factor designs (O'Brien and Kaiser 1985). Windows versions are available for all these programs. Statistica, SYSTAT, and JMP are other statistical packages that run on personal computers and do both univariate and multivariate repeated-measures analyses.

8.8 Conclusions

8.8.1 Texts

Many of the texts and much of the statistical research on repeated-measures analysis over the past several decades have been associated with the psychological literature. The current texts span the spectrum in terms of their coverage of the topic: Winer et al. (1991) present primarily the univariate approach and Maxwell and Delaney (1990) give an extensive discussion of both the univariate and the multivariate approaches. Harris (1985) and Stevens (1996) concentrate on the multivariate approach, but give summaries of the univariate. Hand and Taylor (1987) introduce the multivariate approach with an extensive discussion of contrasts and detailed analyses of real examples. Crowder and Hand (1990) present a statistician's review of the topic at a more technical level than the others and cover techniques not covered in the other texts. Hand and Crowder (1997) focus on the regression (general linear model and general linear mixed model) approaches to repeated-measures analysis. It is mathematically friendlier than Crowder and Hand (1990) but still is intended for statisticians and uses matrix algebra notation. The introductions to the topic in the articles by O'Brien and Kaiser (1985), Gurevitch and Chester (1986), and Potvin et al. (1990b) have been described previously. See also Koch et al. (1980). With the exception of Winer et al. (1991), the trend in the psychologically oriented texts is toward greater reliance on the multivariate approach. Crowder and Hand (1990) and Hand and Crowder (1997), however, do not show this same preference. Zar (1996) has a clear introduction to univariate repeated-measures ANOVA. Milliken and Johnson (1992) have a more extensive discussion, including the effects of unbalanced and missing data, and a brief introduction to the multivariate approach. Hatcher and Stepanski (1994) discuss one-way and two-factor ANOVA repeated-measures designs in detail, as do Cody and Smith (1997), who also include three-factor designs. Both provide SAS code and output for their examples.

8.8.2 Recommendations: Repeated-measures ANOVA
Versus Profile Analysis

There is no unanimity among statisticians in recommending which approach to take in repeated-measures analyses. The alternative views are summarized here.

1. The general opinion seems to be not to use Mauchly's sphericity test as the basis for whether to use univariate repeated measures (O'Brien and Kaiser 1985; Winer et al. 1991; Stevens 1996).
2. If you agree with Mead's (1988) criticism of using time as the within-subject factor, the multivariate approach seems preferable (profile analysis) (O'Brien and Kaiser 1985; Maxwell and Delaney (1990):
   a. Maxwell and Delaney (1990) recommend using MANOVA only when ($N - M > k + 9$), so the test has sufficient power. These conditions may be hard to meet in some ecological systems.

b. If you cannot meet these criteria and cannot increase replication or reduce the number of levels of the within-subject factor, you can follow the recommendation of Maxwell and Delaney (1990) and set up individual multivariate contrasts.

c. Alternatively, the multivariate analysis can be done with low replication, but with α set at 0.1 or 0.2 to compensate for the low power, especially if it is an exploratory study (Stevens 1996).

d. If neither of these is feasible, then a nonparametric approach (Potvin et al. 1990b) can be used.

e. As alternatives to repeated-measures ANOVA when time is the within-subject factor, Mead (1988) suggests individual contrasts on the different time intervals, MANOVA, or curve fitting.

3. If one does not accept Mead's criticism and there is sufficient replication to use the multivariate approach, then there are three different opinions on what to do. The recommendations depend primarily on the authors' views of the respective overall power of the univariate and multivariate approaches for different experimental designs and tests:

a. O'Brien and Kaiser (1985) prefer, and Maxwell and Delaney (1990) give a "slight edge" to, the multivariate approach over the adjusted univariate approach, including performing individual multivariate contrasts if there is trouble meeting the criteria for MANOVA.

b. Stevens (1996) recommends doing both the multivariate and the adjusted univariate tests (α = 0.025 for each test), especially if it is an exploratory study.

c. Finally, Crowder and Hand (1990) say to use either if it is possible to do MANOVA.

Given this variation in opinion, Stevens' (1996) recommendation seems a safe option.

4. There also are several options when you do the univariate repeated-measures analysis:

a. The conservative approach is to use the Greenhouse–Geisser corrected probability. Maxwell and Delaney (1990) recommend this because the Huynh–Feldt adjustment occasionally can fail to properly control Type I errors (overestimates ε).

b. The more liberal approach is to use the Huynh–Feldt adjustment.

c. Stevens (1996) suggests averaging the ε-values of the two corrections.

5. Because there are limitations of both the standard univariate and multivariate techniques, you should consider fitting response curves, nonparametric analysis, and especially the mixed linear model approach described previously (section 8.6). Continued advancements in the theory and applications of the general mixed linear model, along with the development of appropriate software, are expected to make this an important statistical methodology for ecologists in the future.

Acknowledgments   I thank Jessica Gurevitch and Sam Scheiner for their indulgence and especially helpful editorial comments, Ann Hedrick for first introducing me to O'Brien and Kaiser (1985), an anonymous reviewer for highlighting inconsistencies and providing the Muir (1986) reference, Stephen F. O'Brien and Donald R. Brown for answering my naive statistical questions, and Mohsen Pourahmadi for trying to educate me with respect to the mixed linear model. Any mistakes remain my own.

Appendix. SAS Program Code for ANOVA and MANOVA (Profile Analysis) Repeated-measures Analyses

A. Analyses of Data in table 8.1;
   Output in tables 8.2, 8.3, 8.5, 8.6, 8.7

```
PROC ANOVA;                          /use procedure ANOVA for balanced designs/
   CLASS NUTRIENT;
   MODEL WKI WK2 WK3 WK4 WK5                 /NOUNI suppresses ANOVAs
      = NUTRIENT/NOUNI;                          of dependent variables/
   REPEATED TIME 5 (I 2 3 4 5)         /REPEATED = repeated-measures factor label
      PROFILE/SUMMARY;                  TIME. 5 indicates levels of TIME. (I 2 3 4 5) are
                                     intervals of TIME, which can be unequally spaced, default
                                   is equally spaced intervals. PROFILE = profile transformation.
                                     SUMMARY prints the ANOVA for each contrast/
   REPEATED TIME 5 (I 2 3 4 5)         /POLYNOMIAL = polynomial transformation/
      POLYNOMIAL/SUMMARY;
```

B. Repeated-measures Analysis Data in table 8.8
   (Two Within-subject Factors): Output in table 8.9

```
PROC ANOVA;
   CLASS NUTRIENT;
   MODEL YIWI YIW2 YIW3            /YIWI ... YIW5, Y2WI ... Y2W5 identifies the
      YIW4 YIW5 Y2WI Y2W2           I0 dependent variables for the five weeks
      Y2W3 Y2W4 Y2W5 =                 in the two years/
      NUTRIENT/NOUNI;
   REPEATED YEAR 2, WEEK 5         /YEAR = first within-subject factor with 2 levels,
      POLYNOMIAL/SUMMARY;          WEEK = second within-subject factor with 5 levels.
                                     See SAS Institute Inc. (1989b, p. 926) for an
                                   explanation of how to define dependent variables
                                     when there are two within-subject factors./
```

9

Time Series Intervention Analysis: Unreplicated Large-scale Experiments

PAUL W. RASMUSSEN

DENNIS M. HEISEY

ERIK V. NORDHEIM

THOMAS M. FROST

9.1 Unreplicated Studies

Some important ecological questions, especially those operating on large or unique scales, defy replication and randomization (Schindler 1987; Frost et al. 1988; Carpenter et al. 1995). For example, suppose we are interested in the consequences of lake acidification on rotifer populations. Experimental acidification of a single small lake is a major undertaking, so it may be possible to manipulate only one lake. Even with baseline data from before the application of the acid, such a study would have no replication, and thus could not be analyzed with classical statistical methods, such as analysis of variance (chapter 4).

An alternative approach might be to use some biological or physical model of the system of interest that allows for replication. For example, we could construct small replicated enclosures in lakes, and acidify some of them. Although this would (with proper execution) permit valid statistical analysis of differences among units the size of the enclosures, it is questionable whether this model allows valid ecological generalization to the lake ecosystem (Schindler 1998). For large-scale phenomena, experiments on small-scale models may not be a trustworthy ecological substitute for large-scale studies (although they may provide valuable supplementary information).

In this chapter, we examine how certain types of unreplicated studies can be analyzed with techniques developed for time series data. Time series are repeated measurements, or subsamples, taken on the same experimental unit through time. Time series analysis techniques include methods for determining whether nonrandom changes in the mean level of a series have occurred at prespecified times.

The results of such analyses can help determine whether other changes or manipulations occurring at those prespecified times may have caused changes in the observed series.

The methods we describe take advantage of the long-term data available for many large-scale studies (see also Jassby and Powell 1990; Stow et al. 1998). Previous authors have proposed a variety of techniques for assessing such studies, ranging from graphical approaches (Bormann and Likens 1979) to more sophisticated statistical analyses (Matson and Carpenter 1990). These techniques share an emphasis on time series data and usually involve comparing a series of pre- and posttreatment measurements on a treatment and a reference system (e.g., Stewart-Oaten et al. 1986; Carpenter et al. 1989). We will confine our discussion of time series methods to the use of ARIMA modeling techniques, although many other time series methods are available.

We first consider how time series data affects the use and interpretation of classical statistical analysis methods (section 9.2). Section 9.3 presents some examples of unreplicated time series designs. Section 9.4 introduces some of the key ideas of time series analysis. Section 9.5 describes intervention analysis, an extension of time series analysis useful for examining the impact of a treatment or natural perturbation. Section 9.6 illustrates the application of intervention analysis to data from a whole-lake acidification experiment. (If you are not an aquatic ecologist, mentally replace "rotifer" and "lake" with your favorite organism and ecosystem, respectively, in our examples. The principles and potential applications are general.) We discuss some general issues in section 9.7.

9.2 Replication and Experimental Error

We first return briefly to the lake acidification example. The prescription in classical experimental design for testing the effect of lake acidification on rotifer densities would be similar to the following:

1. Identify all lakes to which we want our inferences to apply.
2. Select, say, four lakes at random from this population.
3. Randomly select two of the four lakes to be acidified.
4. Use the other two lakes as reference, or control, lakes.

Suppose estimates of pre- and posttreatment rotifer densities are available for all four lakes; for each lake, we can compute the change (difference) between the pre- and posttreatment periods for further analysis. Classical analysis then proceeds to examine the question of whether the changes in the acidified lakes are consistently different from the changes in the reference lakes, judged in light of the typical lake variability. This typical lake variability, or experimental error, can be estimated only by having replicate lakes within each group. (Two lakes per group is the minimum that permits estimation of experimental error, although a design with only two lakes per group would normally return a rather poor estimate of experimental error [Carpenter 1989].) An important feature of the classical approach is that we can specify the probability of mistakenly declaring an acidification effect when none exists.

Although it may be difficult to acidify more than one lake, getting a long time series of pre- and posttreatment data from a single acid-treated lake may be feasible. With such a series, we can examine the related question of whether there was an unusual change in the series at the time of the treatment, judged in light of normal variation in the series through time. The hypothesis tested here is weaker than the classical hypothesis. A test of this hypothesis can answer only the question of whether a change occurred at the time of the treatment; it cannot resolve the issue of whether the change was due to the treatment rather than to some other coincidental event (Frost et al. 1988; Carpenter et al. 1998). Making the case for the change being due to the treatment in the absence of replication is as much an ecological issue as a statistical one; such arguments can often be supported by corroborating data, such as from enclosure experiments. In the next section, we discuss time series designs that to some extent guard against the detection of spurious changes.

Determining the normal variation through time of a time series is a more difficult issue than determining experimental error in a replicated experiment. In a replicated, randomized, experiment, the experimental units are assumed to act independently of one another, which greatly simplifies the statistical models. On the other hand, measurements in time series are usually serially dependent or autocorrelated; the future can (at least in part) be predicted from the past. The variation of a system through time depends on its autocorrelation structure, and much of time series analysis focuses on constructing and evaluating models for such structure.

We caution the reader to guard against treating subsampling through time as genuine replication and then analyzing the unreplicated experiment as if it were replicated. In some cases, it may turn out that a classical test, such as a t-test, can be applied to a time series. Such a test could be used if analysis indicated that the measurements were not autocorrelated (Stewart-Oaten et al. 1986). Even then it must be remembered that the hypothesis tested has to do with a change in level at the time of the treatment application, and that such a change may not be due to the treatment. A basic understanding of these issues is necessary for conducting effective ecological experimentation; see the discussions in Hurlburt (1984) and Stewart-Oaten et al. (1992).

We do not advocate unreplicated designs as a matter of course, but they are sometimes necessary. Even with the detailed machinery of time series, some caution will be required in the interpretation of results. Without replication, we cannot be sure that we have an adequate understanding of the underlying error against which treatment effects should be judged. Increasing confidence in making such an interpretation can be developed in several ways, however. These include (1) considering information on the natural variability that is characteristic of the types of systems being manipulated, (2) integrating smaller scale experiments within the large-scale manipulations, and (3) developing realistic, mechanistic models of the process being evaluated (Frost et al. 1988).

9.3 Unreplicated Time Series Designs

Suppose an ecologist monitors rotifer densities in a lake biweekly for 5 years, then acidifies it and continues to monitor the lake for an additional 5 years. In this example, there is a sequence of observations over time (a time series) on a single experimental unit (the lake). We refer to such a single series with an intervention as a before–after design. (Acidification is the intervention).

Sometimes nature, perhaps unexpectedly, applies a perturbation to a system. This can be called a *natural experiment*. In such a case, there may be no choice but to use a simple before–after design. Such designs are most prone to having coincidental "jumps" or trends spuriously detected as intervention effects. When the treatment is under the investigator's control, improvements in the before–after design are possible. One simple modification is to have multiple interventions or to switch back and forth between the treatment and control conditions. If each intervention is followed by a consistent response, the likelihood that the observed responses are merely coincidental is reduced.

A design with paired units is also useful. One unit receives a treatment (intervention) during the course of the study, and the other serves as a reference, or baseline, and is never treated. Both units are sampled repeatedly and at the same times before and after the treatment is applied. This design has been discussed by Stewart-Oaten et al. (1986), and earlier by Eberhardt (1976), although it had been in use among field biologists before either of these discussions (see Hunt 1976). We will refer to this design as the before-after-control-impact (BACI) design, following Stewart-Oaten et al. (1986). In a BACI design, the reference unit provides a standard against which to compare the treatment unit. This helps to determine whether changes seen in the treatment unit are due to the treatment itself, long-term environmental trends, natural variation in time, or some other factor.

Two important decisions involved in implementing a BACI design are the selection of experimental units, and the timing and number of samples per unit. The treatment and reference units should be representative of the systems to which an observed treatment effect is to be generalized, and they should be similar to one another in their physical and biological characteristics. Typically, samples are evenly spaced, with the time between samples dependent on the rate of change in the population or phenomenon studied (Frost et al. 1988). For instance, rapidly changing populations of insects or zooplankton are often sampled many times a year, whereas slowly changing vertebrate or plant populations are usually sampled only once a year. The sampling frequency will depend on both the system studied and the specific questions asked. The duration of each evaluation period (before or after treatment) is determined by the periodicity of natural cycles, the magnitude of random temporal fluctuations, and the life span of the organisms studied. Longer sampling periods are necessary when random fluctuations are greater, cycles have longer periods, or study organisms are longer-lived.

We will use an example of a BACI design with multiple interventions to demonstrate in some detail the analysis and interpretation of a "real" unreplicated design. This concerns the effect of acidification on Little Rock Lake, a small, oligotrophic seepage lake in northern Wisconsin (Brezonik et al. 1986). Since 1983, it has been the site of an ecosystem-level experiment to investigate the responses of a seepage lake to gradual acidification (Watras and Frost 1989). Little Rock Lake was divided into a treatment and a reference basin with an impermeable vinyl curtain. After a baseline period (August 1983 through April 1985), the treatment basin was acidified with sulfuric acid in a stepwise fashion to three target pH levels, 5.6, 5.1, 4.7, each of which was maintained for 2 years. The reference basin had an average pH of 6.1 throughout the experiment. Details on the lake's limnological features are provided in Brezonik et al. (1986). Our analysis here will focus on populations of the rotifer *Keratella taurocephala*, which exhibited marked shifts with acidification (Gonzalez et al. 1990; Gonzalez and Frost 1994).

Creating good unreplicated designs requires some creativity. So does the analysis of the resulting data. The analytical methods for most studies with genuine replication (e.g., analysis of variance) are quite straightforward. However, much of the analysis of unreplicated experiments is less clear-cut and tends to be more subjective. With time series data of sufficient length (at least 50 observations), it may be possible to use statistical techniques that have a rigorous theoretical underfooting. We discuss some of these techniques in sections 9.4 and 9.5. The general goal of such statistical modeling is to find a parsimonious model, a simple model that fits the data well. By simple, we mean a biologically and physically reasonable model with a small number of parameters to be estimated. We will use analytical techniques appropriate to time series with an intervention. As will be seen in the next two sections, proper use of these techniques requires mastery of certain fundamentals and considerable care. Unless you are quite skilled in the area, we recommend consultation with a statistician when conducting the analysis. (Such consultation at the design phase is also strongly recommended.)

9.4 Time Series

Time series analysis encompasses a large body of statistical techniques for analyzing time-ordered sequences of observations with serial dependence among the observations. We will focus on a subset of time series techniques developed for autoregressive integrated moving average (ARIMA) models. The classic reference for ARIMA modeling is Box and Jenkins (1976). Other references include McCleary and Hay (1980), Cryer (1986), Diggle (1990), and Wei (1990).

ARIMA models are valuable because they can describe a wide range of processes (sequences of events generated by a random model) using only a few parameters. In addition, methods for identifying, estimating, and checking the fit of ARIMA models have been well studied. These models are appropriate for modeling time series where observations are available at discrete, (usually) evenly spaced, intervals.

ARIMA models include two basic classes of models: autoregressive (AR) and moving average (MA) models. AR models are similar to regression models in which the observation at time *t* is regressed on observations at earlier times. First, we introduce some notation. Let y_t be the original time series, say, rotifer density at time t. It is convenient to work with the series centered at its mean, say, $z_t = y_t - \mu$; z_t is the deviation of the density at time t from the long term mean μ. An AR model of order $p = 2$ [denoted AR(2)] for the centered rotifer density is of the following form:

$$z_t = \phi_1 z_{t-1} + \phi_2 z_{t-2} + \varepsilon_t \tag{9.1}$$

where the ϕ's are coefficients (like regression coefficients) and ε_t is a random error at time t (usually assumed to be uncorrelated with mean 0 and variance σ^2). This model states that the present rotifer density is a linear function of the densities at the previous two sampling times plus random error.

An MA model relates an observation at time t to current and past values of the random error ε_t. For example, an MA model of order $q = 2$ [denoted MA(2)] has the form:

$$z_t = \varepsilon_t + \theta_1 \varepsilon_{t-1} + \theta_2 \varepsilon_{t-2} \tag{9.2}$$

where the θ's are coefficients.

More general models can have both AR and MA terms, so-called ARMA models (note that ARIMA models include ARMA models, as described subsequently). The goal of fitting an ARMA model is to describe all of the serial dependence with autoregressive and moving average terms so that the residuals, or estimated error terms, look like uncorrelated random errors, or "white noise."

ARMA processes have the same mean level, the same variance, and the same autocorrelation patterns, over whatever time interval they are observed. This is an intuitive description of a fundamental property known as stationarity. (More rigorous definitions can be found in the references). Stationarity allows valid statistical inference and estimation despite a lack of genuine replication. A stationary process exhibits the same kind of behavior over different time intervals, as if replication were built into the process. Because the autocorrelation between observations depends on the number of time steps between them, the observations usually must be equally spaced. For some biological processes, however, it is not clear that this is necessary. For instance, the autocorrelation among zooplankton abundance estimates 2 weeks apart may be larger in the winter when the population is changing slowly than in the summer when it changes more quickly.

Because many processes observed in nature are clearly not stationary, it is desirable to find a way to modify an observed time series so that the modified series is stationary; then ARMA models with few parameters can be fit to the modified data. There are two primary ways to do this. One is to incorporate deterministic functions into the model, such as a linear trend over time, a step increase at a known time, or periodic functions, such as sines and cosines, to represent seasonal behavior. The other way is to difference the series, that is, to compute new observations such as $z_t - z_{t-1}$ (first difference) or $z_t - z_{t-12}$ (seasonal difference with period 12). Differencing is a more general approach because it can represent both deterministic and random trends. On the other hand, when deterministic trends are of special interest, it may make more sense to fit them than simply to remove them by differencing. Because of its greater generality, we will focus on achieving stationarity by differencing.

Describing some processes may require both AR and MA parameters as well as differencing. This leads to ARIMA models with p AR parameters, q MA parameters, and d differences [denoted ARIMA(p,d,q)]. For instance, an ARIMA(1,1,1) model is of the form

$$x_t = \phi_1 x_{t-1} + \varepsilon_t + \theta_1 \varepsilon_{t-1} \qquad (9.3)$$

where $x_t = z_t - z_{t-1}$ and ε_t is uncorrelated error as previously stated.

Box and Jenkins (1976) developed methods for identifying appropriate forms for ARIMA models from data. Their methods require computing sample autocorrelations and related functions from the data and using the properties of these functions to identify possible ARIMA models. Each sample autocorrelation, r_k, at lag k (i.e., the correlation among observations k time steps apart) is computed for lags 1, 2, . . . to obtain the sample autocorrelation function (ACF). A related function, the sample partial autocorrelation function (PACF), represents the sample autocorrelation among observations k time steps apart, when adjusted for autocorrelations at intermediate lags.

The model identification process involves identifying a subset of ARIMA models for fitting from the ACF, the PACF, and plots of the original series. The original series may show a long-term trend or seasonal behavior and thus suggest nonstationarity. The ACF for nonstationary series also declines very slowly to zero. If the series appears nonstationary, it should be differenced until it is stationary before examining the ACF and PACF further.

The theoretical ACF and PACF give distinctive signatures for any particular ARIMA model. In ARIMA modeling, the sample ACF and PACF are estimated from the (perhaps differenced) data with the hope of recognizing such a signature, and thereby identifying an appropriate ARIMA parameterization. The theoretical ACF for pure MA models has large values only at low lags, whereas the PACF declines more slowly. The number of lags with autocorrelations different from zero suggests q, the order of the MA model. The opposite is true for pure AR models, where the PACF has large values at low lags only and the ACF declines more slowly. In this case, the order p of the AR model is suggested by the number of partial autocorrelations different from zero. Often AR or MA models of low order (1 or 2) provide good fits to observed series, but models with both AR and MA terms may be required. The adequacy of fit of the model can be checked by examining the residuals. The sample ACF and PACF of the residuals should not have large values at any lags and should not show strong patterns of any sort. Box and Pierce (1970) suggested an overall test for lack of fit, which has since been refined (Ljung and Box 1978).

9.5 Intervention Models

Intervention analysis extends ARIMA modeling methods to investigate the effects of known events, or interventions, on a time series. The response to a treatment, or intervention, can have a number of forms. The two simplest forms are a permanent jump to a new level after the intervention and a temporary pulse, or spike, at the intervention. For example, a step change in an ARIMA model can be represented by a model of the form

$$z_t = \omega\, S_t + N_t \tag{9.4}$$

where $S_t = 0$ before intervention and $S_t = 1$ at and after intervention, ω is a coefficient, and N_t is an ARIMA model. In this representation, z_t is the suitably differenced series if the original series was not stationary. To model a spike or pulse response, $S_t = 1$ at the intervention and 0 otherwise. Box and Tiao (1975) discuss more complicated models such as linear or nonlinear increases to new levels. Examples are illustrated in McCleary and Hay (1980). The lag between the intervention and the response can also be examined by fitting models with different lags.

There is usually no direct way to identify the ARIMA form of the intervention model from the observed time series itself. Typically, a plausible ARIMA model is developed for the series as a whole or separately for the series before and after the intervention. The form of the response to intervention may be suggested by examination of residuals from the ARIMA model without the intervention or from theoretical knowledge about the expected response. Usually a number of plausible models must be examined and the best chosen from among them. If the intervention form is unknown, McCleary and Hay (1980) suggest first trying the spike model, then the gradual increase model, and finally the step model. Box and Jenkins (1976) have emphasized the iterative approach to model building, where the goal is to obtain a model that is simple but fits the data adequately and that has a reasonable scientific interpretation. In following this iterative model-building philosophy, a series of decisions is made, thus analyses of different data sets will proceed in different ways. Although we can give a general outline of the steps involved in intervention analysis, we cannot exactly prescribe a method that will work for all data sets.

Although each analysis is unique, the following sequence of steps is useful in carrying out intervention analysis:

1. Plot the time series.
2. Choose a transformation of the original series, if necessary.
3. Determine whether the series must be differenced.
4. Examine the sample ACF and PACF of the (possibly) transformed and differenced series to identify plausible models.
5. Iterate between fitting models and assessing their goodness of fit until a good model is obtained.

We emphasize that the way an analysis proceeds depends on the data set. The following example should not be viewed as a prescription for intervention analysis, but as an example of a general process. Other examples of the use of intervention analysis in ecology and environmental monitoring have been discussed by Pallesen et al. (1985), Madenjian et al. (1986), van Latesteijn and Lambeck (1986), Noakes (1986), Bautista et al. (1992), and Rudstam et al. (1993).

9.6 Example

9.6.1 Data

We will analyze abundance data (animals/liter) for the rotifer *Keratella tauro-cephala* from Little Rock Lake, Wisconsin. Sampling methods are described in Gonzalez et al. (1990). Data were collected every 2 weeks when the lake was ice free, and every 5 weeks when the lake was ice covered. There were slight variations from this schedule during the first years of monitoring; we have dropped extra observations or used an observation for two consecutive sampling dates when necessary to produce the same schedule during all years (we dropped a total of eight observations, and used six observations twice). We will discuss the effects of spacing the observations, as well as ways to obtain equal spacing, in section 9.6.5. The example data set includes 19 observations per year (106 total), with observations on approximately the same dates during all years. We will examine the effect of two interventions: the pH drop from 6.1 to 5.6 on 29 April 1985 and from 5.6 to 5.1 on 27 April 1987. We have not used data collected after the third intervention on 9 May 1989.

One of the early steps in any data analysis should be careful examination of plots of the data. The plot of the abundance series for both basins (figure 9.1A) suggests that there is little change in *K. taurocephala* abundance in the reference basin over time, but there is an increase in both the level and variability of abundance in the treatment basin after the interventions.

9.6.2 Derived Series

The first decision is whether to analyze the series for the two basins separately or to analyze a single series derived from the two original series. There are advantages to both approaches. The single derived series may be simpler to analyze: not only is there only one series, but the derived series may have less serial autocorrelation and less pronounced seasonal behavior than the original series (Stewart-Oaten et al. 1986). In addition, the analysis of the derived series may lead to a single direct test of the intervention effect, whereas separate analyses of the treatment and reference series would have to be considered together to evaluate the effect of the intervention. On the other hand, the derived series is more remote from the observed data, and we may wish to model the serial dependence, seasonal behavior, and effects of interventions in the original series themselves. If temporal patterns in the two units are quite different, it may be essential to analyze the two series separately. In many cases, both approaches may be appropriate. We will discuss both here.

Two forms of derived series are possible: (1) the series of differences between the treatment and reference observations at each sampling date and (2) the series of ratios of the treatment to reference observations at each date. Most discussions of this topic in the ecological literature have argued that the log-transformed ratio series (equivalent to the difference of the two log series) best meets the assumptions of statistical tests (Stewart-Oaten et al. 1986, Carpenter et al. 1989; Eber-

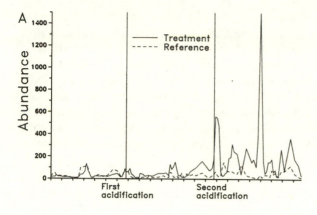

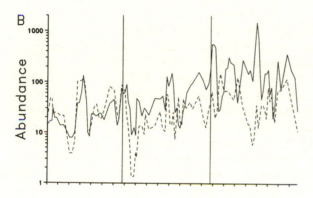

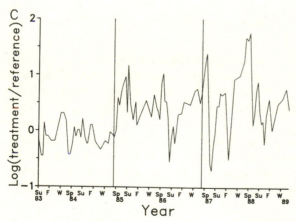

Figure 9.1 Time series for *Keratella taurocephala* in Little Rock Lake, Wisconsin, during acidification treatments. (A) The abundance series for both the treatment and reference basins shown on the original scale, animals/liter. (B) The same series shown on a $\log_{10}$ scale. (C) The series derived as a $\log_{10}$ of the ratio of the treatment series over the reference series.

hardt and Thomas 1991). The issue in part revolves around the assessment of whether factors leading to change in the response are additive or multiplicative. Because factors that affect populations often have multiplicative effects, the series of ratios may work best for abundance data. It is usually best to examine both derived series. For the *K. taurocephala* data, both derived series showed similar patterns, with the mean level and variability of the series increasing after each intervention. The increase in variability was much greater for the series of differences; we chose to work with the ratio series because we could find a transformation that more nearly made the variance of that series constant. Note that we added one to each value to avoid dividing by zero.

9.6.3 Transformations

A fundamental assumption of ARIMA models, as well as of many standard statistical models, is that the variance of the observations is constant. The most common violation of this assumption involves a relationship between the variance and the mean of the observations. We will use a simple method for determining a variance-stabilizing transformation when the variance in the original series is proportional to some power of the mean (Box et al. 1978). Poole (1978) and Jenkins (1979) have described the use of this procedure in time series analysis. Jenkins (1979) recommends dividing the series into subsets of size 4 to 12, related to the length of the seasonal period (we used 6-month intervals). Then the mean and variance are calculated for each subset, and the log standard deviation is regressed on the log mean. The value of the slope (*b*) from this regression suggests the appropriate transformation (see SAS commands in appendix 9.1). The assumption that the variance is proportional to some power of the mean determines the form of the transformation:

$$z = y^{1-b} \qquad (1 - b) \neq 0 \qquad\qquad (9.5)$$
$$= \log y \qquad b = 1$$

For the ratio series, the slope from the regression of log standard deviation on log mean was 1.33. The estimate of 1.33 has some associated error; we traditionally use a value near the estimate that corresponds to a more interpretable transformation rather than the exact estimate. Thus, we might consider either the reciprocal square-root transformation (the reciprocal of the square root of the original value; appropriate for a slope of 1.5) or the log transformation (for a slope of 1.0). We carried out analyses on both scales and found that conclusions were similar, although the reciprocal square-root scale did a slightly better job of stabilizing the variance. We will report analyses of the log ratio series (we used logs to the base 10), since that transformation is more familiar and somewhat easier to interpret.

9.6.4 ARIMA Models

The plot of the abundance series on a $\log_{10}$ scale for both basins (figure 9.1B) presents a very different appearance than that of the abundance series (figure

9.1A). The large increase in both level and variability of the treatment basin series on the original scale is much reduced on the log scale. The reference basin series may show a slight increasing trend on the log scale. The log ratio series (figure 9.1.C) appears to have increased in level after the first intervention and perhaps in variability after the second intervention. These subjective impressions should be kept in mind while carrying out the statistical analyses.

Methods for identifying ARIMA models require that the series be stationary. If an intervention changes the level of a series, that series is no longer stationary, and standard model identification procedures may be difficult to apply. One approach to determining the ARIMA form for an intervention series is to fit models to the segments of the series between interventions. Before discussing the results of this process for the log ratio series, we will first demonstrate model identification procedures on the longest stationary series available, the log reference series.

The first step in ARIMA model identification is to examine the sample ACF and PACF of the series (appendix 9.2A). The sample ACF and PACF are usually presented as plots (see SAS output in table 9.1) in which each autocorrelation or partial autocorrelation is a line extending from zero, with 2 standard error (SE) limits (approximate 95% confidence limits) indicated on the plot. The sample ACF for the log reference series has large values for lags 1, 2, and 3, and values less than the 2 SE limits elsewhere; the sample PACF has a single large value at lag 1. This pattern suggests an AR(1) process, since the PACF "cuts off" at lag 1, whereas the ACF declines more slowly to zero. The PACF for an AR(2) process would have large values at both lags 1 and 2, for instance. Both the sample ACF and PACF have values at lags 18 or 19 that almost reach the 2 SE limits; this suggests that there may be a weak annual cycle.

Once a model form has tentatively been identified, the next step is to fit that model to the series (appendix 9.2B) and examine the residuals. In this case, the residuals from fitting an AR(1) model have the characteristics of random, uncorrelated error (white noise): there are no large values or strong patterns in the sample ACF or PACF, and a test of the null hypothesis that the residuals are random and uncorrelated is not rejected [SAS procedure ARIMA carries out this test whenever the ESTIMATE statement is used to fit a model (SAS Institute Inc. 1988)]. The AR(1) model produced estimates for the mean of 1.44 (SE = 0.08), an autoregressive coefficient ($\hat{\phi}$) of 0.65 (SE = 0.07), and a residual variance of 0.084 (these estimates and standard errors are included in the output produced by SAS procedure ARIMA when the ESTIMATE statement is used).

Carrying out the same procedure on the three segments of the log ratio series between interventions suggests that all three can be fit by AR(1) models. This leads to the following parameter estimates (SE values are in parentheses):

Segment	$\hat{\mu}$	$\hat{\phi}$	Variance
1: pH 6.1	−0.10 (0.05)	0.28 (0.18)	0.04
2: pH 5.6	0.37 (0.09)	0.47 (0.15)	0.09
3: pH 5.1	0.54 (0.19)	0.62 (0.13)	0.21

Table 9.1 Plots of the autocorrelation function (ACF) and partial autocorrelation function (PACF) from the program listed in appendix 7.2A

```
                        ARIMA Procedure

                 Name of variable = LOGREF.

                 Mean of working series = 1.443589
                 Standard deviation      = 0.379975
                 Number of observations =      106

                        Autocorrelations

Lag Covariance Correlation -1 9 8 7 6 5 4 3 2 1 0 1 2 3 4 5 6 7 8 9 1        Std
  0   0.144381    1.00000  |                     |********************|          0
  1   0.093730    0.64918  |                   . |*************        |   0.097129
  2   0.049058    0.33978  |                   . |*******             |   0.131855
  3   0.028400    0.19670  |                   . |****  .             |   0.139871
  4   0.010254    0.07102  |                   . |*     .             |   0.142457
  5  0.0034776    0.02409  |                   . |      .             |   0.142791
  6  0.0010632    0.00736  |                   . |      .             |   0.142829
  7 -0.0067425   -0.04670  |                   . *|      .             |   0.142833
  8 -0.0084151   -0.05828  |                   . *|      .             |   0.142977
  9 -0.0005594   -0.00387  |                   . |      .             |   0.143201
 10 -0.0014308   -0.00991  |                   . |      .             |   0.143202
 11 -0.0089326   -0.06187  |                   . *|      .             |   0.143208
 12 -0.0090990   -0.06302  |                   . *|      .             |   0.143460
 13 -0.0028932   -0.02004  |                   . |      .             |   0.143721
 14  0.00078262   0.00542  |                   . |      .             |   0.143747
 15  0.0053034    0.03673  |                   . |*     .             |   0.143749
 16 -0.0010200   -0.00706  |                   . |      .             |   0.143838
 17  0.00089464   0.00620  |                   . |      .             |   0.143841
 18   0.022174    0.15358  |                   . |***   .             |   0.143844
 19   0.026635    0.18448  |                   . |****  .             |   0.145382
 20   0.010681    0.07398  |                   . |*     .             |   0.147574
 21 -0.0084435   -0.05848  |                   . *|      .             |   0.147924
 22  -0.026879   -0.18617  |                   . ****|     .             |   0.148142
 23  -0.026095   -0.18074  |                   . ****|     .             |   0.150333
 24  -0.023388   -0.16199  |                   . ***|     .             |   0.152369
                                  "." marks two standard errors
                        Partial Autocorrelations

     Lag Correlation -1 9 8 7 6 5 4 3 2 1 0 1 2 3 4 5 6 7 8 9 1
       1    0.64918  |                   . |*************        |
       2   -0.14114  |                  .***|      .             |
       3    0.06457  |                   . |*     .             |
       4   -0.09598  |                  . **|      .             |
       5    0.04833  |                   . |*     .             |
       6   -0.01962  |                   . |      .             |
       7   -0.07093  |                   . *|      .             |
       8    0.01676  |                   . |      .             |
       9    0.07100  |                   . |*     .             |
      10   -0.06325  |                   . *|      .             |
      11   -0.07167  |                   . *|      .             |
      12    0.02016  |                   . |      .             |
      13    0.05950  |                   . |*     .             |
      14   -0.00623  |                   . |      .             |
      15    0.02763  |                   . |*     .             |
      16   -0.09466  |                  . **|      .             |
      17    0.10644  |                   . |**    .             |
      18    0.20260  |                   . |****  .             |
      19   -0.05757  |                   . *|      .             |
      20      13497  |                  .***|      .             |
      21   -0.12215  |                  . **|      .             |
      22   -0.11148  |                  . **|      .             |
      23    0.08049  |                   . |**    .             |
      24   -0.09544  |                  . **|      .             |
```

Although all three segments have the same AR(1) form, parameter estimates and variances may differ. These estimates may be poor because the series are short (between 31 and 38 observations), but the differences suggest some caution when interpreting the intervention model. Differences between the mean levels for the three segments provide estimates for the effects of the interventions (see subsequent discussion). Notice that if the reference and treatment series had the same mean level initially, the estimated mean for the log ratio series would be zero at pH 6.1.

The preceding analyses indicate that we can begin fitting intervention models for all series by assuming an AR(1) form. We also must specify the form for the interventions. The acid additions to the treatment basin were step changes, so they may be represented by variables that take the value of zero before, and one after, the intervention. These variables are just like dummy variables used in regression (Draper and Smith 1981). We have constructed two dummy variables, one for each intervention. The first has the value of one between 29 April 1985 and 27 April 1987, and zero elsewhere, the second has the value of one after 27 April 1987 and zero elsewhere. In this parameterization, the parameter for the mean estimates the initial level of the series, and the intervention parameters estimate deviations from that initial level to the levels after the interventions. Alternative dummy variables can be used and will lead to the same predicted values, although the interpretation of the intervention parameters will differ. The dummy variables are treated as input variables when using the ESTIMATE statement in SAS procedure ARIMA to fit the model (appendix 9.3).

We fit intervention models to the log ratio series and also to the log treatment basin and log reference basin series separately (appendix 9.3). In all cases, the simple AR(1) model fit well, although for the log treatment series there was also some evidence for an annual cycle. We also fit models for more complicated responses to intervention, but the additional parameters were not significantly different from zero. Results for the three series are presented here (SE values are in parentheses after each estimate; $\hat{\omega}_1$ and $\hat{\omega}_2$ are coefficients for the first and second intervention parameters):

Series	$\hat{\mu}$	$\hat{\phi}$	$\hat{\omega}_1$	$\hat{\omega}_2$	Variance
Ratio	−0.06 (0.13)	0.57 (0.08)	0.40 (0.17)	0.60 (0.18)	0.12
Treatment	1.40 (0.12)	0.56 (0.08)	0.23 (0.15)	0.69 (0.16)	0.09
Reference	1.46 (0.12)	0.60 (0.08)	−0.16 (0.16)	0.09 (0.16)	0.08

We can construct statistics similar to t-statistics for the parameter estimates by dividing each estimate by its SE, just as in linear regression. Approximate tests can be based on the standard normal distribution; P-values can be determined from the Z-score using a table or software package.

Models for all three series have similar autoregressive coefficients; the value of approximately 0.6 indicates that there is strong serial correlation in all series. The intervention coefficients provide estimates of the shift in level from the initial baseline period. Thus, for the log ratio series, 0.4 is the shift from the initial level of −0.06 to the level after the first intervention, and 0.6 is the shift from the initial level to the level after the second intervention. Both of these coefficients are significantly different from zero ($Z = 2.35$, $P = 0.02$; and $Z = 3.33$, $P = 0.001$). Shifts at the time of the second intervention can be computed as the difference, $\hat{\omega}_2 - \hat{\omega}_1$. The largest shift in level for the log ratio series was at the first intervention, whereas for the log treatment series it was at the second intervention.

Notice that the differences between intervention parameters for the treatment and reference series approximately equal the parameters for the log ratio series [i.e., $0.23 - (-0.16) = .39$ and $0.69 - 0.09 = 0.60$]. The larger shift at the time of the first intervention for the log ratio series is due to an increase for the log treatment series and a decrease for the log reference series, whereas the smaller shift in the log ratio series at the time of the second intervention is due to increases in both the log reference and treatment series. [Such additivity of the individual series and the derived, transformed series should not be expected in general. It occurs with the log ratio because $\log(x/y) = \log(x) - \log(y)$.]

The intervention parameters are more easily interpreted when converted back to the original abundance scale. For the log ratio series, the initial predicted level is $\log_{10}(t/r) = -0.06$, which implies $t = 0.87r$ (t is the treatment basin value and r is the reference basin value). Thus, the treatment abundance is 0.87 as large as the reference abundance before the first intervention. After the first intervention, the level of $-0.06 + 0.40 = 0.34$ implies that $t = 2.19r$, and after the second intervention, the level of $-0.06 + 0.60 = 0.54$ implies that $t = 3.47r$. The shift in level of 0.23 at the time of the first intervention in the log treatment series corresponds to a multiplication of $10^{0.23} = 1.70$ in the abundance of the rotifer. The shift at the second intervention represents a multiplication of $10^{0.69} = 4.9$ times the initial abundance.

9.6.5 Spacing of Observations

We analyzed our series as though the observations were equally spaced when in fact the spacing differed between summer and winter. To investigate how the spacing of observations may have affected our analyses, we constructed two series with equally spaced observations from the log ratio series and fit intervention models to them. Some approaches to constructing series with equally spaced observations are as follows:

1. Interpolate to fill in missing observations.
2. Aggregate data to a longer sampling interval (Wei 1990).
3. Use a statistical model to fill in missing observations (Jones 1980).

We used method 1 to construct a series with 24 observations per year by interpolating between winter observations taken at 5-week intervals. We also used method 2 to construct a series with 13 observations per year by averaging all observations within each 4-week time period.

Both of these constructed series were adequately fit with AR(1) models, and estimates of intervention effects were very similar to those from the model for the original series. The variance was smaller for the two constructed series because the process of constructing them involved smoothing. The size of the auto-regressive parameter increased as the frequency of observation increased. The important point here is that the intervention parameter estimates were not much affected by the spacing of observations. Thus, we can feel comfortable that the unequal spacing has not distorted our estimation of the intervention parameters. We chose to describe analyses of the data set that corresponded most closely to the actual observations.

9.6.6 Comparison with Other Analytical Procedures

If we ignore the serial dependence among observations over time, we can compare the postintervention data with the preintervention data using t-tests. Recall that a fundamental assumption of the t-test and similar classical statistical procedures is that the observations are independent. The *K. taurocephala* series we analyzed had autocorrelations of about 0.57 for observations 1 time unit apart, with the autocorrelations decreasing exponentially as the number of time units between observations increased. The t-statistic is positively inflated when observations are positively correlated, and the degree of inflation increases as the autocorrelation increases (Box et al. 1978; Stewart-Oaten et al. 1986). For the log ratio series, the Z-statistics calculated for the intervention coefficients were one-third to one-half the size of the t-statistics calculated by standard t-tests comparing the preintervention period to either of the postintervention periods. In this case, all tests are significant, but in situations where the intervention has a smaller effect, the t-test could lead to misleading conclusions.

Randomized intervention analysis (RIA) has been proposed as an alternative approach for analyzing data from experimental designs, such as we have discussed here (Carpenter et al. 1989). Randomization tests involve comparing an observed statistic with a distribution derived by calculating that statistic for each of a large number of random permutations of the original observations (chapter 7). Although randomization tests do not require the assumption of normality, they may still be adversely affected by the lack of independence among observations (chapter 14). Empirical evaluations suggest that RIA works well even when observations are moderately autocorrelated (Carpenter et al. 1989), although Stewart-Oaten et al. (1992) question the generality of these evaluations. Conclusions drawn from RIA for the Little Rock Lake *K. taurocephala* data parallel those we found (Gonzalez et al. 1990).

Although his approach is not possible when there is only one reference unit, Schindler et al. (1985) used conventional statistical methods to compare a single manipulated system with many reference systems. This may be useful when standard monitoring can provide data from a number of reference systems (Underwood 1994).

9.6.7 Ecological Conclusions

Based on our intervention analysis, we conclude that there have been nonrandom shifts in the population of *K. taurocephala* in Little Rock Lake's treatment basin with acidification. However, the question is whether these shifts can be attributed to the acidification or whether they reflect natural population shifts coincident with the pH manipulations.

Information on the ecology of *K. taurocephala* provides strong support for the conclusion that the increases in acidification in Little Rock Lake reflect a response to the pH manipulation. In surveys across a broad sample of lakes, *K. taurocephala* showed a strong affinity with low pH conditions (MacIssac et al. 1987). Such previous information on the distribution of *K. taurocephala* led to the prediction, before any acidification, that its population would increase at lower pH levels in Little Rock Lake (Brezonik et al. 1986). Detailed mechanistic analyses of the response of *K. taurocephala* to conditions in Little Rock Lake's treatment basin revealed that its increased population levels were linked with reductions in the population levels of its predators under more acid conditions (Gonzalez and Frost 1994).

9.7 Discussion

Intervention analysis provides three major advantages over other analytical methods that have been used to analyze similar time series data sets. First, it provides good estimates of intervention effects and standard errors even in the presence of autocorrelation among observations. Second, intervention analysis requires that the autocorrelation structure of the data is modeled explicitly, and this may provide additional information about underlying processes. Third, the iterative model-building approach that is inherent in intervention analysis is itself a way of learning about characteristics of the data that may otherwise be difficult to discover. It is certainly a very different approach from that of simply carrying out a standard test to compute a *P*-value. The disadvantages of intervention analysis are that it requires long series of observations (usually 50 or more) to determine the pattern of serial dependence and it involves a large investment of time on the part of the data analyst, both in becoming familiar with the methodology and in carrying out each analysis.

We have not covered all of the complexities of intervention analysis. We analyzed time series with weak seasonal behavior, where nonseasonal models were adequate. In many data sets, seasonal behavior may be pronounced and thus will have to be incorporated into ARIMA intervention models. We have used methods appropriate for univariate time series. Our series was initially bivariate with one series for the treatment and another for the reference. We transformed the problem to a univariate one by taking the ratio of the two series at each sampling period. In general, such a combination of series requires that the series be synchronous, that is, one series does not lag or lead the other. Such a combination also requires that the two processes are operating on the same scale; Stewart-Oaten et al. (1986) discuss this in some detail within the context of achieving additivity. Bivariate approaches allow these assumptions to be relaxed to some extent but at the expense of ease of model identification and interpretation. One such approach is to use a general transfer function model (Box and Jenkins 1976), of which intervention models are a special type. The methods of identifying and estimating such models can be quite involved and are well beyond the scope of this chapter, but they are worth considering if the assumptions required for the univariate approach are suspect.

Despite our emphasis on unreplicated studies, replication should always be employed when possible. Replication is the only way of reducing the probability of detecting a spurious treatment effect in a manner that is readily quantifiable. A detailed treatment of replication is beyond the scope of this chapter, but we will mention some basics. A design with replicated time series is referred to as a repeated-measures design; repeated, or sequential, measurements are taken over time within each experimental unit (in an unreplicated time series, there is just one experimental unit). The usual approach to analyzing repeated-measures designs relies on applying a standard multivariate analysis of variance (chapter 8). Such an approach requires estimating a moderate number of parameters, which might be a problem if the amount of replication is not fairly substantial. This problem can be reduced to some extent by using repeated-measures models with structured covariance matrices; such models are hybrids of ARIMA models and multivariate analysis of variance models (Laird and Ware 1982; Jennrich and Schluchter 1986). The advantage of such repeated-measures models over standard multivariate ANOVA models are (1) they allow hypothesis-testing about the nature of the serial dependencies within units just as in ARIMA modeling and (2) they require the estimation of fewer parameters and hence demand less data for estimation. Software implementations of such procedures are relatively recent (e.g., SAS procedure MIXED, Littell et al. 1996).

Time series methods have broad potential applicability in ecology; intervention analysis is a relatively small facet. Poole (1978) gives an interesting review and summary of time series techniques in ecology, including transfer function models. Because of their underlying statistical similarities, time series models are also useful in understanding spatial statistical techniques. Time series techniques typically receive little or no attention in the basic statistics courses that most graduate ecology programs require. Many universities offer applied time series courses that are accessible to students with an introductory statistics background. Considering the frequency with which ecologists deal with time series data, graduate ecology programs should encourage such formal training. We also encourage ecologists who are designing or analyzing time series studies to seek the collaboration of their statistical colleagues.

Acknowledgments   We thank Pamela Montz for compiling the rotifer data evaluated here. Comments by Steve Carpenter on a previous version of this manuscript substantially improved it. This work was supported in part by funds from the U.S. Environmental Protection Agency and the National Science Foundation, Long-term Ecological Research Program.

Appendix 9.1 SAS Program Code for Data Input and Determination of Value for Transformation

```
DATA ABUN;                              /input data/
   INPUT MONTH                       /month sampled/
      DAY                              /day sampled/
      YEAR                            /year sampled/
      PERIOD          /six-month period dating from the start of monitoring/
      S1                 /dummy variable—has value of 1 while treatment
                                   basin pH is 5.6, value of 0 otherwise/
      S2                 /dummy variable—has value of 1 while treatment
                                   basin is pH 5.1, value of 0 otherwise/
      REF              /abundance of K. taurocephala in reference basin/
      TRT;             /abundance of K. taurocephala in treatment basin/
   RATIO = (TRT + 1)/(REF + 1);   /compute ratio of treatment to reference abundance/
   LOGRATIO = LOG10(RATIO);            /compute log transformation of ratio/
   LOGREF = LOG10(REF);           /compute log transformation of original data/
PROC SUMMARY NWAY;             /calculate mean and SD of ratios for each year/
   CLASS YEAR; VAR RATIO;
   OUTPUT OUT = SUMRY                      /output SDs/
      MEAN = MRATIO STD = SD;
DATA SUMRY; SET SUMRY;              /calculate log of mean and SD/
   LOGRATIO = LOG10(MRATIO);
   LOGSD = LOG10(SD);
PROC PLOT;                          /plot log SD vs log mean/
   PLOT LOGSD*LOGRATIO;


PROC REG;                           /regress log SD on log mean/
   MODEL LOGSD = LOGRATIO;
```

Appendix 9.2 SAS Program Code for Fitting the ARIMA Model to the Log Reference Series. Note That the Data Must Be Sorted by Sample Date

A. Initial Analysis for Examining the ACF and PACF Ratios

```
PROC ARIMA;                         /compute the ACF and PACF/
   IDENTIFY VAR = LOGREF;      /ACF and PACF plots produced automatically/
```

B. Based on the Initial Analysis, a Model of Order $p = 1$ is Fit to the Data

```
PROC ARIMA;                         /fit an AR(1) model/
   IDENTIFY VAR = LOGREF;
   ESTIMATE P = 1                   /P = order of model/
      MAXIT = 30              /maximum number of iterations = 30/
```

```
    METHOD = ML                              /use maximum likelihood/
    PLOT;                    /plot ACF and PACF residuals to test for model fit/
*** Similar analyses of reference and treatment basins not shown for brevity ***
```

Appendix 9.3 SAS Program Code for Fitting an Intervention Model to the Log Ratio Series

```
PROC ARIMA;                                      /fit intervention model/
    IDENTIFY VAR = LOGRATIO
        CROSSCOR = (S1 S2);                  /indicate intervention variables/
    ESTIMATE P = 1 INPUT = (S1 S2)                /list input variables/
        MAXIT = 30 METHOD = ML
        PLOT;
```

10

Nonlinear Curve Fitting

Predation and Functional Response Curves

STEVEN A. JULIANO

10.1 Ecological Issues

10.1.1 Predation and Prey Density

The number of prey that an individual predator kills (or the number of hosts a parasitoid attacks) is a function of prey density and is known as the functional response (Holling 1966). In general, the number of prey killed in a fixed time approaches an asymptote as prey density increases (Holling 1966). There are at least three types of curves that can be used to model the functional response (Holling 1966; Taylor 1984; Trexler et al. 1988) that represent differences in the proportion of available prey killed in a fixed time. In type I functional responses, the number of prey killed increases linearly to a maximum then remains constant as prey density increases (figure 10.1A). This corresponds to a constant proportion of available prey killed to the maximum (density independence), followed by a declining proportion of prey killed (figure 10.1B). In type II functional responses, the number of prey killed approaches the asymptote hyperbolically as prey density increases (figure 10.1C). This corresponds to an asymptotically declining proportion of prey killed (inverse density dependence; figure 10.1D). In type III functional responses, the number of prey killed approaches the asymptote as a sigmoid function (figure 10.1E). This corresponds to an increase in the proportion of prey killed (density dependence) to the inflection point of the sigmoid curve, followed by a decrease in the proportion of prey killed (figure 10.1F).

Most ecological interest in functional responses has involved types II and III. At a purely descriptive level, it is often desirable to be able to predict the number

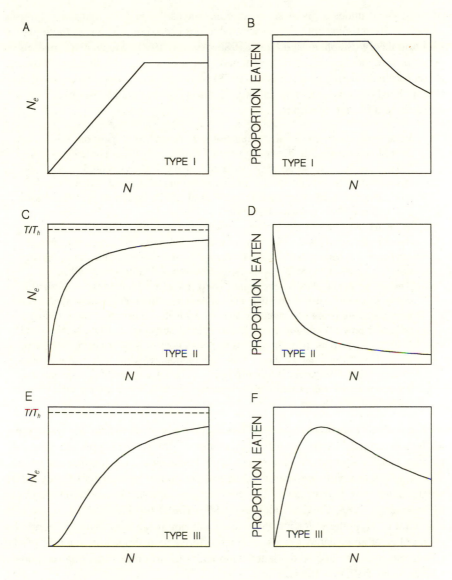

Figure 10.1 Three types of functional responses. The relationships between number of prey eaten (N_e) and number of prey present (N) are depicted in parts A, C, and E. The corresponding relationships between proportion eaten (N_e/N) and number of prey present (N) are depicted in parts B, D, and F.

of prey killed under a given set of circumstances. Type II functional responses especially have figured prominently in behavioral ecology, serving as the basis for foraging theory (Stephens and Krebs 1986; Abrams 1990). Mechanistic models of population dynamics of resource-limited consumers (e.g., Williams 1980) and predation-limited prey (e.g., Hassell 1978) have also made extensive use of type II functional responses. Biological investigations of functional responses may address these different questions:

1. What is the shape (type) of the functional response? This question is often of interest in attempting to determine whether density-dependent predation is a stabilizing factor in predator–prey population dynamics, and should also be answered before fitting any specific mathematical model to the functional response (van Lenteren and Bakker 1976; Hassell et al. 1977; Trexler et al. 1988; Casas and Hulliger 1994; Kabissa et al. 1996).
2. What are the best estimates of parameters of a mechanistic model of the functional response? This question must be answered by investigators who wish to use functional responses in mechanistic models of resource competition or predation (e.g., Hassell 1978; Williams 1980; Lovvorn and Gillingham 1996).
3. Are parameters of models describing two (or more) functional responses significantly different? Such a question may arise when different predator species or age classes are being compared to determine which is most effective at killing a particular prey (Thompson 1975; Russo 1986; Kabissa et al. 1996; Nannini and Juliano 1998), when different species or populations of prey are being compared to test for differential predator–prey coevolution (Livdahl 1979; Houck and Strauss 1985; Juliano and Williams 1985), or when different environments are being compared for predator effectiveness (Smith 1994; Messina and Hanks 1998; Song and Heong 1997).

These three questions require different methods of statistical analysis; these analytical methods are the subject of this chapter. The general approach described in this chapter comprises two steps: model selection and hypothesis testing. Model selection involves using a logistic regression of proportion of prey killed versus number of prey to determine the general shape of the functional response (Trexler et al. 1988; Trexler and Travis 1993; Casas and Hulliger 1994; see also chapter 11). Hypothesis testing involves using nonlinear least-squares regression of number of prey eaten versus number offered to estimate parameters of the functional response and to compare parameters of different functional responses (Juliano and Williams 1987).

Both of these statistical methods are related to more typical regression models, and I will assume that readers have some familiarity with regression analyses and assumptions (see Neter and Wasserman 1974; Glantz and Slinker 1990). Although the methods described in this chapter may have other applications in ecology (section 10.4), the descriptions of procedures given in this chapter are fairly specific to analyses of functional responses. General descriptions of methods for logistic regressions are given by Neter and Wasserman (1974) and Trexler and Travis (1993) and in chapter 11, and descriptions of methods for nonlinear regressions are given by Glantz and Slinker (1990) and Bard (1974).

10.1.2 Mathematical Models of Functional Responses

Because most interest has centered on types II and III functional responses, the remainder of this chapter will concentrate on these. Numerous mechanistic and phenomenological models have been used to describe functional responses. Holling's (1966) original model assumed that asymptotes were determined by time limitation, with predators catching and eating prey as rapidly as they could. Holling also assumed that (1) encounter rates with prey were linearly related to prey density, (2) while predators were handling prey they could not make additional captures, and (3) prey density was constant. He modeled type II functional responses using the Disc Equation:

$$N_e = \frac{aNT}{1 + aNT_h}$$

(10.1)

where N_e = number eaten, a = attack constant (or instantaneous search rate), which relates encounter rate with prey $(=aN)$ to prey density, N = prey density, T = total time available, and T_h = handling time per prey, which includes all time spent occupied with the prey and unable to attack other prey. This model is the most widely used representation of the type II functional response. Although the ecological literature contains numerous citations of this model with alternative symbols, Holling's symbols are used throughout this chapter.

Type III functional responses can be modeled using equation 10.1 if the attack constant a is made a function of prey density N (Hassell 1978). In the most general useful form, a is a hyperbolic function of N:

$$a = \frac{d + bN}{1 + cN}$$

(10.2)

where b, c, and d are constants. This is a more general form of the hyperbolic relationship postulated by Hassell (1978), which was equivalent to equation 10.2 with $d = 0$. Substituting equation 10.2 into equation 10.1 and rearranging yields

$$N_e = \frac{dNT + bN^2T}{1 + cN + dNT_h + bN^2T_h}$$

(10.3)

In general, type III functional responses can arise whenever a is an increasing function of N. Numerous mathematical forms could describe the relationship between a and N, and it is difficult to know beforehand which form will be best for a given data set. The form given in equation 10.2 has the advantage that if certain parameters are equal to 0, the expression reduces to produce other likely relationships of a and N (table 10.1). However, if both b and d are 0, then a is 0 and there is no functional response.

These models describe predation occurring at a constant prey density N. For experimental studies of predation, this assumption is often not realistic. In typical functional response experiments, individual predators in a standardized state of hunger are offered a number of prey (initial number = N_0) under standardized

Table 10.1 Various forms of the relationships between a and N produced by equation 10.2[a]

Parameters equal to 0	Resulting equation for a	Relationship of a to N
None	$\dfrac{d+bN}{1+cN}$	Increasing to asymptote; intercept $\neq 0$
c	$d+bN$	Linear increase; intercept $\neq 0$
d	$\dfrac{bN}{1+cN}$	Increasing to asymptote; intercept $= 0$
d and c	bN	Linear increase; intercept $= 0$

[a]It is assumed that the functional response is known to be type III; hence at least $b > 0$.

conditions of temperature, light, container size, and shape. Usually prey are not replaced as they are eaten, or they are only replenished at some interval. Although it would be desirable for experimenters to replace prey as they are eaten (Houck and Strauss 1985), this would obviously become prohibitive because it would require continuous observation. After a set time (T), the number of prey eaten (N_e) is determined. This process is repeated across a range of numbers of prey, usually with replication at each number of prey. The resulting data set contains paired quantitative values, usually (N_0, N_e). Investigators then attempt to answer one or more of the three questions outlined in the first section by statistical analysis of these data. Thus, in typical functional response experiments, prey density declines as the experiment proceeds. Under such circumstances, equations 10.1 and 10.3 do not accurately describe the functional response, and parameters estimated using these models will be subject to errors that are dependent on the degree to which prey are depleted (Rogers 1972; Cock 1977; Juliano and Williams 1987; Williams and Juliano 1996). When prey depletion occurs, the appropriate model describing the functional response is the integral of equation 10.1 or 10.3 over time to account for changing prey density (Rogers 1972). For the type II functional response (equation 10.3), integrating results in the random predator equation,

$$N_e = N_0\{1 - \exp[a(T_h N_e - T)]\} \tag{10.4}$$

where $N_0 =$ the initial density of prey.

For type III functional responses, the precise form of the model incorporating depletion depends on whether the attack constant a is a function of initial density (N_0) or current prey density (N) (Hassell et al. 1977; Hassell 1978). The simplest form arises when a is a function of initial density, as in equation 10.2:

$$N_e = N_0\{1 - \exp[(d + bN_0)(T_h N_e - T)/(1 + cN_0)]\} \tag{10.5}$$

We will use this form throughout this chapter.

Other mathematical forms have been used to model functional responses. Ivlev (1961) used an exponential model based on the assumption of hunger limitation of the asymptotic maximum number eaten to generate type II functional responses. Several authors (e.g., Taylor 1984; Williams and Juliano 1985) have used phenomenological models derived from Michaelis–Menten enzyme kinetics to model type II functional responses (eq. 10.2 is such a model). Trexler et al. (1988) showed that several phenomenological models could generate type II and III functional responses. Although different models may be employed, the general issues involved in statistical analyses of functional responses are similar. In this chapter, I will concentrate on models derived from Holling's case of limited handling time.

10.2 Statistical Issues

Many investigators have analyzed experiments done without replacement of prey, but they have used models appropriate for predation with constant prey density (equations 10.1 and 10.3; e.g., Livdahl 1979; Juliano and Williams 1985; Russo 1986). Such analyses are of questionable value because parameter estimates and comparisons are sure to be biased and the shape of the functional response may be incorrectly determined (Rogers 1972; Cock 1977). Although the biases associated with fitting equations 10.2 and 10.3 to data that incorporate prey depletion have been well known since 1972 (Rogers 1972), this inappropriate approach continues to be used and advocated (Fan and Pettit 1994, 1997) primarily because fitting models such as equations 10.2 and 10.3 to data is a simpler task than fitting models such as equations 10.4 and 10.5 (see Houck and Strauss 1985; Williams and Juliano 1985; Glantz and Slinker 1990). Such models involving no prey depletion have the desirable property of giving the correct expectation for stochastic predation (Sjoberg 1980; Houck and Strauss 1985). However, if experiments are conducted with prey depletion, prey depletion must be incorporated into the statistical analysis, using equations 10.4 and 10.5, to reach valid conclusions (Rogers 1972; Cock 1977). The wide availability of powerful statistical packages makes fitting impicit functions like equations 10.4 and 10.5 rather simple.

The most common method used to analyze functional responses is some form of least-squares regression involving N_0 and N_e (e.g., Livdahl 1979; Livdahl and Stiven 1983; Juliano and Williams 1985, 1987; Williams and Juliano 1985; Trexler et al. 1988). Some authors have used linear regressions, especially on linearized versions of equations 10.1 and 10.4 (Rogers 1972; Livdahl and Stiven 1983). However, these linearizations often produce biased parameter estimates, and comparisons of parameters may not be reliable (Cock 1977; Williams and Juliano 1985; Juliano and Williams 1987). Linearized expressions are also unsuitable for distinguishing type II from type III functional responses.

Nonlinear least squares has proved effective for parameter estimation and comparison (Cock 1977; Juliano and Williams 1987), but it seems to be a less desirable choice for distinguishing types II and III functional responses (Trexler et al. 1988; Casas and Hulliger 1994). Distinguishing the curves in figure 10.1C and E using nonlinear least squares could be done by testing for a significant lack of fit for models using equations 10.4 and 10.5 (Trexler et al. 1988). Significant lack of fit for equation 10.4 but not for equation 10.5 would indicate a type III functional response. These two types could also be distinguished by fitting a nonlinear model using equation 10.5 and by testing H_0: $b = 0$ and $c = 0$. Rejecting the null hypothesis that $b \leq 0$ is sufficient to conclude that the functional response is of a type III form. However, comparison of figure 10.1C and E illustrates why nonlinear regressions involving N_0 and N_e are unlikely to distinguish the two types. In many type III functional responses, the region of increasing slope is quite short, hence the difference between the sigmoid curve of figure 10.1E and the hyperbolic curve of figure 10.1C is relatively small and will be difficult to detect in variable experimental data (Porter et al. 1982; Trexler et al. 1988; Peters 1991).

Another difficulty in using equation 10.5 to distinguish type II and III functional responses is that the parameter estimates for b, c, and d may all be nonsignificant when incorporated into a model together despite the fact that a model involving only one or two of these parameters may yield estimates significantly different from 0. This may occur because of reduced degrees of freedom for error in more complex models and inevitable correlations among the parameter estimates.

10.3 Statistical Solution

10.3.1 Experimental Analysis

Analyzing functional responses and answering questions 1, 2, and 3 posed in section 10.1.1 require two distinct steps. The methods best suited for answering question 1 are distinct from those best suited for answering questions 2 and 3. Further, question 1 must be answered before questions 2 or 3 can be answered.

Model selection: Determining the shape of the functional response. Trexler et al. (1988) demonstrated that the most effective way to distinguish type II and III functional responses involves logistic regression (Neter and Wasserman 1974; Glantz and Slinker 1990) of proportion of prey eaten versus number of prey present. Casas and Hulliger (1994) reached a similar conclusion using a more complex approach to logistic regression. Logistic regression focuses the comparison on the curves in figures 10.1D and F, which are clearly much more distinct than their counterparts in figures 10.1C and E. Logistic regression involves fitting a phenomenological function that predicts proportion of individuals in a group responding (in this case, being eaten) from one or more continuous variables (Neter and Wasserman 1974; SAS Institute Inc. 1989b; Glantz and Slinker 1990). The dependent variable Y is dichotomous, representing individual prey alive ($Y = 0$) or dead ($Y = 1$) at the end of the trial. In this case, the continuous variable is N_0. The general form of such a function is

$$\frac{N_e}{N_0} = \text{Prob}\{Y = 1\} = \frac{\exp(P_0 + P_1 N_0 + P_2 N_0^2 + P_3 N_0^3 + \cdots + P_z N_0^z)}{1 + \exp(P_0 + P_1 N_0 + P_2 N_0^2 + P_3 N_0^3 + \cdots + P_z N_0^z)} \tag{10.6}$$

where P_0, P_1, P_2, P_3, $\cdots$, P_z are parameters to be estimated. In most statistical packages, the parameters are estimated using the method of maximum likelihood rather than least squares (e.g., SAS Institute Inc. 1989a, procedure CATMOD; see also chapter 15). In this method, rather than minimizing sums of squared deviations of observed from expected, the probability of the observed values arising from the parameter estimates is maximized (Glantz and Slinker 1990). This involves an iterative search for the parameter values that maximize the likelihood function L:

$$L = \prod_k \frac{[\exp(P_0 + P_1 N_{0k} + P_2 N_{0k}^2 + P_3 N_{0k}^3 + \cdots + P_z N_{0k}^z)]^{Y_k}}{1 + \exp(P_0 + P_1 N_{0k} + P_2 N_{0k}^2 + P_3 N_{0k}^3 + \cdots + P_z N_{0k}^z)} \tag{10.7}$$

where k is the subscript designating individual observations in the data set, and Y_k is the value of the dichotomous variable (1 for prey eaten, 0 for prey surviving). For further details, consult Glantz and Slinker (1990, pp. 520–522).

The strategy is to find the polynomial function of N_0 that describes the relationship of N_e/N_0 versus N_0. The curves depicted in figures 10.1D and F may both be fit by quadratic (or higher order) expressions. However, in figure 10.1D, the linear term would be negative (initially decreasing), whereas in figure 10.1F the linear term would be positive (initially increasing). Thus, one criterion for separating type II and III functional responses by analyzing the proportion of prey eaten is to test for significant positive or negative linear coefficients in the expression fit by the method of maximum likelihood to data on proportion eaten versus N_0. A cubic expression will often provide a good fit to a type III functional response (Trexler et al. 1988), and cubic expressions provide a good starting point for fitting a logistic regression (Trexler and Travis 1993). Higher order expressions will, of course, fit even better, but this improved fit is usually the result of a better fit to points at higher values of N_0 (see subsequent discussion). It is likely that the initial slope of the curve will have the same sign. Whatever the order of the expression that is fit, plotting observed and predicted values versus N_0 and observing the slope near the origin is desirable for distinguishing type II and type III functional responses (see chapter 3). At the model selection stage, significance testing is not as important as is obtaining a good description of the relationship of proportion eaten versus N_0.

Distinguishing between type II and III functional responses requires determining whether the slope near $N_0 = 0$ is positive or negative (figure 10.1). In designing experiments, this is an important consideration in the choice of values for N_0. If values of N_0 are too large, any region of density-dependent predation may be undetected. Because the dependent variable is a proportion, the relative variability of observations will necessarily increase at low N_0. Replication of observations at very low N_0 should therefore be greater than replication of observations at higher N_0.

One alternative method for determining the shape of the functional response curve is LOWESS (LOcally WEighted regrESSion, chapter 3). This computer-intensive smoothing technique has the advantage of less restrictive assumptions than regression techniques.

Hypothesis testing: Estimating and comparing functional response parameters. Although logistic regression is the most useful technique for distinguishing the types of functional responses, many investigators want to fit the mechanistic models given in equations 10.1, 10.3, 10.4, or 10.5 to data to obtain estimates of the parameters. Nonlinear least squares is the preferred method for obtaining such estimates (Cock 1977; Williams and Juliano 1985, 1996; Juliano and Williams 1987). Because most experiments are done without prey replacement, equations 10.4 and 10.5 are appropriate. Both of these equations give only an implicit function relating N_e to N_0 (with N_e on both sides of the expressions). Thus, iterative solutions are necessary to find the predicted values of $\hat{N}_e$ for any set of parameters. This can be done using Newton's method for finding the root of an implicit equation (Turner 1989). Equation 10.4 can be rewritten as

$$0 = N_0 - N_0 \exp[a(T_h N_e - T)] - N_e = f(N_e) \tag{10.8}$$

The problem is to find the value of $\hat{N}_e$ that satisfies equation 10.8. Graphically, this problem is represented in figure 10.2, in which the value of $\hat{N}_e$ at which $f(N_e) = 0$ is the value that satisfies equation 10.8. Newton's method approaches this value iteratively, by using the first derivative of $f(N_e)$ [$=f'(N_e)$]. This derivative defines the slope of a line that crosses the horizontal axis at some point (figure 10.2). The slope of this line can also be estimated from any two points (i, $i + 1$) on the line by calculating the difference in the vertical coordinates over the difference in the horizontal coordinates, or in this case

$$f(N_{ei}) = \frac{0 - f(N_{ei})}{N_{ei+1} - N_{ei}} \tag{10.9}$$

where N_{ei+1} is the next value of N_e that serves to start an iteration. Each successive step in the iterative process brings N_{ei+1} closer to the true value that makes equation 10.8 true (figure 10.2). $\hat{N}_e$ is then the value of N_{ei+1} at which the value of the function $f(N_e)$ is sufficiently close to 0.

Using this procedure to run a nonlinear regression on an implicit function with a software package such as SAS requires first substituting initial parameter esti-

Figure 10.2 Graphic representation of how Newton's method is used to find N_e, the root of the implicit equation describing the functional response with prey depletion (eqs. 10.4 or 10.5). The point at which the function $f(N_e)$ crosses the horizontal axis is the true solution. Starting from N_{e1}, new approximations of N_e are made using the derivative of $f(N_{ei})$: $N_{ei+1} = N_{ei} - \{f(N_{ei})/ f'(N_{ei})\}$. The process is repeated at each new approximate N_e until the approximation is arbitrarily close to the true value [i.e., $f(N_e) = 0$]. See text for details.

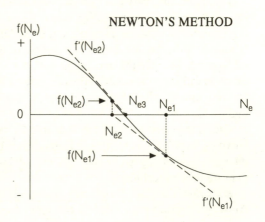

mates into equation 10.8 and obtaining an initial estimate of $\hat{N}_e$. The nonlinear least-squares routine then computes sums of squares and modifies parameter estimates so that the residual sum of squares is minimized. With each new set of parameter estimates, Newton's method is reemployed to determine $\hat{N}_e$.

Using nonlinear least squares assumes normally distributed, homogeneous error variance over the range of N_0 (Bard 1974; Glantz and Slinker 1990). Most, if not all, real data fail to meet these assumptions because variance typically increases with N_0 (Houck and Strauss 1985; Juliano and Williams 1987; Trexler et al. 1988). Violating these assumptions in analyses of functional responses may result in confidence intervals that miss the true parameter values more often than expected (Juliano and Williams 1987). It appears, however, that even fairly severe departures from these assumptions distort the results only slightly and nonlinear least squares is thus relatively robust (Juliano and Williams 1987). Residual plots (Neter and Wasserman 1974) can be used to examine these assumptions graphically. In extreme cases, other remedial measures may be necessary, such as using a gamma distribution of residuals (Pacala and Silander 1990), nonparametric regressions (Juliano and Williams 1987), or weighted nonlinear least squares, which may be implemented in the SAS procedure NLIN (SAS Institute Inc. 1989b).

10.3.2 Examples and Interpretation

The following two examples illustrate how to go about answering the three questions described in section 10.1.1. Both are based on published functional response data.

Example 1. Notonecta preying on Asellus. Notonecta glauca is a predatory aquatic hemipteran. *Asellus aquaticus* is an aquatic isopod crustacean. This original experiment was described by Hassell et al. (1977). Raw data are given by Trexler et al. (1988) and Casas and Hulliger (1994). Individual *Notonecta* were exposed to a range of densities of *Asellus* from 5 to 100, and allowed to feed for 72 hours, with prey replaced every 24 hours. Substantial prey depletion occurred over the 24-hour periods. At least eight replicates at each density were run. The questions of interest were questions 1 and 2 in section 10.1.1: What is the shape of the functional response? and, What are the best parameter estimates for the function describing the functional response?

The first step in analyzing the data is to use logistic regression to determine whether the functional response is type II or type III. Data input for this analysis using the SAS procedure CATMOD features the following: (1) two data lines for each replicate are given, one for prey eaten (FATE = 0) and one for prey alive (FATE = 1); (2) when FATE = 0, the variable NE is equivalent to N_e (prey eaten); and (3) for a given replicate, NE sums to N0. This data structure allows both analysis of proportions using logistic regression in CATMOD and, with minor modification, analysis of number eaten using nonlinear least squares.

The first step in evaluating the shape of the functional response is to fit a polynomial logistic model. Cubic models are likely to be complex enough to describe most experimental data, hence the specific form of the logistic equation is

$$\frac{N_e}{N_0} = \frac{\exp(P_0 + P_1 N_0 + P_2 N_0^2 + P_3 N_0^3)}{1 + \exp(P_0 + P_1 N_0 + P_2 N_0^2 + P_3 N_0^3)}$$ (10.10)

where P_0, P_1, P_2, and P_3 are parameters to be estimated. Program steps to fit this model to data are given in appendix 10.1A at the book's companion Website [see http://www.oup-usa.org/sc/0195131878/]. Once estimates of parameters from the logistic regression have been obtained, observed and predicted mean proportions eaten can be plotted (figure 10.3).

After information concerning the data set are input and iterative steps toward the solution are taken (not shown), CATMOD produces maximum likelihood tests of hypotheses that the parameters are zero, along with a likelihood ratio test, which tests the overall fit of the model (table 10.2A). A significant test indicates lack of fit. In this example, all parameters are significantly different from 0, and even this cubic model still shows significant lack of fit. Next, the output shows the actual parameter estimates with their standard errors (table 10.2B). The intercept is not particularly informative. Because the linear parameter (labeled N0 in the output) is positive and the quadratic parameter (labeled N02 in the output) is negative, these results indicate a type III functional response, with proportion eaten initially increasing then decreasing as N_0 increases, as in figure 10.1F. A simple verification of this can be obtained by plotting observed mean proportions eaten along with predicted proportions eaten (program steps not shown), which clearly indicates a type III functional response (figure 10.3). The scatter of the observed means around the predicted curve is consistent with the significant lack-of-fit test. A better fit would likely be obtained by fitting quartic or higher order equations to the data. Fitting a quartic equation (results not shown) to this data

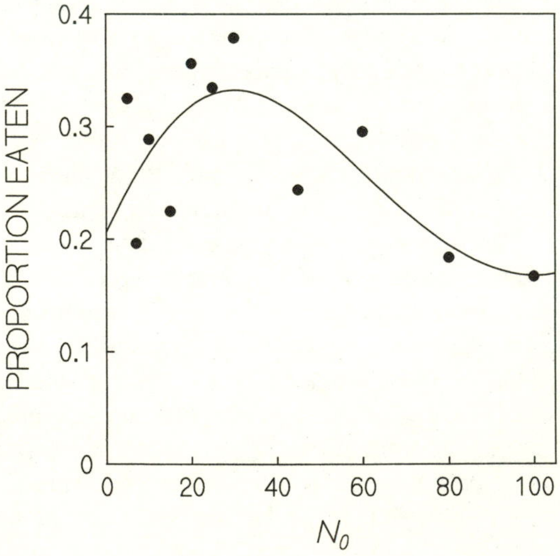

Figure 10.3 Observed mean proportions of prey eaten at each initial prey density in the *Notonecta–Asellus* example, and the fitted relationship produced by logistic regression. Parameter estimates for the logistic regression are given in table 10.2B.

Table 10.2 SAS Output for example 1, analysis of *Notonecta–Asellus* data[a]

A. Procedure CATMOD: Maximum–likelihood analysis of variance table

Source	DF	Chi-Square	Prob.
INTERCEPT	1	28.68	0.0000
N0	1	6.66	0.0098
N02	1	7.52	0.0061
N03	1	5.67	0.0172
LIKELIHOOD RATIO	85	206.82	0.0000

B. Procedure CATMOD: Analysis of maximum-likelihood estimates

Effect	Parameter	Estimate	Standard Error	Chi-Square	Prob.
INTERCEPT	1	−1.3457	0.2513	28.68	0.0000
N0	2	0.0478	0.0185	6.66	0.0098
N02	3	−0.00103	0.000376	7.52	0.0061
N03	4	5.267E-6	2.211E-6	5.67	0.0172

C. Procedure NLIN: Full model

Nonlinear Least-Squares Summary Statistics
Dependent Variable NE

Source	DF	Sum of Squares	Mean Square
Regression	4	9493.754140	2373.438535
Residual	85	1766.245860	20.779363
Uncorrected Total	89	11260.000000	
(Corrected Total)	88	4528.808989	

Parameter	Estimate	Asymptotic Std. Error	Asymptotic 95% Confidence Interval	
			Lower	Upper
BHAT	0.000608186	0.0246696938	−0.0484419866	0.0496583578
CHAT	0.041999309	2.6609454345	−5.2486959321	5.332694549
DHAT	0.003199973	0.0591303799	−0.1143675708	0.1207675172
THHAT	3.157757984	1.7560019268	−0.3336593588	6.6491753258

D. Procedure NLIN: Reduced model 1

Nonlinear Least-Squares Summary Statistics
Dependent Variable NE

Source	DF	Sum of Squares	Mean Square
Regression	3	9526.797240	3175.599080
Residual	86	1733.202760	20.153520
Uncorrected Total	89	11260.000000	
(Corrected Total)	88	4528.808989	

Parameter	Estimate	Asymptotic Std. Error	Asymptotic 95% Confidence Interval	
			Lower	Upper
BHAT	0.000415285	0.00022766485	−0.0000373001	0.0008678694
DHAT	0.000835213	0.00380836014	−0.0067355902	0.0084060158
THHAT	4.063648373	0.35698844155	3.3539756864	4.7733210598

Table 10.2 Continued

E. Procedure NLIN: Reduced model 2

Nonlinear Least-Squares Summary Statistics
Dependent Variable NE

Source	DF	Sum of Squares	Mean Square
Regression	2	9525.757196	4762.878598
Residual	87	1734.242804	19.933825
Uncorrected Total	89	11260.000000	
(Corrected Total)	88	4528.808989	

Parameter	Estimate	Asymptotic Std. Error	Asymptotic 95% Confidence Interval Lower	Upper
BHAT	0.000460058	0.00010357262	0.0002541949	0.0006659204
THHAT	4.104328418	0.28920091748	3.5295076540	4.6791491819

ªOnly the analysis and final parameter estimates are shown; output from the iterative phases of NLIN is omitted for brevity.

set yields a significant fourth-order term, but does not alter the general shape of the curve and does not alter the conclusion that *Notonecta* shows a type III functional response to *Asellus*. Independent logistic analysis using more complex methods (Casas and Hulliger 1994) and analysis using LOWESS (chapter 3) also yield the conclusion that *Notonecta* shows a type III functional response.

If a cubic equation yields a nonsignificant cubic parameter (labeled NO3 in the output), then even if all other parameters are nonsignificant, it is desirable to reduce the model by eliminating the cubic term from equation 10.10 and to retest the other parameters. Models fit with excess nonsignificant parameters may be misleading.

The utility of determining by logistic regression whether a functional response is type II or type III is apparent when attempting to answer the second question and determining values of the parameters. If logistic regression indicates a type II functional response, a simpler model can be fit. In this case, the more complex type III model is known to be necessary.

To fit a mechanistic model to the data and to estimate parameters, nonlinear least-squares regression is used (SAS procedure NLIN, SAS Institute Inc. 1989b). Several methods are available in this procedure, some of which require the partial derivatives of the nonlinear functions with respect to the parameters of the model. A simpler derivative-free procedure (referred to as DUD in SAS) is also available, and because of its simplicity, this procedure will be used. In many cases, different methods within the procedure NLIN will yield similar results. However, investigators may wish to try two or more methods on the same data set to determine whether the solution is independent of the method used. NLIN allows programming steps to be incorporated into the procedure, and this makes it possible to use Newton's method to solve the implicit functions given in equations 10.4 or 10.5.

If the experiment had been done with prey replacement (constant N), NLIN could be used to fit one of the explicit nonlinear relationships given in equations 10.1 or 10.3. For this circumstance, the lines from the "Define implicit function" comment through the MODEL statement (appendix 10.1A on Website) would be eliminated and replaced with a simple model statement representing equation 10.1 or 10.3.

In this example, because logistic analysis indicates a type III functional response, the general function given in equation 10.5 is the starting point. This initial model involves the estimation of four parameters. Three parameters (b, c, and d), which are defined in equation 10.2, also define the relationship between the attack constant, a, and N_0. Because the functional response is type III, at least the parameter b in equations 10.2 and 10.5 must be greater than 0. This will prove important in defining reduced models if the full model is not satisfactory. The fourth parameter is T_h.

The original data set must be modified to use nonlinear least squares. Data lines enumerating prey alive (FATE = 1) are deleted, resulting in one paired observation, with values (N_0, N_e).

NLIN, like other nonlinear procedures, is iterative and requires initial estimates of the parameters. In many cases, the success of nonlinear regression depends on having reasonably accurate estimates of the parameters. In functional response models, handling time (T_h) can be approximated by T/N_e at the highest N_0. In this model of type III functional responses, initial estimates of the other parameters are less obvious. The parameter d is likely to be small in many cases, hence an initial estimate of 0 usually works. For b and c, it is usually best to select several initial estimates varying over as much as two orders of magnitude and to let the procedure find the best initial estimate. This is illustrated in appendix 10.1A (on Website) for initial values of the parameter b. Multiple initial parameter values may be useful for T_h and d as well; however, using multiple initial values for all four parameters may require long computation time. Values of all parameters depend on the time units chosen.

NLIN allows bounds to be placed on parameters, and this step should always be used in the analysis of functional responses. This will help to avoid nonsensical results, such as negative handling times. In type III functional responses, the parameters T_h and b must be greater than 0, and the parameters c and d must be nonnegative. In appendix 10.1A and B, a BOUNDS statement is used to set these limits.

Use of Newton's method to find $\hat{N}_e$ also requires an initial estimate of $\hat{N}_e$. For simplicity, observed values of N_e may be used, as in appendix 10.1. Mean N_e at each value of N_0 could also be used and would likely be closer to the final value of $\hat{N}_e$, hence it would reduce the number of iterations necessary to reach a solution.

The strategy for obtaining parameter estimates is to begin with the full model (four parameters) and to eliminate from the model the parameters c and d (one parameter at a time) if these are not significantly different from 0 (appendix 10.1A, B). For a type III functional response, the minimal model includes T_h and b. In the output from NLIN, significant parameters have asymptotic 95% confidence intervals that do not include 0 (table 10.2C). For the full model (input in Appendix 10.1A), estimates of b, c, and d were not significantly different from 0 (table 10.2C). The next step is to eliminate the parameter c, resulting in a linear relationship of a to N_0, and rerun the model (program steps not shown). Again, all parameters except T_h were not significantly different from 0 (table 10.2D). The final step is to eliminate the parameter d, resulting in the minimal model for a type III functional response, and rerun the model (appendix 10.1B on Website). In this case, the parameters b and T_h were both significant (table 10.2E), indicating that the relationship of a to N_0 is linear with a slope ($\pm$SE) of 0.00046 ± 0.00010 and the line passes through the origin. Handling time is estimated to be 4.104 ± 0.289 h.

Graphic presentation of observed and predicted values (figure 10.4) indicates a very short region of density dependence and a reasonably good fit of the model to the data. A residual plot (not shown) indicates that variance increases with N_0, as is common in functional response data sets. Weighted least squares may be needed to remedy this problem.

Because this type III response is described by a two-parameter model, it can be compared to the alternative two-parameter model for a type II functional response (constant a). Fitting a type II functional response results in parameter estimates $\pm$ SE of $\hat{a} = 0.0240 \pm 0.0043$ and $\hat{T}_h = 0.908 \pm 0.161$. Both parameters are significantly different from 0. However, residual sum of squares for the type II model is 1813.5—considerably greater than that for the minimal type III model (table 10.2E). Thus, a type III functional response fits the data better than a type II functional response, even though the number of parameters is the same. This example also illustrates the value of using logistic regression to establish the type of functional response, even if the primary interest is in estimating the parameters. Once it is known that this is a type III functional response, the decision of which

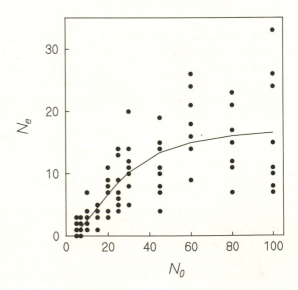

Figure 10.4 Observed numbers of prey eaten in the *Notonecta–Asellus* example, and the functional response curve fit by nonlinear least squares. Parameter estimates for the functional response are given in table 10.2E.

two-parameter model to choose is simplified. Finally, this example illustrates how models with excess parameters (the four- and three-parameter models) may give misleading conclusions.

Example 2. Toxorhynchites preying on two strains of Aedes triseriatus. Aedes triseriatus is a tree-hole dwelling mosquito from eastern North America. In the southern part of its range, larval *A. triseriatus* are preyed upon by larvae of *Toxorhynchites rutilus*, a large predatory mosquito. In the northern part of its range, larval *A. triseriatus* do not encounter this predator. Livdahl (1979) conducted functional response experiments in which single first-instar *T. rutilus* were offered 3 to 48 first-instar larvae of one of two strains of *A. triseriatus*, one from northern Illinois, where *T. rutilus* is absent, and one from North Carolina, where *T. rutilus* is abundant. Predation proceeded for 24 hours, with prey not replenished. The primary question of interest was number 3 in section 10.1.1: Do the functional responses of the predator to the two prey strains differ, and, if so, which parameters differ?

Analyses of these functional response data have already appeared in the literature (Livdahl 1979; Juliano and Williams 1985). Both of these analyses were conducted assuming a type II functional response and using equation 10.1, the model appropriate for experiments done at constant prey density (no depletion). However, the experiments were conducted with prey depletion (Livdahl 1979), hence equation 10.4 is in fact the appropriate model. It also appears that explicit tests of the form of the functional responses have never been done (Livdahl 1979; Juliano and Williams 1985). Hence, the conclusion from prior analyses (Juliano and Williams 1985)—that the two populations do produce significantly different functional responses, with attack constants (a), but not the handling times (T_h), significantly different—may be based on inadequate analyses.

To test for differences in the parameters of the functional response, nonlinear least squares will again be used. However, we will use indicator variables (Neter and Wasserman 1974; Juliano and Williams 1985; Juliano and Williams 1987) to test the null hypotheses that the parameters for the two groups differ. Such a test could also be implemented by fitting two separate regressions to the data and then comparing parameter estimates (see Glantz and Slinker 1990, pp. 500–505). The indicator variable approach has the advantage of producing an explicit test of the null hypothesis of equal parameters (Neter and Wasserman 1974). The indicator variable approach is analogous to the computational method employed by SAS in the procedure GLM for linear models with categorical and continuous variables, such as analysis of covariance models (chapter 5).

To compare type II functional responses of two groups, the implicit function with indicator variables is

$$0 = N_0 - N_0 \exp\{[a + D_a(j)]\{[T_h + D_{T_h}(j)](N_e) - T\}\} - N_e \qquad (10.11)$$

where j is an indicator variable that takes on the value 0 for population 1 and the value 1 for population 2. The parameters D_a and D_{Th} estimate the differences between the populations in the values of the parameters a and T_h, respectively. If these parameters are significantly different from 0, then the two populations differ significantly in the corresponding parameters. For population 1, $\hat{a}$ and $\hat{T}_h$ are the estimates of the population parameters a_1 and T_{h1}. For population 2, $\hat{a} + \hat{D}_a$ and $\hat{T}_h + \hat{D}_{T_h}$ are the estimates of the population parameters a_2 and T_{h2}.

To employ this model, the variable j must be present in the data set. The model is implemented in NLIN in the same way that the previous models for a single population were implemented, using Newton's method. Initial estimates of $D_a = 0$ and $D_{T_h} = 0$ usually are adequate. The estimates of a and T_h for the two populations take on the same values that would be obtained from separate regressions on the two populations (Juliano and Williams 1987).

In analyzing this experiment, it is, again, first necessary to determine the shape of the functional response. CATMOD was used to fit equation 10.10 to the data for each of the two populations (not shown). For both populations, the cubic coefficient was not significantly different from 0. Fitting a quadratic model to the data resulted in quadratic and linear coefficients that were significantly different from 0 in both populations (not shown). The estimates of the linear terms were significantly less than 0, whereas the estimates of the quadratic terms were significantly greater than 0, indicating a shape similar to that illustrated in figure 10.1D, as well as type II functional responses. Plots of observed mean proportions eaten and predicted values (not shown) support this conclusion. These results indicate that fitting the indicator variable model in equation 10.11 is appropriate.

Input for fitting the indicator variable model is given in appendix 10.2 at the Website, and partial output is in table 10.3. In this example, the North Carolina population is arbitrarily designated as population 1 ($j = 0$) and the Illinois population is designated as population 2 ($j = 1$). The populations do not differ significantly in values of a or T_h (asymptotic 95% confidence intervals for D_a and D_{T_h} include 0). From this output, t-tests can be conducted for the null hypotheses that D_a and D_{T_h} are 0 (t = parameter/asymptotic standard error, degrees freedom = DF

Table 10.3 SAS output from procedure NLIN for example 2, analysis of *Toxorhynchites—Aedes* data[a]

		Nonlinear Least-Squares Summary Statistics Dependent Variable NE	
Source	DF	Sum of Squares	Mean Square
Regression	4	5362.8970908	1340.7242727
Residual	79	293.1029092	3.7101634
Uncorrected Total	83	5656.0000000	
(Corrected Total)	82	964.7228916	

			Asymptotic 95% Confidence Interval	
Parameter	Estimate	Asymptotic Std. Error	Lower	Upper
AHAT	0.057857039	0.01527262547	0.0274575315	0.0882565457
THHAT	2.040486509	0.25143619643	1.5400135252	2.5409594925
DA	0.186433867	0.11875924855	−0.0499513330	0.4228190660
DTH	−0.016849199	0.29005690323	−0.5941950459	0.5604966480

[a]The output from procedure CATMOD is omitted.

RESIDUAL). For the North Carolina population, $\hat{a}_{NC} = 0.058$ and for the Illinois population, $\hat{a}_{IL} = 0.244$ ($=\hat{a}_{NC} + D_a = 0.058 + 0.186$). $D_a = 0.186$, with $t_{79} = 1.57$, indicating a nonsignificant difference at $P = 0.10$. Thus the attack constants do not differ significantly. This conclusion differs from that reached by Juliano and Williams (1985), who used an indicator variable model that did not incorporate prey depletion (equation 10.1) and who concluded that the two populations had significantly different values of a. The results of the analysis in table 10.3 actually produce values of $\hat{a}$ that differ in the same direction ($\hat{a}_{NC} < \hat{a}_{IL}$) and by a greater absolute amount ($\hat{a}_{NC} = 0.058$, $\hat{a}_{IL} = 0.244$) than the values of $\hat{a}$ from the analysis by Juliano and Williams (1985) ($\hat{a}_{NC} = 0.040$, $\hat{a}_{IL} = 0.068$). The conclusions based on the model incorporating prey depletion are more realistic, and the earlier results of Juliano and Williams (1985) appear to have been biased by the use of a model that did not incorporate prey depletion.

Using this indicator variable approach, it is possible to fit models with some parameters in common between the populations (e.g., by dropping the parameter D_{T_h}, a model with common T_h would result). In this example, such a model yields the same conclusion regarding the difference between a_{NC} and a_{IL} (not shown).

Although the results of this analytical approach are often clear, some experiments can yield analyses that are harder to interpret. Functional response curves may not reach asymptotes (Nannini and Juliano 1998) either because predator behavior changes to partial consumption of prey at high prey density or because prey densities are simply not sufficiently high. Direct comparisons of parameters may be difficult when the form of the functional responses (type II versus type III) differs between predators (Kabissa et al. 1996) or between conditions (Messina and Hart 1998). Nonetheless, application of these methods has been successful in several cases (Kabissa et al. 1996; Messina and Hanks 1998), enabling investigators to estimate and to compare functional responses.

10.4 Related Techniques and Applications

A nonparametric method for estimating parameters for the implicit form of the type II functional response has been described (Juliano and Williams 1987). This method is based on earlier work describing nonparametric methods for estimating nonlinear enzyme kinetics relationships (Porter and Trager 1977). This nonparametric method has the advantage of not requiring the assumption of normally distributed error. In simulated data sets, parameter estimates from the nonparametric method are comparable in variability and low bias to those produced by nonlinear least squares (Juliano and Williams 1987). The nonparametric method has several disadvantages: (1) direct comparisons of parameters between experimental groups is difficult; (2) few, if any, statistical packages implement this procedure directly; and (3) computational time and programming effort required are considerably greater than those associated with nonlinear least squares.

The general approaches outlined in this chapter have other applications in ecological studies. Logistic regression has been used to test for density-dependent survival in field experiments that manipulate density (e.g., Juliano and Lawton 1990) and for comparing survivorship curves under different, naturally occurring conditions (e.g., Juliano 1988). Nonlinear least squares can be used to fit growth curves (e.g., Sedinger and Flint 1991; see also chapter 8) and to estimate parameters for lognormal distributions of species abundances (e.g., Ludwig and Reynolds 1988).

Acknowledgments   I thank Sam Scheiner and Jessica Gurevitch for the invitation to contribute this chapter, Vickie Borowicz, Sam Scheiner, and Jessica Gurevitch for helpful comments on the manuscript, F. M. Williams, Brian Dennis, Joel Trexler, Frank Messina, and Joe Kabissa for valuable discussions of the analysis of functional response experiments, and Aaron Ellison, Jay Ver Hoef, and Noel Cressie for providing drafts of their chapters. Computer resources necessary for this work were provided by grants from NIH (1 R15 AI29629-01), NSF (BSR90-06452), and Illinois State University.

11

Logit Modeling and Logistic Regression

Aphids, Ants, and Plants

TED FLOYD

11.1 Ecological Issues

11.1.1 Example

Ecological data are often discontinuous or categorical. This means that measurement variables tend to have values like "present vs. absent" or "treatment vs. control." The example that I consider in detail in this chapter involves categorical data at nearly every level of analysis: patterns of association among predatory ants in the genus *Pheidole*, herbivorous aphids in the genus *Aphis*, and host plants in the genus *Piper*. Categorical measurements include experimental manipulation of ant abundance (ants present vs. ants removed), experimental manipulation of host plant quality (fertilized plants vs. control plants), observations of aphid abundance (colony present vs. colony absent), characterization of aphid diet breadth (specialist vs. generalist), year of study (year 1 vs. year 2), and so forth.

An important topic in categorical data analysis is causality. In some cases, the direction of causality is obvious. For example, year of study cannot be influenced by aphid abundance, but the reverse is theoretically possible. In other cases, however, causal order is far from obvious, and the ant–aphid–plant system presents a good case study for distinguishing among alternative causal models. I will focus primarily on the hypothesis that aphid abundance is influenced by the other factors in the system, but a proper justification of this view awaits further discussion.

11.1.2 Definitions

In the broadest sense, categorical data analysis seems to refer to any situation where one or more variables are discontinuous. Such situations are ubiquitous in nature. However, I will restrict this discussion primarily to cases were the distribution of a categorical response variable is influenced by one or more predictor variables that can be either categorical or continuous. In the ant–aphid–plant system, for example, we might wish to know whether the presence or absence of aphid colonies (a categorical response variable) can be explained by the presence or absence of predatory ants (a categorical predictor variable) and/or by the total phenolic content of the aphid host plant (a continuous predictor variable). Traditional ANOVA, in which a continuous response variable is modeled on a categorical predictor variable, is not usually viewed as a type of categorical data analysis. A summary of widely used names for major families of causal analyses of categorical and continuous data is given in table 11.1. In this chapter, I discuss two major related topics (logistic regression and logit modeling) that fall under the general rubric of categorical data analysis.

The distinction between categorical and continuous variables can be blurred—sometimes by semantics, but sometimes by legitimate mathematical considerations. Truly continuous and unbounded random variables (of the sort that ordinary least-squares (OLS) regression technically requires) are rather uncommon in ecology. Conversely, many discrete measurement classifications represent underlying continuous variables that are difficult or impossible to measure (e.g., deterrent vs. neutral vs. attractant). In other cases, we may know the underlying continuous distribution, but find it necessary to create artificial categories (e.g., higher than median vs. lower than median); this may be due to a paucity of data or to nonrandom sampling across the domain of the distribution. In any case, it is desirable to view the problem along a broad spectrum, rather than as a clear dichotomy.

This spectrum of random variable distributions along the continuous–categorical axis should be apparent to anyone who has ever examined an aphid (table 11.2). The midgut redox potential of an aphid is an example of a *continuous* random variable that is unbounded, because it can theoretically take all values between $+\infty$ and $-\infty$. Body mass is also continuous, because it has a lower bound of zero. Crawling speed, another continuous random variable, is likewise bounded from below by zero, but it also bounded from above by the speed of light (and, practically, at lower velocities by other factors). The number of setae is an *inter-*

Table 11.1 Commonly used names of analyses of categorical and continuous data

		Predictor Variables		
		Continuous	Categorical	Mixed
Response Variable	Continuous	OLS regression	analysis of variance	analysis of covariance
	Categorical	logistic regression	logit modeling	logistic regression or logit modeling

Table 11.2 Characteristics of an aphid, showing a broad range of random variable distributions

Distribution	Example
Continuous, unbounded	midgut redox potential
Continuous, upper or lower bound	body mass
Continuous, upper and lower bounds	crawling speed
Categorical, many regular intervals	seta number
Categorical, few regular intervals	generation
Categorical, ordinal with fairly evenly distributed rankings	instar
Categorical, ordinal with unevenly distributed rankings	diet breadth
Categorical, nominal with no apparent ranking	species

val variable with so many regular divisions as to resemble a continuously distributed variable. However, the numbers of generations (say, in a population study), although also divisible into regular intervals, are typically so few that the continuous approximation is risky. The instars (first, second, third, and so forth) of an individual aphid are ordinally distributed, in rankings that are not perfectly regular, even though the biological "distances" between adjacent instars may be roughly equivalent. However, the categorization of aphid diet breadth (monophagous, oligophagous, polyphagous, and so on) implies no such equivalency between adjacent rankings. Finally, the classification of aphids as different species is truly nominal and imparts no information whatsoever about relative ranking.

There are pedagogical and analytical advantages to be gained from viewing this spectrum of random variables as a hierarchy (Agresti 1990) in which unbounded continuous distributions are highest and pure nominal distributions are lowest. Analytical methods are nested within this hierarchical framework, such that statistical procedures for any given situation can be used lower in the hierarchy, but not higher in the hierarchy. For example, methods for ordinal variables can be used on interval variables, but not vice versa.

I will focus on causal categorical models that are best suited to the analysis of purely nominal data where there is no intrinsic ranking of the variable levels (as in the case of the different species of aphids). Fortunately, these methods are fairly robust and can accommodate many classes of ordinal categorical variables. Moreover, there has been considerable recent progress toward the development of analyses that are explicitly designed for ordinal data (Ishii-Kuntz 1994). But unless otherwise indicated, my approach throughout will be to focus on nominal data, which is at the extreme categorical end of the spectrum of random variable distributions (table 11.2). The approach is idealized, but it is conceptually straightforward and it is robust.

11.1.3 Causality

Categorical data analyses typically distinguish between cause and effect (section 11.1.2, table 11.1). For example, I have assumed (thus far without justification) that the presence or absence of an herbivore is determined by two factors: (1) the presence or absence of natural enemies, and (2) total phenolic content of the host plant. We might represent causal direction in this system as

$$(\text{phenolics, predators}) \rightarrow (\text{herbivores})$$

There is no a priori reason to believe that the preceding model is correct. Perhaps the system is controlled by bottom-up forces,

$$(\text{phenolics}) \rightarrow (\text{herbivores}) \rightarrow (\text{predators})$$

or perhaps it is controlled by top-down forces,

$$(\text{predators}) \rightarrow (\text{herbivores}) \rightarrow (\text{phenolics})$$

Any of the preceding scenarios, or a combination of them, is plausible.

How do we know which model is correct? There are three approaches to the problem of determining causal order, and the best idea is to draw from the strengths of all three, described here:

1. The first approach is to rely on statistical methods, such as path analysis (chapter 12) or graph theory (Christiansen 1997), that partition complex models into explanatory and response factors. This approach is especially successful in restrospective and observational studies.
2. The second approach is to conduct an experiment. Experiments have the powerful advantage of controlling the direction of causality, but it is important to realize that experiments do not necessarily prove that a particular causal hypothesis is correct. The advantages and disadvantages of ecological experiments are discussed by Diamond (1986), Hilborn and Mandel (1997), and Houston (1997).
3. The third approach is to rely on knowledge and familiarity of the study organisms. Sometimes ecological theory can guide us toward one causal model or another; but in many cases, there are well-established theories for more than one causal model. Thoughtful perspective on the importance of "a feeling for the organism" is given by Keeler (1983).

Intrinsic causal direction is probably far less straightforward than is often imagined (Bungs 1959; Davis 1985), and we will pay careful attention to causality in the development of categorical data models. In turn, we will see that modern methods of categorical data analysis have very handy properties vis-à-vis the problem of causality.

11.2 Statistical Issues

11.2.1 Overview

Consider two straightforward sets of data (table 11.3). In the first case (table 11.3A), we might test the association between host chemistry and herbivore preference using a χ^2 goodness-of-fit test. In the second case (table 11.3B), we would probably use OLS regression to test the relationship between caloric intake and body mass. Typically, these two procedures are presented as mathematically unrelated. However, we will see that the two methods are, in fact, closely related, and that similar methods can be used for either type of data.

Table 11.3 Relationship between two variables

A. Categorical variables

Host Chemistry	Herbivore Preference
toxic	rejects
palatable	accepts
toxic	accepts
toxic	rejects
toxic	rejects
palatable	accepts
palatable	accepts
toxic	rejects
palatable	rejects
toxic	rejects

B. Continuous variables

Caloric Intake (Kcal)	Body Mass (kg)
901	72
844	64
910	80
1005	73
970	71
726	55
906	76
1029	89
930	72
986	78

11.2.2 OLS Regression Revisited

It may seem odd to review OLS regression—the workhorse of continuous data analysis—in a chapter on categorical data analysis. However, the basic OLS regression model provides the fundamental statistical framework on which categorical data modeling is based. The regression model, $Y_i = \Sigma\beta_k X_{ik} + u_i$, places no restrictions on the X values (i.e., the predictor variables), except that they be noncollinear. Otherwise, the values of the X values are totally unconstrained: they can be continuous or categorical, or anything in between. One consequence of this is the fact that ANOVA (X's are categorical) and regression (X's are continuous) are really the same statistical procedure. A good treatment of this subject can be found in Hamilton's (1992) text.

The central problem in OLS regression is to model, or predict, the response variable Y in terms of k X's, by estimating the unknown k β-coefficients. In other words, the goal is to make the predicted value of Y, as represented by the sum of the βX's, as close to the observed value of Y as possible. No other manipulation or transformation is required. Therefore, we say that the response variable and the predictor variables are connected, or related, via the identity link.

The identity link imposes certain restrictions on the distribution of the Y values. Since the X's are unconstrained, the Y's must be unconstrained, too. That is, the Y's must be continuously distributed along the entire domain of X's. But ecology is full of cases where the Y's are strongly constrained. For example, we have already asked whether the probability of occurrence (i.e., presence/absence data) can be related to one or more predictor variables. The problem is that probability, p, is a highly constrained response variable: $0 \le p \le 1$. This constraint introduces several severe violations of the assumptions of OLS regression (Aldrich and Nelson 1984), and it is unadvisable to attempt to transform the data to meet OLS regression assumptions (Trexler and Travis 1993). In particular, the popular angular transformation (i.e., $\sin^{-1}\sqrt{p}$) is recommended against, especially in cases where p is close to 0 or 1 (J. C. Trexler, written commun., 1999).

It is therefore necessary to remove the constraints on the Y's—or in the present case, the p's. We can remove the upper bound by dividing p by $1 - p$. This manipulation is well known to gamblers and is termed the odds transformation, O. The odds are unbounded from above and span this range of values: $0 \le O \le +\infty$. To remove the lower bound, we take the logarithm of the odds and call the result the logit transformation (whence "logit modeling" and "logistic regression" in table 11.1). The logit transformation has the very useful property of being continuous, unbounded, and symmetrical around zero. We can now rewrite the transformed regression function as

$$\ln[p/(1-p)] = \Sigma\beta_k X_{ik} + u_i$$

which relates the Y's and X's according to the logit link. The logit link restores the assumptions of the regression model, and it powers categorical data analysis.

11.2.3 Odds Ratios

The simplest possible categorical data model is a 2×2 array. For example, the data in table 11.3A could be conveniently reconfigured as a 2×2 array (table 11.4) in which host chemistry (palatable vs. toxic) would be plotted against herbivore preference (accepts vs. rejects). The marginal odds that an herbivore will accept a host are 4 to 6 (0.67). The conditional odds that an herbivore will accept a palatable host are 3 to 1 (3.00), and the conditional odds that an herbivore will accept a toxic host are 1 to 5 (0.20). In direct comparison, it is 15 times more likely that an herbivore will accept a palatable host than a toxic host. The ratio of two conditional odds (here, 3.00/0.20) is called the odds ratio. The odds ratio is a very important quantity with some rather unusual properties.

The odds ratio is important, because it provides a measure of association within a table; if there is no association (i.e., if there is independence), the odds ratio is equal to one. A test of significance is given by dividing the logarithmic odds ratio by its asymptotic standard error and comparing the result against the Z-distribution (standard normal). The asymptotic standard error is equal to the square root of the sum of the reciprocals of the cells in the 2×2 array. In the present example, the Z-statistic is given by

$$\ln 15/[(1/3) + (1/1) + (1/1) + (1/5)]^{1/2} = 1.70$$

Table 11.4 Relationship between host chemistry (palatable vs. toxic) and herbivore preference (accepts vs. rejects)

Host Chemistry	Herbivore Preference		Total
	Accepts	Rejects	
Palatable	3	1	$\Sigma = 4$
Toxic	1	5	$\Sigma = 6$
Total	$\Sigma = 4$	$\Sigma = 6$	$\Sigma\Sigma = 10$

which is not statistically significant ($P = 0.0889$). The same result can be obtained by computing a χ^2-statistic for the association, but comparison against the Z-distribution has a powerful pedagogical advantage: it disabuses us of the notion that contingency table analysis is somehow nonparametric. In fact, contingency table analysis is based on the standard normal distribution.

The odds ratio is unusual because it is invariant under marginal transposition (rows and columns can be interchanged) and because it is invariant under marginal multiplication (each entry in a row or column can be multiplied by a constant). These are special properties of odds and odds ratios; probabilities do not exhibit these properties. A more complete treatment is given by Edwards (1963). In addition to whatever intrinsic mathematical appeal these properties may hold, they also have practical consequences for statistical analysis and interpretation, which can be summarized as follows:

1. Marginal transposition results from switching the axes of a 2×2 array, as would happen if we reversed the causal order of a model. The odds ratio, and all other test statistics and parameter estimates are unaffected, whether the model is $(x) \rightarrow (y)$ or $(y) \rightarrow (x)$. However, probability ratios and differences usually vary between alternative causal models.
2. Marginal multiplication arises when we sample disproportionately on one level of a variable. Such a scheme is often desirable or necessary because of the cost, difficulty, or feasibility of sampling. Odds ratios are unaffected by marginal multiplication, but probability ratios and differences usually are.

Typically, these features are more useful in observational studies than in experimental studies because a good experiment controls causal direction and samples adequately across all levels of each variable. However, it is important to remember that careful observations often precede experimental studies and inform experimental design. An observational study, with ambiguous causal direction and unbalanced sampling, is discussed in section 11.4.2. First, we return to our experiment on aphids, ants, and plants.

11.3 Statistical Solution

11.3.1 Experimental Design

A good first step in experimental design is to determine which variables are predictors and which variables are responses. The variables of interest are herbivore distribution, predator distribution, host plant chemistry, and year of study. As we have already seen, it is unlikely that year of study would be a response variable; for example, it is hard to imagine how host plant chemistry could determine what year it is. However, the causal relationships among the remaining three variables are not necessarily obvious. At this point, it is appropriate to turn to ecological theory as a guide for experimental design. One particularly powerful view of tritrophic interactions is that populations are influenced simultaneously (and sometimes interactively) by top-down and bottom-up forces (Hunter and Price 1992). This view has the very handy property of automatically ascribing cause and effect in our system: predators are a top-down influence, plant chemistry is a bottom-up influence, and herbivores are caught in the middle as a response variable. In the simplest possible form, the model could be written as

(predator distribution, plant chemistry, year of study) → (herbivore distribution)

It is important to recognize that this model is based both on ecological theory (with regard to the three biotic variables) and on physical fact (time is an independent variable). It is equally important to remind ourselves that the experimental approach is designed to control causal direction, not to prove that the causal model is the right one.

The second step is to design the experiment. We can manipulate predator abundance by removing or excluding ants from experimental plants, we can manipulate host chemistry by fertilizing plants, but we cannot do a thing about what year it is. A similar experiment was performed by Krebs et al. (1995).

The third step is to describe the random variable distributions in this system and to characterize the analytical design of the experiment. In the ant–aphid–plant system, all three predictor variables are categorical. Predator occurrence is a nominal categorical variable (present vs. absent) and so is host chemistry (fertilized plants vs. control plants). Study year could be an ordinal categorical variable (year 1 vs. year 2 vs. year 3); however, it simplifies matters to restrict the analysis to only 2 years of study. The response variable (herbivore occurrence) is also a nominal categorical variable: aphid colonies present versus aphid colonies absent. The data constitute a $2 \times 2 \times 2 \times 2$ array, which is difficult to depict graphically. A more typical presentation is given in table 11.5. There are approximately 40 replicates for each combination of treatment effects. The number of replicates is not always 40 because a few plants may be lost or destroyed during the course of the experiment.

11.3.2 Data Analysis: Log-linear Modeling

Categorical data are usually analyzed by a family of methods called log-linear models. In their most general form, log-linear models do not distinguish between predictor variables and response variables. When a model contains both categorical response variables and predictor variables, it is usually called a *logit model*. When a model contains a categorical response variable and continuous predictor variables, it is usually called *logistic regression*. These distinctions are somewhat semantic; logit modeling and logistic regression utilize similar analytical proce-

Table 11.5 Four-way association among year of study (1 vs. 2), fertilizer treatment (fertilized plants vs. control plants), predator occurrence (ants present vs. ants removed), and herbivore occurrence (aphid colonies present vs. aphid colonies absent)

Year	Plants	Ants	Aphids	N
1	control	absent	absent	31
1	control	absent	present	9
1	control	present	absent	22
1	control	present	present	18
1	fertilized	absent	absent	27
1	fertilized	absent	present	12
1	fertilized	present	absent	17
1	fertilized	present	present	23
2	control	absent	absent	31
2	control	absent	present	8
2	control	present	absent	22
2	control	present	present	19
2	fertilized	absent	absent	26
2	fertilized	absent	present	13
2	fertilized	present	absent	15
2	fertilized	present	present	24

dures and are based on similar assumptions, just as regression and ANOVA for continuous response variables really are equivalent procedures (see Nelder and Wedderburn 1972; Dobson 1990). In some treatments (e.g., Agresti 1990), the difference among types of log-linear models has to do primarily with sampling distributions. In other treatments (e.g., DeMaris 1992), the role of causality is emphasized. We will develop an understanding of analytical methods for categorical data by first examining simple two-factor tables and then working our way up to three-factor and higher factor tables.

Two-way tables. As we have already seen, the simplest possible categorical data model is a 2×2 array. In this situation, the null hypothesis of no relationship is tantamount to an assumption of independence between two variables. The expected cell counts under the null hypothesis can be estimated from the marginal proportions, in the following manner:

$$m_{ij} = (\Sigma\Sigma m_{++}) \times (\Sigma\Sigma m_{++}\Sigma m_{i+}) \times (\Sigma\Sigma m_{++}\Sigma m_{+j})$$

where m_{ij} denotes the expected count in cell m_{ij}, $\Sigma\Sigma m_{++}$ denotes the total sample size, Σm_{i+} denotes the subtotal in row i, and Σm_{+j} denotes the subtotal in column j. To simplify matters, we can refer to the three terms on the right-hand side of the equation as n (total sample size), p_{i+} (the marginal proportion in row i), and p_{+j} (the marginal proportion in column j), respectively, as follows:

$$m_{ij} = n \times p_{i+} \times p_{+j}$$

We can now represent m_{ij} as a linear function of the new terms on the right-hand side of the equation, via logarithmic transformation, as follows:

$$\ln(m_{ij}) = \ln(n) + \ln(p_{i+}) + \ln(p_{+j})$$

Intuitively, the preceding equation tells us that the logarithm of the expected cell count can be partitioned into three additive components: an effect associated with location in row i, an effect associated with location in column j, and a constant term. The logarithms of the variables have been expressed as a linear function of each other; this is why we refer to the result as a log-linear model. The similarities among the present derivation, the logit transformation (section 11.2.2), and the OLS regression model should be apparent.

In addition to the model of independence, there is a slightly more complex model that admits the possibility of interaction between the rows and columns. Using an even more simplified terminology, we can write this equation as

$$\ln(m_{ij}) = [i] + [j] + [ij]$$

where $[i]$ denotes an effect of variable i, $[j]$ denotes the effect of variable j, and $[ij]$ denotes the effect of variable $[ij]$. This model, with all terms present (both main effects plus their interaction) is called the saturated model. By definition, the saturated model provides a perfect fit to the data. But it is not necessarily the simplest or most correct description of association. In fact, it may be unnecessarily complex and quite incorrect. Instead, a lower order model may be superior. In fitting log-linear models, an important objective is to identify the most parsimonious model that fits the data. Model-fitting procedures are sometimes confusing to newcomers to categorical data analysis, but the basic approach is straightforward: a model "fits" if the model χ^2-statistic is greater than the critical α-value (usually 0.05 or 0.10). Practically speaking, a model with a good fit is one that yields estimated cell frequencies that are statistically indistinguishable from the null hypothesis of no association among variables. In the case of a two-way relationship, model selection is trivial: either the most parsimonious model—the model of independence, $\ln(m_{ij}) = [i] + [j]$ fits or it does not. In the case of higher order relationships, model selection is not nearly so trivial.

Three-way tables. In the case of a three-way analysis, the saturated model is

$$\ln(m_{ijk}) = [i] + [j] + [k] + [ij] + [ik] + [jk] + [ijk]$$

Although there were only two possible two-way log-linear models (the saturated model, $\ln(m_{ij}) = [i] + [j] + [ij]$, and the model of independence, $\ln(m_{ij}) = [i] + [j]$), there are nine possible three-way log-linear models (table 11.6). The nine different log-linear models refer to at least five different general patterns of association among three factors. Model 1 is called complete independence, and it means that everything is independent of everything else. In model 1, the observed cell counts do not differ significantly from the expected cell counts. Models 2, 3, and 4 are the three models of joint, or marginal, independence. For example, the model $\ln(m_{ijk}) = [i] + [j] + [k] + [jk]$ tells us that factor i is independent of factors j and k. Models 5, 6, and 7 are the three models of conditional independence. For example, if factors j and k are independent of each other at each level of factor i, then this relationship is given by the model $\ln(m_{ijk}) = [i] + [j] + [k] + [ij] + [ik]$. Model 8 does not have a name; it has posed something of a pedagogical chal-

Table 11.6 Hierarchical log-linear models for a three-way association

Model	Interpretation
1. $\ln(m_{ijk}) = [i] + [j] + [k]$	i, j, k completely independent
2. $\ln(m_{ijk}) = [i] + [j] + [k] + [jk]$	i marginally independent of j and k
3. $\ln(m_{ijk}) = [i] + [j] + [k] + [ik]$	j marginally independent of i and k
4. $\ln(m_{ijk}) = [i] + [j] + [k] + [ij]$	k marginally independent of i and j
5. $\ln(m_{ijk}) = [i] + [j] + [k] + [ik] + [jk]$	i and j conditionally independent of k
6. $\ln(m_{ijk}) = [i] + [j] + [k] + [ij] + [jk]$	i and k conditionally independent of j
7. $\ln(m_{ijk}) = [i] + [j] + [k] + [ij] + [ik]$	j and k conditionally independent of i
8. $\ln(m_{ijk}) = [i] + [j] + [k] + [ij] + [ik] + [jk]$	see text
9. $\ln(m_{ijk}) = [i] + [j] + [k] + [ij] + [ik] + [jk] + [ijk]$	saturated model

lenge, ever since it was first brought to light by Bartlett (1935). Tautologically, it is the model with no three-way interaction term; substantively, it is the model in which no pair of variables is conditionally independent. Finally, model 9 is the saturated model, which we have already considered.

Four-way and higher way tables. As we have just seen, three-way tables are considerably more complex than two-way tables. And the jump to four-way tables is vastly more complex: whereas a three-factor analysis contains 9 log-linear models, a four-factor analysis contains 114 log-linear models. A five-factor analysis contains thousands of log-linear models, and nobody has even bothered to determine the number of log-linear models in a six-factor analysis. This is sobering news for ecologists, especially in light of Kareiva's (1994) view that higher order interactions are the rule, not the exception, in ecology.

Recall that an important goal in higher order categorical data analysis is to find the most parsimonious model that fits the data. To do so, it is not necessary to diagnose the hundreds or thousands of possible models. A particularly elegant solution to the problem of higher order model selection comes from graph theory, which is discussed in a general context by Fienberg and Kim (1999) and in an ecological context by Boecklen and Niemi (1994). Readers who are unfamiliar with graph theory can instead use stepwise procedures, which are conceptually straightforward. We can either add terms in stepwise fashion, via stepwise forward selection, or delete terms, via stepwise backward elimination. Stepwise methods are highly sensitive to the cutoff values (i.e., critical α's) for inclusion or exclusion of a term, so it is advisable to compare methods from different selection procedures (Christiansen 1997). Further details are given in SAS (1989a).

11.3.3 Worked Example: Logit Modeling

Let's return to the experiment on aphids, and see how the principles of model selection can be put into action. It is tempting simply to report a four-way model with all terms and interactions present. But as we will see, the saturated model can be unnecessarily complex and misleading. Instead, let's work through the data, moving from simple associations to more complex ones. We will use logit modeling, because our experiment on aphids involves (1) only categorical variables and (2) an assumption about causality. There are strong similarities between a logit analysis and a log-linear analysis of these data, but the logit analysis is simpler, shorter, and easier to understand. An excellent exercise would be to compare the results between a logit analysis (see subsequent discussion) and a log-linear analysis (the SAS code in appendix 11.2 at the Website can be used for this purpose). DeMaris (1992) provides a comparison between the two methods.

We begin with an analysis of the effect of ant removal on aphid occurrence, restricting our attention to ants and aphids on control plants in year 1. The odds ratio is $(31/9)/(22/18) = 2.82$. Note the direction of this relationship: aphid colonies are 2.82 times more likely to be present on plants from which ants were not removed—the opposite of what was predicted. Perhaps the ants are beneficial mutualists, instead of predators on the aphids. According to the output from the SAS program in appendix 11.1, the relationship is statistically significant ($\chi^2 = 4.39$, df = 1, $P = 0.0361$). The preceding results can also be generated by a two-factor log-linear model or with a hand calculator. Recall that a test of significance can be obtained by dividing the logarithmic odds ratio by its asymptotic standard error and comparing the result against the standard normal distribution ($Z = 2.10$, df = 1, $P = 0.0361$). The output from the SAS program in appendix 11.1 yields the following logit model:

$$[H] = 0.5180[P] + 0.7187$$

where [H] denotes the occurrence of aphid colonies and [P] denotes the effect of ant removal. A precise interpretation of the numerical coefficients will be taken up in section 11.3.4.

We can now expand the year 1 analysis to include the effect of fertilizer addition. The saturated model for this association is

$$[H] = 0.5373[P] + 0.2322[F] - 0.0193[PF] + 0.4865$$

Note that the only lower order model,

$$[H] = 0.5383[P] + 0.2347[F] + 0.4876,$$

provides an excellent fit to the data (model $P = 0.9102$). (Recall that a model fits if its χ^2-value is greater than critical α). In the present case, the lower order model is a simpler and better description of the effects of ant removal and fertilizer addition on the presence or absence of aphid colonies. From the SAS output, we see a positive effect of having ants present ($P = 0.0016$) and no effect of adding fertilizer ($P = 0.1660$).

Finally, we conduct a four-factor analysis, involving one response variable (presence or absence of aphid colonies) and all three predictor variables (ant removal, fertilizer addition, and year of study). We assume that the organisms in year 1 are not the same as the organisms used in year 2; if this were not the case, we would need to use a repeated-measures approach (section 11.4.6). The rather ungainly saturated model is

$$[\mathbf{H}] = 0.5650[\mathbf{P}] + 0.2758[\mathbf{F}] + 0.0277[\mathbf{Y}] - 0.0040[\mathbf{PF}] - 0.0277[\mathbf{PY}] -$$
$$0.0437[\mathbf{FY}] - 0.0152[\mathbf{PFY}] + 0.4588$$

Fortunately, all eight of the lower order models fit the data, so it is best to report the simplest of these. This is the model of complete independence, which can be written as

$$[\mathbf{H}] = 0.5644[\mathbf{P}] + 0.2758[\mathbf{F}] + 0.0338[\mathbf{Y}] + 0.4589$$

This model fits the data quite well ($P = 0.9964$), and it tells us that the occurrence of aphid colonies is dependent on having ants present ($P < 0.0001$) and fertilizing plants ($P = 0.0220$), but independent of year of study ($P = 0.7779$). An obvious implication of this result is that an even simpler final model, $[\mathbf{H}] = [\mathbf{P}] + [\mathbf{F}]$, is appropriate, but see section 11.4.4 for a cautionary note.

11.3.4 Interpretation of the Results of the Analysis: Parameter Estimates

One of the strengths of the log-linear model approach is its interpretability. In particular, log-linear modeling emphasizes the selection of simple models with high interpretative power. At this point, then, we should have a good grasp of the qualitative nature of the relationships in our model; specifically, the presence of aphid colonies is positively affected by the main effects of having ants present and adding fertilizer to plants, but unaffected by year of study. Now it is appropriate to consider the quantitative nature of association in categorical data analysis.

Parameter estimates provide quantitative measures of association among variables in a log-linear model and serve as clear analogs to regression (β) coefficients in OLS regression. We begin by rewriting the original two-way log-linear model, $\ln(m_{ij}) = [i] + [j] + [ij]$, as

$$\ln(m_{ij}) = \lambda_0 + \lambda_i[i] + \lambda_j[j] + \lambda_{ij}[ij]$$

The λ-coefficients that provide unique estimators of the expected cell frequencies are given by

$$\lambda_0 = [\Sigma\ln(m_{ij})]/IJ$$
$$\lambda_i = [\Sigma\ln(m_{i+})]/I - \lambda_0$$
$$\lambda_j = [\Sigma\ln(m_{+j})]/J - \lambda_0$$
$$\lambda_{ij} = \ln(m_{ij}) - \lambda_0 - \lambda_i - \lambda_j$$

Using the data in table 11.5, we proceed to obtain parameter estimates for the case of aphid colonies absent, given ant removal. To keep matters simple, we will restrict our analysis to control plants in year 1. The parameter estimates are as follows:

$$\lambda_0 = (\ln31 + \ln9 + \ln22 + \ln18)/(2 \cdot 2) = +2.9032$$
$$\lambda_j = (\ln31 + \ln9)/(2) - 2.9032 = -0.0876$$
$$\lambda_i = (\ln31 + \ln22)/(2) - 2.9032 = +0.3593$$
$$\lambda_{ij} = \ln31 - 2.9032 - (-0.0876) - (0.3593) = +0.2591$$

Substituting the λ-coefficients into the saturated log-linear model yields

$$\ln(m_{ij}) = 2.9032 - 0.0876 + 0.3593 + 0.2591 = 3.4340$$

Taking the exponent gives

$$\exp(3.4340) = 31.0000$$

which, conveniently, is the actual cell count.

We can now expand our treatment of parameter estimates to encompass the logit model. The response variable, **[H]**, is linked to the predictor variable, **[P]**, as follows:

$$\ln[\mathbf{H}] = \lambda_1 \ln[\mathbf{P}] + \lambda_0$$

Note the analogy with the regression equation:

$$\mathbf{Y} = \beta_1 \mathbf{X} + \beta_0$$

We obtain parameter estimates as we did previously, for factor **[H]** conditional on either level of factor **[P]**. For example, the parameter estimates for factor **[H]**, given ant removal, were shown in section 11.3.3 to be

$$\lambda_1 = 0.5180$$
$$\lambda_0 = 0.7187$$

Substituting the λ-coefficients into the equation yields

$$\ln[\mathbf{H}] = 0.5180 + 0.7187 = 1.2367$$

which is identical to the logarithm of the conditional odds of the absence of aphids, given the removal of ants:

$$\ln(31/9) = 1.2367$$

Finally, it can be shown that the odds ratio (2.82) is the ratio of the antilogarithms of the two expected logics: $\exp(1.2367)/\exp(0.2007)$.

11.3.5 Other Issues: Sample Size and Sampling Distribution

In any cross-classified data analysis, the fundamental unit of analysis is the *cell count*, which is the number of observations of a particular combination of levels of the variables in a study. The higher the cell count, the better. Of course, many readers will wish to know just how low they can go, especially when dealing with large and expensive organisms or study subjects. Fienberg (1979) provides some basic guidelines that will be welcomed by those accustomed to the arbitrary rule that each expected cell count must be ≥ 5; Molenberghs et al. (1999) give a more general overview of the problem.

A particularly thorny problem is that of cell counts of zero, which can arise for two reasons. The most common situation is the sampling (or random) zero, in which one or more cells contain no observations because of insufficient sampling. For example, an observational study of the associations among butterfly coloration (bright vs. cryptic), toxicity (toxic vs. palatable), and response by predators (attacked vs. ignored) may fail to turn up any observations of brightly colored toxic butterflies that are attacked by birds—even though this combination may occur in the wild. Unfortunately, maximum-likelihood estimates cannot always be obtained for sampling zeros (Christiansen 1997). One approach is to transform the sampling zeros by adding a small constant to each cell (as suggested by Haldane 1955 and SAS 1989a), but this can result in significant departures from the correct asymptotic estimates (Knoke and Burke 1980). An alternative approach, which may seem drastic but which sometimes makes good intuitive (not to mention statistical) sense, is simply to exclude effects and interactions with sampling zeros. In the previous hypothetical case, we might exclude toxicity from the analysis and concentrate only on the association between butterfly coloration and the response of predators.

The other type of empty cell is called the structural (or fixed) zero. In this case, a cell contains no observations because it is impossible for it to do so. There are two ways in which structural zeros can occur. First, a particular cell may represent an impossible combination of variable levels. For example, our study of butterflies and predators may fail to turn up any observations of cryptic butterflies that happen to be toxic—because this combination of factors is impossible in the particular system we are studying. Second, a particular cell may contain no observations because it was lost or destroyed during the course of an experiment. Recall, for example, that several of the plants were lost during the experiment on aphids, ants, and plants (table 11.5). When analyzing cross-classified data with structural zeros, it is usually necessary to exclude all empty cells from the analysis. When SAS is used to analyze cross-classified data with empty cells, SAS assumes that all empty cells refer to structural zeros unless instructed to do otherwise.

The various analytical methods for cross-classified categorical data are broadly equivalent to one another, but they may assume different sampling distributions. In most cases, the sampling distributions are either Poisson, if the total sample size is random, or multinomial, if the total sample size is fixed. Technically, log-linear models are appropriate for Poisson sampling, whereas logit models are used for multinomial samples. Random sample sizes arise from situations where the total sample size is not fixed by the observer; for example, our sample might include all the butterflies that we were able to catch. In contrast, we might determine beforehand that our experiment is to involve 50 toxic butterflies and 50 palatable butterflies; this would be an example of a fixed sample size. In the aphid data (table 11.5), the total sample size is fixed, which provides another justification for having used a logit analysis. Advanced discussion of fixed and random samples is provided by Sokal and Rohlf (1995) and Underwood (1997). Although these authors' conclusions refer explicitly to OLS regression, they are broadly applicable to categorical situations as well. Note that the sampling distribution (which refers to how the data were collected) is not the same thing as the intrinsic distribution of the different levels of the study variables (nominal, ordinal, continuous, and so on).

11.4 Related Issues and Techniques

11.4.1 Logistic Regression

Ecologists use the term *logistic regression* to refer to situations where we wish to model a categorical response variable on a continuous predictor variable or variables, or on a mixture of continuous and categorical predictor variables. However, this usage is not universal. Regardless of what the procedure is called, it is important point to know how to regress a categorical response variable on a continuous predictor variable. Menard (1995) provides a good introduction to logistic regression, and the sections on standardized regression coefficients and indices of predictive efficiency are especially useful. For a review with specific relevance to ecology, see Trexler and Travis (1993).

We can explore the properties and advantages of logistic regression by reexamining our experiment on aphids. Until now, we have viewed fertilizer addition as a categorical predictor variable: either plants receive fertilizer or they do not. Another approach, however, would be to measure leaf nitrogen (presumably higher in fertilized plants) and to use this information to predict the presence or absence of aphid colonies. Data are given in table 11.7, and a SAS program is provided on-line in appendix C. Leaf nitrogen is a continuous variable, so logistic regression is appropriate. Hypothesis tests and parameter estimates are generated using the same procedures we used previously, and the logistic regression is given by

$$[\mathbf{H}] = 0.3108[\mathbf{F}] - 5.2564$$

where $[\mathbf{H}]$ denotes the logarithm of the odds that aphid colonies are present, as previously, and $[\mathbf{F}]$ denotes the amount of nitrogen in a leaf sample. Thus, the odds that an aphid colony is present on a plant with 13 g N/kg leaf dry mass are given by

$$\exp(0.3108 \cdot 13 - 5.2564) = \exp(-1.216) = 0.30$$

Note that the λ_1-coefficient (here, 0.3108) is a measure of unit change. Specifically, the odds that an aphid colony are present increase by $\exp(0.3108)$ for each unit increase in $[\mathbf{F}]$. This unit increase (1.36) is the odds ratio for logistic regression.

Table 11.7 Effect of leaf nitrogen content on aphid occurrence

Leaf nitrogen content (g N/kg leaf dry mass)	Aphid colonies
11	absent
11	absent
12	absent
14	present
17	absent
16	present
20	absent
22	present
23	present
24	present

11.4.2 Simpson's Paradox

In our previous treatment of higher way tables, we saw that the saturated model and other complex models can be unnecessarily complex. Now we will see that the saturated model can be plain wrong. Consider an observational study of the association among herbivore occurrence (present vs. absent), parasitoid occurrence (present vs. absent), and host plant quality (low vs. high). Data are presented in table 11.8, and further reading on this interesting subject can be found in Turlings et al. (1990).

This is an observational study with random sampling and no assumption of causal order, so a log-linear model is appropriate. In the saturated three-way model, the association between occurrence of herbivores and parasitoids is not statistically significant ($\chi^2 = 0.08$, df $= 1$, $P = 0.7718$). But when we collapse across the effect of host plant quality and consider the simpler two-way log-linear model, we do find a statistically significant association between the occurrence of herbivores and parasitoids ($\chi^2 = 7.58$, df $= 1$, $P = 0.0059$). We can look at this another way: the two-way model without the interaction term does not fit ($\chi^2 = 7.77$, df $= 1$, $P = 0.0053$). Examination of the odds ratio of the association between parasitoids and hosts also confirms this result:

$$(48/36)/(28/51) = 2.43$$

The association is statistically significant ($Z = 2.75$, df $= 1$, $P = 0.0059$).

In this case, parasitoid occurrence and herbivore occurrence exhibit conditional independence but marginal association. Parasitoids and herbivores occur independently of one another within either host plant category, even though the overall pattern of occurrence is nonrandom. This phenomenon is termed Simpson's paradox, and it frequently arises in situations where marginal sample sizes are unequal. Simpson's paradox has received considerable attention from workers outside the biological sciences, and an interesting example from the ecological literature is given by Horvitz and Schemske (1994).

Table 11.8 Three-way association among host plant quality (high vs. low), occurrence of parasitoids (present vs. absent), and occurrence of herbivores (present vs. absent)

Plant Quality	Parasitoids	Herbivores	N
low	absent	absent	48
low	absent	present	17
low	present	absent	27
low	present	present	11
high	absent	absent	3
high	absent	present	11
high	present	absent	9
high	present	present	37

11.4.3 Hierarchical Log-linear Models

How can we resolve this paradox? The answer lies in making use of the fact that log-linear models are nested, or hierarchical: χ^2-values and df values for higher order models are composed of the additive χ^2-values and df values of lower order models. In the present case, we wish to compare the χ^2 of the log-linear model that contains the herbivore $\times$ parasitoid interaction term with the χ^2 of the log-linear model that does not contain the term. The relevant models correspond to model 1 and model 2 for the full log-linear model of association (appendix B at the Website) among parasitoid occurrence **[P]**, herbivore occurrence **[H]**, and plant quality **[Q]**:

$$\chi_{(1)}^2 = 69.76, \; df_{(1)} = 4$$
$$\chi_{(2)}^2 = 61.99, \; df_{(2)} = 3$$

By subtracting the model 2 parameters from the model 1 parameters, we get

$$\chi_{(diff)}^2 = 7.77, \; df_{(diff)} = 1$$

These parameters are identical to the model parameters for the two-factor log-linear analysis.

The take-home lesson is this: the saturated model (main effects plus interactions) does not necessarily include all the statistically significant associations in a system. Indeed, the saturated model masks a potentially interesting and statistically significant association between herbivores and parasitoids. This is not to say the three-factor approach is wrong in this situation; rather, the best three-factor model happens to be something other than the saturated model. In the present case, it is necessary to consult one of the three models of marginal independence (table 11.6). The preceding caveats apply to OLS regression and ANOVA as well.

11.4.4 Nonhierarchical Log-linear Models

We have just seen that it is advantageous to view log-linear models as hierarchical, but it is also possible to test hypotheses about nonhierarchical log-linear models. Any log-linear model that does not contain all the terms nested within a higher order interaction term is nonhierarchical. An example of a nonhierarchical log-linear model is

$$\ln(m_{ijkl}) = [i] + [j] + [k] + [l] + [ij] + [kl] + [ijk]$$

because it does not contain the $[jk]$ term that is nested within $[ijk]$. Nonhierarchical log-linear models have peculiar interpretations (Knoke and Burke 1980), and they should be used with caution. However, nonhierarchical models seem to hold some promise for ecologists, and the interested reader would do well to consult the recent work of Dyer (e.g., Dyer 1995) on this matter.

11.4.5 Polytomous Response Variables

The examples we have encountered to this point involved dichotomous categorical variables. Fortunately, all of the methods discussed thus far extend readily to situations where a categorical response variable is polytomous. If, however, a polytomous variable is ordinal, then we can increase the power of statistical tests by making use of our knowledge of the known or suspected ranking of the various levels of the variable. Although most methods for ordinal data are fairly new, it is interesting that the problem dates back at least to Yule (1912). For a more recent treatment, see Ishii-Kuntz (1994), who shows that, under many circumstances, ordinal log-linear models can be surprisingly powerful and highly flexible.

11.4.6 Repeated-measures Analysis

Another area of recent activity by applied statisticians has been the case of *repeated-measures* categorical data modeling, in which multiple measurements are made on the same individual or subject. The discussion in Vermunt (1997) provides a good introduction to the topic. In a typical application, a categorical response variable might be measured at different points in time on the same individual or at different locations on the same individual. Recent inroads against the problem of repeated-measures categorical data modeling have been admittedly modest, in comparison with progress on analogous problems for continuous data (chapter 8). However, it seems likely that the methods of generalized linear modeling theory will prove useful in the development of a comprehensive and powerful approach to repeated-measures categorical data modeling.

11.4.7 Computational Issues

Parameter estimates and hypothesis tests for all but the simplest of categorical data analyses cannot be calculated except with computer-intensive techniques. In particular, iterative fitting procedures are required to obtain maximum-likelihood estimates in log-linear models, and, more generally, in generalized linear models. Fortunately, the major statistical packages can handle most problems in categorical data analysis. For example, GLIM, BMDP, SPSS, MINITAB, and SAS can be used for any of the topics discussed in this chapter.

The CATMOD procedure in SAS has gained some popularity among ecologists, and most of the SAS programs in the online appendix to this chapter use CATMOD. CATMOD is satisfactory for most problems, but beware of the parameter estimates; curiously, they are reported in alphabetical order (as opposed to the order in which they were entered), and all redundant parameters are omitted from the standard report. The LOGISTIC procedure is useful for, not surprisingly, logistic regression; what is surprising, however, is that parameter estimates and logarithmic odds ratios are reported with the wrong signs, unless the DESCENDING option is used in the PROC statement. The newer GENMOD procedure seems to hold considerable promise for a wide variety of GLMs, including all of the methods handled by the CATMOD and LOGISTIC procedures. In particular, the GEE (generalized estimating equations) macro in GENMOD is highly flexible and can handle repeated-measures and other problems that other categorical data procedures in SAS cannot (SAS 1989b; SAS 1996). For an application with specific relevance to ecology, consult Crawley's (1993) text on GLIM.

Acknowledgments   I am grateful to Joel Trexler and an anonymous referee for their thorough reviews of this chapter, but I accept sole responsibility for any errors or omissions. I would also like to thank Heidi Appel, who provided information on aphid natural history. Finally, I would like to acknowledge the contributions of Lee Dyer and Kei Sochi, who have guided me toward a better understanding of categorical data analysis over the years.

12

Path Analysis

Pollination

RANDALL J. MITCHELL

12.1 Ecological Issues

Naturalists have recognized the importance of pollination in plant reproduction for more than 2500 years (Baker 1983). However, modern attempts to understand the details of this interaction have faced numerous difficulties. The major complication is that plant reproduction involves a number of sequential and relatively distinct stages, and experimental investigation of all the stages simultaneously is not feasible. Consider, for example, reproduction of the herbaceous monocarp (dies after reproducing) scarlet gilia (*Ipomopsis aggregata*, Polemoniaceae), found in mountains of western North America. Reproduction for this self-incompatible plant depends on pollination by broad-tailed and rufous hummingbirds (*Selasphorus platycercus* and *S. rufus*) that probe the red tubular flowers to extract the nectar produced within.

Because of the timing of events and the basic biology involved, a logical first hypothesis about the factors influencing plant reproduction might be summarized as follows:

$$\text{plant traits} \rightarrow \text{visitation} \rightarrow \text{pollination} \rightarrow \text{reproduction}$$

Each arrow indicates an effect of one class of traits on another. Other interactions are possible (e.g., direct effects of plant traits on reproduction, or of traits such as density or location on visitation), but I will ignore them for this initial example.

One way of determining whether this general hypothesis is correct is to mechanistically (and especially experimentally) assess the effects of each component in isolation from the others. Assuming that the components can be combined linearly and additively, the expected overall effect of each trait can be calculated using the mean effects from different experiments. For example, three experiments on three different groups of plants might indicate that (1) plants producing more nectar receive more hummingbird visits, (2) plants receiving more visits receive more pollen per flower, and (3) plants receiving more pollen per flower tend to mature more seeds per flower. Multiplication of the numeric estimates for these components predicts the effect of nectar production on seed production. For example, (visits/mg nectar sugar) × (pollen deposited/visit) × (seeds/pollen deposited) = (seeds/mg nectar sugar). Combining these estimates is intuitively satisfying but depends on an untested assumption: each step in the pathway is independent of (uncorrelated with) those preceding it, except for the causal linkages (Welsh et al. 1988). When this assumption is violated, estimated effect sizes can be misleading. Unfortunately, violation may be common; for example, large plants may produce both more nectar and more seeds/fruit (even with equivalent pollen/flower), so that predictions based on the separate estimates would be inaccurate. Such correlations among traits are common (e.g., Campbell et al. 1991), and current understanding of the factors causing the correlations is poor, so the assumption of independence is probably not justified in many cases.

12.2 Statistical Issues

Field biologists generally use two major approaches to deal with intertrait correlations and complicated causal relationships: experimental manipulation and statistical control. Experimental manipulation of suspected causes is a very effective way to draw strong inferences about causality (e.g., Hairston 1989), primarily because experiments eliminate correlations among traits through randomization (but see Bollen 1989, p. 73, for some qualifications). However, experiments are not always logistically possible, ethical, or even a reasonable place to start (Wright 1921). So, as effective as the experimental approach can be, it is not always appropriate.

Statistical control is another way to deal with correlations among traits (Pedhazur 1982). A common example is multiple regression, which accounts for correlations among causal variables to estimate the effect of a particular trait with all else statistically held constant. Statistical control is especially suited for observational data, where we can capitalize on natural variation in the putative causal variables. Multiple regression as a method of statistical control has one disadvantage: it cannot directly handle complicated causal schemes, such as those outlined previously for pollination, because it only deals with one dependent variable and does not allow effects of dependent variables on one another. This same problem applies to factor analysis and similar multivariate approaches.

Path analysis.   Path analysis, the focus of this chapter, is a more general form of multiple regression that allows consideration of complicated causal schemes with more than one dependent variable and effects of dependent variables on one another. Path analysis deals with intertrait correlations in the same way as does multiple regression, thus providing statistical control. Though not a substitute for experimental research, path analysis (and other methods of statistical control) can be especially valuable when used in combination with judiciously chosen experiments, or as a way to analyze experimental data when some variables cannot be randomized (e.g., sex) or when experiments are not possible (Hayduk 1987, p. xii), provided that treatment levels can be ordered (e.g., amounts of fertilizer) or that there are only two levels per treatment (e.g., pollinated vs. not; Schemske and Horvitz 1988). The main difference is that the experimental method minimizes correlations between the manipulated variables, so no correlation among treatments need be hypothesized (see subsequent discussion).

Path analysis was originally developed by Sewall Wright (1920, 1921, 1934) as a way to partition variation from observational data into causal and noncausal components, according to a particular hypothesis. Along with the required mathematics, Wright introduced the valuable concept of a path diagram, which summarizes an overall hypothesis about how different variables affect one another (as in the previous simple hypothesis). Before conducting a path analysis, you must have at least one specific diagram in mind. In proposing the basic causal hypotheses that make up the path diagram, use all the information available. As Wright (1921, p. 559) noted, "There are usually a priori or experimental grounds for believing that certain factors are direct causes of variation in others or that other pairs are related as effects of a common cause." The diagram reflects natural history, temporal sequence of events, intuition, and experience with the system. Having clear causal hypotheses in mind before beginning the analysis is a critical part of the process. After the fact, it is remarkably easy to change your ideas in response to the data, but this compromises significance tests, as do a posteriori comparisons in ANOVA. This is not to say that you should not have several alternative hypotheses in mind; indeed, this is to be encouraged (Breckler 1990). Methods for comparing different models will be discussed subsequently.

12.3 Statistical Solution

To demonstrate the implementation of path analysis, I use data from a study of *Ipomopsis aggregata*. The original choice of traits and structure of the path diagram were made a priori, but here I analyze a subset of traits chosen for instructive value. To test the very general hypothesis about pollination described previously, I have first customized the hypothesis to more specifically fit the biology of *Ipomopsis*. Flowers of this species are hermaphroditic, so a full accounting of reproductive success would consider both seeds mothered and seeds sired by a plant. Budgetary and logistic considerations prevented estimation of the number of seeds sired, so I only consider a measure of female reproductive success. The customized path diagram is shown in figure 12.1. In path diagrams, a one-headed arrow represents a causal effect of one variable on another (e.g., nectar production affects approach rate), a curved arrow represents a correlation, and "U" represents unexplained causes.

Three dependent variables (approaches, probes/flower, and proportion fruit set) are shown in figure 12.1, and I will justify the causal relationships for each. In

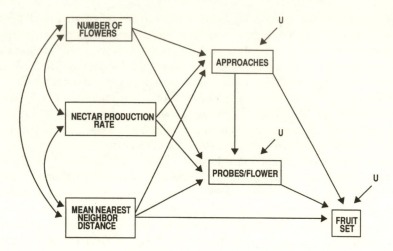

Figure 12.1 Hypothesized relationships determining hummingbird visitation and fruit production for *Ipomopsis aggregata* (model 1).

some cases, several alternative causal hypotheses are possible; one major goal of my analysis is to use the observed data to compare the explanatory value of those alternatives.

Approach rate. Hummingbirds may respond to a wide variety of cues when deciding which plants to visit. In this example, I consider two plant traits and one indicator of local density of conspecifics. Based on foraging theory, on the work of others, and on observation in the wild (Pyke et al. 1977; Wolf and Hainsworth 1990; Mitchell, unpublished data), I hypothesize that birds will preferentially approach and begin to forage at plants with high nectar production rates and many flowers. For similar reasons, birds may more often approach plants in denser clumps. Because little is known about the effects of clumping for *Ipomopsis*, this path is not as well justified as the others; its usefulness will be considered in section 12.3.4. Thus, the diagram includes causal arrows leading to approach rate from plant traits (flower number and nectar production) and from nearest neighbor distance.

Probes/flower/hour. The proportion of flowers probed might also be affected by the same characters as approach rate, for similar reasons. Furthermore, before a bird can decide how many flowers to probe on a plant, it must choose to approach a plant, so an effect of approach rate is also hypothesized.

Proportion fruit set. Logically, the chance that a flower matures into a fruit should increase with visitation rate, since each visit deposits additional pollen (Mitchell and Waser 1992). I use two estimates of visitation rate, each with slightly different properties. The first is approach rate, which may indicate the ability of plants to attract hummingbirds from a distance. Such plants also might consequently receive more cross-pollen and receive a higher diversity of pollen genotypes, because each approach represents a new opportunity for pollen to arrive from different plants. The second is probes/flower/hour, which represents the chance that a particular flower will be visited in a given time period. More probes/flower/hour should deposit more pollen, which increases fruit (and seed) set for *Ipomopsis* (Kohn and Waser 1986). Thus, I included both causal pathways in the model.

Pollination is not the only factor affecting fruit set. Among many other potential factors, I have chosen to consider the effects of near neighbors. The proximity of neighbors may indicate the extent of competition among conspecific plants for resources (Mitchell-Olds 1987), and competition might influence the ability of plants to mature fruits. I have therefore hypothesized effects of neighbor distance on fruit set. Admittedly, little is known of the effects of neighbors for this species, and I will consider these results with caution.

Correlations. Full analysis of a path diagram requires that the pattern of unanalyzed correlations among variables (the curved arrows) be explicitly stated. Traits such as flower number and nectar production may be correlated with one another for both genetic and environmental reasons. With detailed genealogical information, the genetic components of that correlation could be directly analyzed (Wright 1920; Li 1975). Likewise, detailed knowledge of the microenvironment experienced by each plant might allow a causal interpretation of this correlation, but that information also is unavailable. Consequently, I have hypothesized that the two traits are correlated for unanalyzed reasons. Nearest neighbor distance may be one cause of correlation between the plant traits (through effects on resource availability), but, since other interpretations are possible, I have modeled its relationships with flower number and nectar production as unanalyzed correlations.

12.3.1 Design of Observations

To test the causal hypothesis in figure 12.1, I measured floral traits, visitation, and reproduction for a sample of plants in a natural population. Observation of all these variables on the same individuals is required to apply the technique, partly for reasons mentioned in section 12.1. Measurement methods are described elsewhere (Campbell et al. 1991; Mitchell 1992). After deciding which variables to consider, my primary consideration was sample size. A good rule of thumb for both multiple regression and path analysis is to have, at minimum, 10–20 times as many observations as variables (Harris 1985, p. 64; SAS Institute 1989a, p. 140; Petraitis et al. 1996; section 12.3.5). For this example, there are six variables, so a sample size of >60–120 is reasonable. With this in mind, several assistants and I recorded pollinator visitation simultaneously for five distinct subpopulations of 23–30 plants each, for a total of 139 plants.

For each plant, I counted the number of flowers open on each day of observation and used the mean across days in the analysis. Mean nectar production rate and mean distance to the three nearest neighboring conspecifics are straightforward to measure and calculate (Campbell et al. 1991). For visitation, I calculated number of approaches/hour of observation, then took the mean across seven observation periods (about 70 minutes each). Likewise, for each observation period, I divided the total number of flower probes/hour for each plant by the number of flowers open that day to measure probes/flower/hour, and I used the mean across all observation periods. Proportion fruit set was calculated as fruits/total flowers.

Like many parametric techniques, path analysis assumes that the distribution of residuals is normal (section 12.3.5). Meeting this assumption is more likely if the variables themselves are normally distributed. To this end, I transformed most variables and tested the transformed and untransformed variables for deviation from normality using the SAS procedure UNIVARIATE with the NORMAL option (appendix 12.1 at the book's Website). Transformations were natural logarithms for flower number and nearest neighbor distance, square root for nectar production, and arcsin–square root for proportion fruit set (also see table 12.1).

Path analysis, like other techniques, can be sensitive to missing data. For example, some of my plants were eaten or trampled by animals before I could measure all of the traits mentioned. There are two ways to deal with missing data. First, do nothing (this is the default method for SAS). The result is that for some traits there are observations from all plants and for others the sample size is smaller. However, this method of "pairwise deletion" can occasionally make matrix inversion impossible (SAS Institute 1989a, p. 287; Hayduk 1987, p. 327; matrix inversion is a mathematical operation used in solving the regressions). A second method is to completely ignore individuals with incomplete data by deleting plants for which there are no data for any one of the six variables. Most of us are naturally reticent to discard any data, but such "listwise deletion" is generally the safest and most conservative approach (Hayduk 1987, p. 326). Unfortunately, there also are drawbacks, especially if the missing individuals are not a random sample of the population (Bollen 1989, p. 376). Hayduk (1987, p. 327) suggests a practical way to determine whether the method of deletion matters: do the analysis once using each method. If the results do not differ appreciably, there is no problem; otherwise, listwise deletion is preferable. For the *Ipomopsis* data, I found no substantial difference between results from the two deletion methods, and therefore I used listwise deletion for the nine plants with incomplete data.

The observations of plants and pollinators can be summarized in a correlation matrix (table 12.1). In some respects, this matrix answers some questions: there is a significant positive correlation between nectar production rate and proportion fruit set, which agrees with the idea that nectar production rate is important for reproduction. Unfortunately, these simple correlations alone do not provide direct information on the causal relationships. That is where path analysis comes in.

12.3.2 Data Analysis

A simple example.   Consider the simple path diagram in figure 12.2, which represents a portion of figure 12.1. The diagram is a symbolic representation of the following equations:

$$\text{approaches} = p_{\text{APH,NPR}}\text{NPR} + p_{\text{APH,U}}\ \text{U}$$
$$\text{probes} = p_{\text{PPF,NPR}}\ \text{NPR} + p_{\text{PPF,APH}}\ \text{APH} + p_{\text{PPF,U}}\ \text{U}$$

Table 12.1 Observed and expected correlations[a]

	FLR[b]	NPR	NND	APH	PPF	PROPFS
FLR	1.000					**-0.052**
NPR	0.156	1.000				**0.082**
NND	0.148	-0.130	1.000			
APH	0.248	0.271	0.035	1.000		
PPF	-0.145	0.141	-0.064	0.676	1.000	
PROPFS	-0.102	0.213	-0.298	0.182	0.247	1.000

[a]Observed correlations are below the diagonal. The two expected correlations that deviate from observed values are shown above the diagonal, in boldface (section 12.3.4). The expected correlations are derived from model 1 (figure 12.3), using the SAS procedure CALIS.
[b]Abbreviations: FLR = ln(mean number of open flowers each day), NPR = (nectar production rate)$^{1/2}$, NND = ln(mean of distances to three nearest neighbors), APH = mean approaches/plant/hour, PPF = Mean probes/open flower/hour, PROPFS = Arcsin(Proportion fruit set)$^{1/2}$.

The first equation states that all the variation in approach rate is accounted for by nectar production and all other causes (U). The second equation states that the observed variation in probes per flower is due to effects of nectar production and approach rate. These equations may readily be solved as regressions: in SAS, you simply state

PROC REG; MODEL APH = NPR/STB; MODEL PPF = NPR APH/STB

where APH = approaches/hour, NPR = √nectar production, and PPF = visits/flower/hour; these and other abbreviations are defined in table 12.1. SAS implicitly assumes random errors of estimation due to all other unmeasured causes (U). The option STB requests the standardized regression coefficients in addition to the unstandardized output.

Path coefficients are standardized regression coefficients and, therefore, indicate the number of standard deviations of change in the dependent variable expected from a unit change in the independent variable, with any effect of other independent variables statistically held constant. For more on the interpretation of standardized coefficients and statistical control, see Li (1975, p. 170) and Hayduk (1987, p. 39). In this simple example, SAS provides estimates of 0.27 for the path from NPR to APH, -0.04 for $p_{PPF,NPR}$, and 0.69 for $p_{PPF,APH}$.

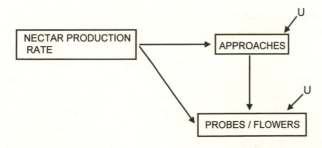

Figure 12.2 Path diagram of a simple multiple regression.

A useful way to think about the analysis so far is to focus on the fact that SAS was effectively presented with a 3×3 matrix of correlations among three variables:

	NPR	APH	PPF
NPR	1.00		
APH	0.27	1.00	
PPF	0.14	0.68	1.00

Consider the correlation between nectar production and probes/flower. Under the hypothesis embodied in figure 12.2, $r_{NPR,PPF} (= 0.14)$ consists of two components: a direct effect and an indirect effect (Li 1975, p. 114). The direct effect is the variation in probes that is uniquely attributable to variation in nectar, and the optimal estimator of this is the path coefficient (partial regression coefficient) from nectar to probes ($p_{PPF,NPR}$; -0.04 in this example). The indirect effect represents the portion of variation in probes jointly determined by nectar and approaches. The indirect effect occurs because nectar affects approaches directly ($p_{APH,NPR} = 0.27$), so that a change in nectar should result in a correlated change in approaches, which itself directly affects probes. A useful feature of path analysis is that the direct and indirect effects can be read directly from the diagram by tracing the two paths connecting nectar to probes [the rules for tracing paths are described by Li (1975, p. 162)]. Thus, the correlation between nectar and probes/flower can be decomposed into a direct effect ($p_{PPF,NPR}$) and an indirect effect mediated through approaches ($p_{APH,NPR} \times p_{PPF,APH}$).

In equations, this is stated as

$$\text{correlation of NPR and PPF} = \text{direct effect} + \text{indirect effect}$$

or, in symbols,

$$r_{NPR,PPF} = p_{PPF,NPR} + p_{APH,NPR} \times p_{PPF,APH}$$

Using the SAS estimates, we get

$$0.14 = -0.04 + 0.27 \times 0.69$$
$$0.14 = 0.14$$

This decomposition indicates that the direct effect of nectar on probes (-0.02) differs in magnitude and sign from the indirect effect ($0.27 \times 0.69 = 0.18$).

A more complicated example.   More complicated path diagrams are similarly treated. Model 1 (figure 12.1) is an elaboration of the simple example, modified to include more variables, as well as effects of dependent variables on one another (e.g., nectar production affects approaches, which in turn affects probes/flower/hour). Just as with the simple example, it is straightforward to write one equation for each dependent variable in the model. Each equation includes a term for every variable affecting that dependent variable, one term for each of the arrows leading to that dependent variable.

$$\text{approaches} = p_{\text{APH,FLR}}\text{FLR} + p_{\text{APH,NPR}}\text{ NPR} + p_{\text{APH,NND}}\text{ NND} + p_{\text{APH,U}}\text{ U}$$

$$\text{probes/flower} = p_{\text{PPF,FLR}}\text{ FLR} + p_{\text{PPF,NPR}}\text{ NPR} + p_{\text{PPF,NND}}\text{ NND}$$
$$+ p_{\text{PPF,APH}}\text{ APH} + p_{\text{PPF,U}}\text{ U}$$

$$\text{proportion fruit set} = p_{\text{PROPFS,APH}}\text{ APH} + p_{\text{PROPFS,PPF}}\text{ PPF}$$
$$+ p_{\text{PROPFS,NND}}\text{ NND} + p_{\text{PROPFS,U}}\text{ U}$$

These equations and the model in figure 12.1 are synonymous, because the correlations among the variables on the far left-hand side are assumed in the mathematics of regression.

One of the advantages of path analysis is that complicated equations can be summarized in fairly simple diagrams. Relationships excluded from the diagram (and therefore forced to equal zero) dictate the structure of the equations as much as do those included. For instance, model 1 specifies that nectar does not directly affect fruit set but may have indirect effects through approaches and probes/flower; this is reflected by the fact that $p_{\text{PROPFS,NPR}}$ is fixed at zero and therefore is not in the diagram or equations.

For three of the dependent variables (approaches/hour, probes/flower/hour, and proportion fruit set), there are no unanalyzed correlations (i.e., no curved arrows lead to them). The implicit assumption is that any correlations involving those variables are completely accounted for by the hypothesized causal scheme. However, there can sometimes be substantial deviations between the correlations implied by the model and the actual correlations, and these deviations can be used to assess the agreement between the model and the observed data (section 12.3.4).

12.3.3 Worked Example

Solving the equations for model 1 is straightforward using SAS procedure REG as outlined in section 12.3.2. The SAS code is on-line in appendix 12.1, and the results are summarized in table 12.2 and figure 12.3. The path coefficients (direct effects) in table 12.2 can be read directly from the SAS regression output (one

Table 12.2 Direct, indirect, and total effects for model 1[a]

	Effect on								
	APH			PPF			PROPFS		
Variable	DE[b]	IE	TE	DE	IE	TE	DE	IE	TE
FLR	0.20	—	0.20	−0.32	0.15	−0.17	—	−0.02	−0.02
NPR	0.24	—	0.24	−0.02	0.18	0.16	—	0.05	0.05
NND	0.04	—	0.04	−0.04	0.02	−0.02	−0.29	−0.00	−0.29
APH	—	—	—	0.76	—	0.76	0.07	0.14	0.21
PPF	—	—	—	—	—	—	0.18	—	0.18
R^2			0.12			0.56			0.14

[a]Direct effects are path coefficients.
[b]Abbreviations: DE = direct effect, IE = indirect effect, TE = total effect; other abbreviations as defined in table 12.1.

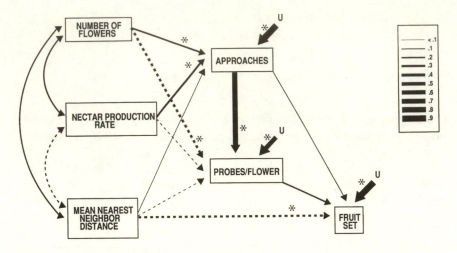

Figure 12.3 Solved path diagram for model 1 (figure 12.1). Solid lines denote positive effects, dashed lines denote negative effects. Width of each line is proportional to the strength of the relationship (see legend), and paths significantly greater than 0 ($P < 0.05$) are indicated with an asterisk. Actual values for path coefficients are in table 12.3.

regression for each of the three dependent variables) in the column labeled "standardized estimate." Figure 12.3 also presents the solution, and for ease of interpretation, arrow width represents the magnitude of each path. The simple correlations are provided by the SAS procedure CORR. The magnitude of the arrows from "U" (unanalyzed causes) indicate influences on each dependent variable that are unexplained by the causal diagram. Since the total explained variance in a dependent variable is defined to be R^2 (Li 1975, p. 178), the magnitude of the path from U to any variable ($p_{X,U}$) is calculated as $(1 - R^2)^{1/2}$. Since R^2 for PROPFS = 0.14 (table 12.2), $p_{PROPFS,U} = (1 - 0.14)^{1/2} = 0.93$. Calculation of total and indirect effects is laborious but straightforward (Li 1975, p. 161; Pedhazur 1982, p. 590; section 12.3.4).

12.3.4 Interpretation of Results

Once you have performed the path analysis, you must interpret it. What does the diagram as a whole tell you? Which paths are especially important? Was your original hypothesis correct? Path analysis has many applications, and each of the questions just raised corresponds to one of the major uses of the technique (Sokal and Rohlf 1995, p. 634).

Heuristic description. A path diagram is a compact method of presenting abundant information that may not be easily absorbed from regression tables (compare table 12.2 and figure 12.3, which are basically synonymous). Even a brief glance at a path diagram such as figure 12.3 can give a substantial understanding of the results: nectar production has a strong effect on approach rate; there are correlations among independent variables; approach rate has little direct

effect on fruit set; and so forth. Because path diagrams are easily grasped intuitively, they can be very useful in data presentation (chapter 3).

Estimation of effect sizes.   Sometimes we are primarily interested in the magnitude of the direct effect of a particular variable when the effects of other variables are statistically held constant (Wright 1983). Since path coefficients are
really multiple regression coefficients, they fit this requirement nicely. Of course,
just as with multiple regression, the path solution is contingent on which variables
are included in the model. Inclusion of other independent variables might well
change the significance, size, and even the sign of coefficients (e.g., Mitchell-
Olds and Shaw 1987).

Testing hypotheses.   Wright (1920, 1921, 1934) originally proposed path
analysis as a way to test explicit hypotheses about causal relations, but achieve-
ment of this goal was not generally practical until the advent of powerful computers. It now is possible (1) to actually test the hypothesis that a particular path
diagram is an adequate description of the causal processes generating the observed correlations and (2) to compare the descriptive power of alternative models
(see subsequent discussion). Aside from directly testing hypotheses, path analysis
also can help indicate which experiments might be especially useful. For instance,
if path analysis indicates that pollinators respond strongly and directly to nectar
production, an experiment to verify that conclusion might be warranted.

Data dredging.   This approach is tempting, but it should be used with caution
(Kingsolver and Schemske 1991). Data dredging is the opposite of hypothesis
testing: after the data have been collected, we look for interesting correlations
and assume that the strongest correlations are causal. There is nothing in the
mathematics that prevents us from then solving for a path diagram that allows
direct effects between strongly correlated variables, but the search through the
data has eliminated any element of hypothesis testing. Just as with a posteriori
comparisons in ANOVA, significance levels are not trustworthy for a posteriori
data dredging. Furthermore, the resulting causal interpretation will not necessarily
be accurate; simulation studies of artificial data reveal that data dredging seldom
arrives at the correct causal model (MacCullum 1986).

Some forms of data dredging are justified, however. For example, pilot studies
are often used to generate hypotheses and develop intuition about a system. Further, cross-validation of a dredged model (i.e., attempting to fit the model suggested by the data to an independent data set or to a previously unanalyzed portion of the original data) also is acceptable and useful (Mitchell-Olds and Shaw
1987; Bollen 1989, p. 278; Breckler 1990).

Interpretation of the Ipomopsis model.   If the diagram in figure 12.3 is correct, we can easily see the following: (1) there are strong effects of plant traits
on approach rate; (2) probes/flower is largely determined by approach rate and
(negatively) by number of flowers; and (3) nearest neighbor distance has strong
negative effects on fruit set.

Next, I evaluate those interpretations of the individual coefficients. But what
about the first proviso: is this diagram correct? This question has been a stum-
bling block in other fields (e.g., psychology, sociology, economics), as well as in
biology, and social scientists have made notable progress toward quantitatively

assessing the agreement between the path diagram and the observed data. They do this using the computer-intensive technique of "structural equation modeling," a more general form of path analysis. These analyses involve such complications as latent factors, specification of errors, and correlation of errors, and also allow tests of the goodness of fit of the model to the data. Although some of these improvements are applicable in the simpler causal structures of most biological applications and are incorporated in this chapter, it is not possible to cover this enormous topic in this space. An overview of the field is provided in any of the excellent general texts (e.g., Hayduk 1987; Loehlin 1987; Bollen 1990), and there are a growing number of biologically oriented introductions to some aspects as well (Maddox and Antonovics 1983; Johnson et al. 1991; Mitchell 1992; Pugesek and Tomer 1995, 1996). For now, it is sufficient to discuss the agreement between model and data.

In structural equation modeling, a goodness-of-fit statistic is used to assess the agreement between the correlations actually observed in the data and the correlations that would theoretically occur if the path diagram were correct (i.e., if the causal structure in the path diagram were the actual causal structure). The goodness-of-fit statistic is calculated from the difference between the observed and expected correlations (table 12.1). It is distributed approximately as a χ^2, with degrees of freedom equal to the difference between the number of unique observed correlations and the number of coefficients estimated. In the example, there are 21 observed correlations and 19 estimated coefficients: 10 direct effects, 3 correlations, and 6 variances. That leaves 2 degrees of freedom for comparing the observed and expected correlations.

Several computer programs quantify goodness of fit (e.g., EQS, LISREL; Bentler 1985; Jöreskog and Sörbom 1988), including a new procedure in SAS version 6 known as CALIS (Covariance Analysis of Linear Structural Equations; SAS Institute 1989a). Programming statements to implement CALIS for the example are given on-line in appendix 12.2. The resulting output includes direct, indirect, and total effects, as well as correlations, so it is not necessary to use the REG procedure. However, to use CALIS and structural equation modeling, you must learn the rather confusing jargon (Loehlin 1987; SAS Institute 1989a).

The observed and expected correlation matrices for the *Ipomopsis* data are shown in table 12.1. The observed correlation matrix simply comes from the raw data, and the expected correlation matrix is provided by CALIS in the standard output. There are only two elements that are free to differ between these matrices, corresponding to the 2 degrees of freedom for comparison. CALIS calculated the χ^2 goodness-of-fit statistic for model 1 (figure 12.3) to be 3.71, with 2 df, $P = 0.16$, so the lack of fit of model to data is not significant. In this case, nonsignificance is good for the model; a nonsignificant χ^2 indicates that there is no significant deviation between the observed and expected correlation matrices under this model. In other words, the model in figure 12.3 has survived an attempted disproof and remains as a potential explanation of the interaction. Whether such agreement between model and data also indicates agreement between the model and the actual causal structure is discussed by Bollen (1989, p. 67), but at least the possibility has not been eliminated.

The details of how the goodness-of-fit statistic is calculated are presented elsewhere (e.g., Hayduk 1987, p. 132); the important point is that we now have a way to determine whether the path diagram is reasonable. And for this example, the hypothesis that this is a correct description of the interaction cannot be disproved.

Not all path diagrams can be tested in this manner—the model must be "overidentified." At a minimum, overidentification requires that there be more observed correlations than estimated coefficients (i.e., there must be more knowns than unknowns), but there are other requirements. A full discussion of model identification is beyond the scope of this chapter; see Hayduk (1987), Loehlin (1987), and Bollen (1989).

If the diagram as a whole had a poor fit to the data, it would be premature to place much importance on the values of individual paths, because the model might be of little descriptive value and the magnitude of individual paths might therefore be misleading. But given that the goodness-of-fit test indicates that model 1 might be reasonable, inspection of the individual paths is in order. In many cases, this may suggest relevant experiments, and results from experiments may suggest improvements or modifications to the path diagram. Such feedback among observations, experiments, and hypotheses is an important part of the strong inference approach to science (Platt 1964; chapter 1).

In model 1, nectar production has a significant effect on approach rate by hummingbirds. One interpretation is that hummingbirds somehow identify or remember plants with high nectar production rates, and they preferentially approach them. Because other evidence suggests that identification from a distance is unlikely (Mitchell 1992), I have experimentally investigated the possibility that hummingbirds actually remember the location of individual plants (Mitchell 1993); I found that birds do not seem to remember high nectar plants, although they do respond after arriving and sampling some flowers. Since those results disagree with the hypothesis that birds remember individual plants, my current working hypothesis is that bird use spatial location as a cue to identify clumps of plants with high or low nectar production rates. Note that nearest neighbor information is apparently not used (figure 12.3). Although this seems a plausible explanation after the fact, I would not have favored this hypothesis without the path analysis and subsequent experimentation.

Flower number significantly affected both approach rate and probes/flower. Although plants with many flowers were approached more often, flower number negatively affected probes/flower. This decrease in probes may represent a proportional cost to plants having many flowers if it lowers seed production.

Nearest neighbor distance had no significant effects on visitation behavior, but had strong, negative effects on proportion fruit set. One potential explanation is that short nearest neighbor distances indicate more intense intraspecific competition (Mitchell-Olds 1987; see Brody 1992 for an alternative explanation). As with the previous examples, this prediction is amenable to experimental investigation.

The largest paths in all cases involve the unexplained influences on each dependent variable (U), indicating that much of the variance in dependent variables cannot be explained by this model, even though the goodness of fit is acceptable.

This may be because many potentially important variables have not been included (e.g., plant resources), but it also may mean that chance variation plays a large role.

Comparing alternative path diagrams.   Often, several different models are considered feasible a priori. In the past, there was no quantitative way to choose among models, but the goodness-of-fit test changes this by greatly increasing the usefulness of path analysis. For instance, model 2 (figure 12.4) is an alternative to model 1 (section 12.3). Model 2 hypothesizes that nearest neighbor distance has no direct effect on hummingbird visitation behavior. The observed data do not deviate significantly from what would be expected if model 2 were correct ($\chi^2 = 4.44$, 4 df, $P = 0.35$), so it also is an adequate description of the interaction. Deciding which of these two acceptable models is a better description of the interaction is facilitated by the fact that they are "nested." Nested models are identical except that one does not include some effects hypothesized in the other. Here, model 2 is nested within model 1 because it can be derived from model 1 by constraining the direct effect of neighbors on visitation to zero (by not estimating that path). To compare nested models, the difference in goodness of fit is used, employing the fact that the difference between two χ^2-values also is distributed as a χ^2, with degrees of freedom equal to the difference in degrees of freedom between the two models (Hayduk 1987). There is a nonsignificant difference in χ^2 between model 1 and model 2 (table 12.3). Because model 2 is simpler than model 1 (it estimates fewer paths), we may prefer it based on the principle of parsimony. But, given the fairly strong theoretical justification for estimating the effects of neighbors on visitation, I will retain the more general model 1 for now. Observations or experimental manipulations of nearest neighbor distance would be useful in further evaluating model 2. Although these particular acceptable models did not differ significantly, in general, there may be strong and significant

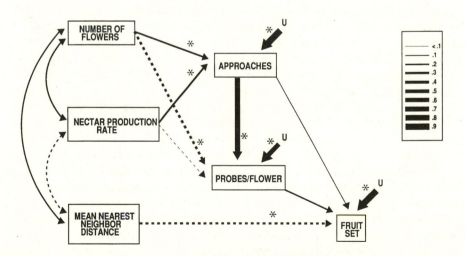

Figure 12.4 Solved path diagram for model 2, modified from model 1 to include no effects of nearest neighbor distance on visitation. Conventions follow figure 12.3.

Table 12.3 Nested comparison of alternative models with model 1 (figure 12.3)

Model number	Description	Goodness of fit			Nested comparison with model 1		
		χ^2	df	P	χ^2	df	P
1	See figure 12.3	3.71	2	0.16	—	—	—
2	Same as model 1, but no effects of neighbors on visitation	4.44	4	0.35	0.73	2	0.7
3	Same as model 2, but no effects of neighbors on fruit set	16.28	5	0.006	12.57	3	0.01

differences among acceptable models. Indeed, social scientists frequently compare nested models to choose among several otherwise acceptable hypotheses (e.g., Loehlin 1987, p. 106).

Model 3 is another nested model, going one step further than model 2 by proposing that nearest neighbor distance does not affect fruit set (figure 12.5). The fit of model 3 is far inferior to that of model 1 (table 12.3), indicating that effects of neighbors on fruit set are very important to an acceptable description of the interaction. As mentioned previously, this may indicate the importance of intraspecific competition (e.g., for nutrients, space, light, or pollination), among other possibilities. Experimental manipulation of one or more of these potential influences could be pursued, perhaps after some are eliminated by further observational study.

Model 3 illustrates an important difference between structural equation modeling and traditional approaches to path analysis. Without the goodness-of-fit test,

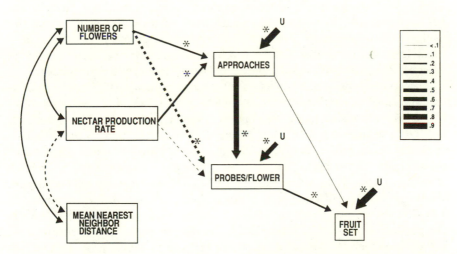

Figure 12.5 Solved path diagram for model 3, modified from model 2 to include no effects of nearest neighbor distance on visitation or fruit production. Conventions follow figure 12.3.

there would be no reason to reject model 3. Just as with model 1, some paths of biological interest are large and R^2-values for some dependent variables are reasonable (for model 3, $R^2 = 0.12$, 0.56, and 0.06 for approaches, probes/flower, and proportion fruit set, respectively; compare to table 12.2). There would be no real clue that a potentially important path is ignored in model 3 and no quantitative method for deciding which is a more reasonable model. Without a way to compare the fit of models, I probably would only have presented results from one model and ignored those I did not consider likely. Because of this, the possibility of alternative models should be carefully considered by both authors and readers, even when goodness of fit is not assessed.

12.3.5 Other Issues

Path analysis relies on a number of assumptions. The assumptions of normal distribution of residuals, additive and linear effects, and inclusion of all important variables are discussed here. Other assumptions commonly made are that residual errors are uncorrelated and that there is no measurement error. The latter assumptions may be relaxed when using structural equation modeling (section 12.4).

First, as with multiple regression, path analysis assumes that the residuals from estimation are normally distributed. Appendix 12.1 (found at the book's Website) demonstrates how to test this assumption using the SAS procedures REG and UNIVARIATE. The same approach can be used to test residuals from GLM and other SAS procedures. For the *Ipomopsis* data, residuals for several variables departed significantly from normality, despite my transformations. Violation of this assumption does not affect the magnitude of the path coefficients (Bollen 1989, p. 126), but significance tests may be untrustworthy. To deal with this, I have compared the traditional significance levels for individual paths with those from a delete-one jackknife test that does not assume normality (Mitchell-Olds 1987; chapter 14). In no case did significance values differ between the two methods. This robustness gives me confidence in traditionally derived significance levels for these data. If there had been disagreement between the two approaches, the jackknifed significance levels would probably be preferable.

The goodness-of-fit test assumes multivariate normality and, further, depends on large sample sizes to utilize the asymptotic χ^2-distribution of the goodness-of-fit statistic. Surprisingly, the assumption of multivariate normality is not very restrictive, because it is a sufficient but not necessary condition for the goodness-of-fit test (Cloninger et al. 1983), and the test is robust to skew and kurtosis (Muthen and Kaplan 1985). Bollen (1989, p. 79) also argues that this assumption is no more restrictive than those made for ANOVA. However, for field ecologists, small sample size will often be a more pressing problem, since observations on 100–200 individuals (sampling units) are usually required to have confidence in the goodness-of-fit test (Tanaka 1987). Because of both these limitations, χ^2-values should generally be used more as an index of fit than as a rigorous statistical test. For example, model 3 does not fit the data well at all, whereas model 1 is reasonable; those conclusions are unlikely to be influenced qualitatively by deviations from multivariate normality or by small sample size.

Second, the analysis assumes that the causal relationships are additive and linear. Wright (1960, 1983) claims that this assumption commonly is met. To test for curvilinearity, inspection of raw data plots or residuals can be useful (Cohen and Cohen 1983, p. 126; chapter 3; see also appendix 12.1). Curvilinear relationships often can be modeled using quadratic and higher terms, as in multiple regression (Sokal and Rohlf 1995, p. 609; Cohen and Cohen 1983, p. 369; Hayduk 1987, p. 219; Scheiner et al. 2000).

Third, the analysis assumes that information on all variables having strong causal effects have been included in the analysis. The reasons for and consequences of violating this assumption are covered in Cohen and Cohen (1983, p. 129) and Mitchell-Olds and Shaw (1987). Note that the assumption is not that all traits are included (an impossibility), but instead that all traits with strong causal effects are included (Cohen and Cohen 1983, p. 354). This assumption should be taken as a warning that the solution may depend on which variables are considered (contrast the significance of the path from probes/flower/hour to fruit set in figures 12.3 and 12.5). Just as with experiments, if important variables are omitted or ignored, the answer may be misleading or wrong.

12.4 Related Issues and Techniques

Multiple regression is covered by Cohen and Cohen (1983), among many others. Good introductions to path analysis and its applications are in Li (1975), Pedhazur (1982), Loehlin (1987), Schemske and Horvitz (1988), Crespi and Bookstein (1989), Crespi (1990), and Kingsolver and Schemske (1991). Currently, the only introduction to CALIS is the SAS manual (SAS Institute Inc. 1989a). Consulting the readable introductions to structural equation modeling and LISREL (a program similar to CALIS) of Hayduk (1987) and Loehlin (1987) will be helpful in deciphering the SAS text, which assumes familiarity with the subject. The advantages and disadvantages of structural equation modeling are discussed in Cloninger et al. (1983), Karlin et al. (1983), Wright (1983), Bookstein (1986), and Breckler (1990). If variables are strongly intercorrelated (multicolinear), inclusion of latent factors may be useful, and these, as well as correlated residuals and measurement error, can be incorporated as an integral part of a structural equation model; see Hayduk (1987), Loehlin (1987), and Bollen (1989), but be sure to read and heed the cautions of Bookstein (1986) and Breckler (1990).

Path analysis has been applied to many other ecological topics, including ecosystem modeling (Johnson et al. 1991), species interactions and community structure (Arnold 1972; Weis et al. 1989; Wootton 1990; Weis and Kapelinski 1994; Grace and Pugesek 1997, 1998; Smith et al. 1997), plant growth rates (Shipley 1995), and phenotypic selection (Maddox and Antonovics 1983; Mitchell-Olds 1987; Schemske and Horvitz 1988; Crespi 1990; Herrera 1993; Mitchell 1994; Jordano 1995; Conner et al. 1996; Murren and Ellison 1996; Sinervo and De-Nardo 1996; Bishop and Schemske 1998; Scheiner and Callahan 1999).

Recent developments in biological usage of path analysis center on two topics. First, path analysis is commonly misused. A review of the ecological literature by Petraitis et al. (1996) revealed that sample sizes are usually too small, colinearity is common, and incomplete information is presented in most published studies. They also point out some useful diagnostics for these problems. Second, new methods of path analysis may help with small data sets. Shipley (1995, 1997) has developed bootstrapping methods for evaluating small data sets, an especially exciting turn of events because of the notoriously small sample sizes in many important ecological applications. Shipley (1997) advocates exploratory uses of path analysis, as opposed to the predictive uses I advocate here, whereas Petraitis et al. (1996) argue against the use of goodness-of-fit tests in path analysis (but see Pugesek and Grace 1998).

Acknowledgments   Alison Brody, Diane Campbell, Ken Halama, Diana Hews, Karen Mitchell, Mary Price, Hadley Renkin, Rich Ring, Peter Scott, and Nick Waser helped in the field. Valuable discussions and comments on the manuscript came from Warren Abrahamson, Alison Brody, Bob Cabin, Ann Evans, Jessica Gurevitch, Ken Halama, Doug Kelt, Diane Marshall, David Reznick, Sam Scheiner, Ruth Shaw, Art Weis, and Nick Waser. Beth Dennis drew the illustrations, Tom Mitchell-Olds kindly provided the jackknife program, and David Reznick demonstrated how to test for normality of residuals. The Rocky Mountain Biological Laboratory provided housing, lab space, and a supportive atmosphere during field work. This research was supported in part by NSF Grant 9001607.

13

Failure-time Analysis

Studying Times to Events and Rates at Which Events Occur

GORDON A. FOX

13.1 Ecological Issues

Do differences among treatments lead to differences in the time until seeds germinate, foragers leave patches, individuals die, or pollinators leave flowers? Do treatment differences lead to differences in the number of events that have occurred by some time or to differences in the rates at which these events occur? Questions of this kind are at the heart of many studies of life history evolution, demography, behavioral ecology, and pollination biology, as well as other ecological subdisciplines. In all of these cases, researchers are concerned either with the time until some event occurs in individual experimental units or with the related problem of the rate at which these events occur.

Ecologists are interested in these types of data for several different reasons. One reason is that time itself is sometimes limiting for organisms because their metabolic clocks are constantly working. For example, animals that do not acquire food quickly enough may become malnourished. Time may also be limiting because of external environmental factors. For example, annual plants that do not flower early enough may be killed by frost or drought without successfully setting seed. Time may also be important because of population-level phenomena. For example, in age-structured populations, individuals that reproduce at a late age will leave fewer descendants on average than those that reproduce at an early age, all else being equal. If resource competition is partly age-dependent, on the other hand, early reproducers may sometimes be poor competitors and therefore may actually have fewer descendants. Finally, there are many situations in which ecologists are interested in knowing how many events have occurred by a particular time, such as the number of flowers that have been visited by insects by the end of the day or the number of animals that survive a particular season.

Many ecologists study failure-time by visually comparing survivorship curves. Although this is often a useful part of any exploratory data analysis (chapter 3), it requires that we make a subjective decision as to whether two survivorship curves are "really" different from one another. I suspect that ecologists do this so frequently because courses in population ecology usually treat survivorship and life tables as fixed features of populations. This makes it easy to forget that failure-time data must be treated statistically.

This chapter discusses data on the timing of events and special methods for their analyses. These methods originated in several diverse fields in which similar statistical problems arise—for example, clinical medicine, human demography, and industrial reliability testing—and consequently are quite well developed.

Recent ecological applications. There are many ecological applications of methods of failure-time analysis. My own research—as reflected in the example developed in subsequent sections—has concerned the timing of life history events, but recent authors have applied failure-time methods to a wide variety of problems. Newman (Dixon and Newman 1991; Newman and McCloskey 1996) made an important contribution to ecotoxicology by showing that studies of time until effect can be much more powerful (and biologically meaningful) than traditional studies of LD_{50}. Petraitis (1998) used failure-time methods to study rates of predation on intertidal mussels. Clark (1989) studied the time between forest fires and how it has changed over the centuries.

Are there costs to sex? In a study that received considerable attention, van Voorhies (1992) used failure-time approaches to show a cost of spermatogenesis in the nematode *Caenorhabditis elegans*. Males with normal spermatogenesis, he showed, have substantially reduced survivorship as compared with males with mutations that stop the chain of events leading to sperm production. To show that dominance rank is of considerable importance in chimpanzees, Pusey et al. (1997) used failure-time methods to show greater survivorship in female chimps with higher rank.

A number of recent studies have focused on problems related to our understanding of aging and senescence. Adult birds (especially Passerines) have long been thought not to senesce, that is, they do not have a classic Type II survivorship curve. By using failure-time approaches, McDonald et al. (1996) were able to show that, in fact, Florida scrub jays do senesce, and Ricklefs (1998) generalized this result to numerous species. Is menopause in mammals an adaptive trait, as suggested by some sociobiological arguments? Using failure-time methods, Packer et al. (1998) were able to show that menopause in baboons and lions (at least) is a senescent trait—postmenopausal females do not increase the fitness of their offspring or grandchildren. Finally, failure-time methods have been important in studies of the factors leading to senescence (e.g., Carey et al. 1998; Vaupel et al. 1998).

Perhaps the most unusual recent application of these methods is to learning and spatial memory. Capaldi and Dyer (1999) studied the ability of honeybees to orient to new landscapes by examining the time it takes until a bee returns to its hive. Naïve bees returned faster than others if they were able to see landmarks near the hive, but when removed to distant sites, resident bees returned more quickly, suggesting an important role for spatial memory.

There are many other potential applications of failure-time analysis to ecological data. We can get an idea of these applications (as well as of the origins of these methods) from the examples in Kalbfleisch and Prentice (1980), Lawless (1982), and Collett (1994). In addition to being well developed statistically, this field of statistics has evolved its own jargon. I introduce this jargon in the following section in the context of a discussion of the peculiar nature of this kind of data.

13.2 Statistical Issues

13.2.1 Nature of Failure-time Data

Ecologists get data on the timing of events by repeatedly observing uniquely identified individuals. As with repeated-measures analysis of variance (ANOVA; see chapter 8), this repeated-observations structure leads to special statistical methods. These observations may take place continuously (as would be necessary in studying giving-up times of foragers in patches) or may be in the form of censuses at intervals (as would be appropriate for studying survival of marked plants). At each observation time, the researcher determines whether the event of interest has occurred for each individual. In simple studies, three general outcomes are possible:

1. The event has not yet occurred, and the individual is still in its original state. In this case, the individual is said to have *survived*. In this sense, *survival* can refer to remaining alive as well as to remaining in a patch, remaining ungerminated, and so on.
2. The event has observably occurred. In this case, statisticians speak of *failure* and of the timing of the event as the individual's *failure time*. This term originated in industrial reliability testing. In an ecological setting, a failure refers to the individual having been observed leaving the patch, being observably dead, or having observably begun to flower.
3. Finally, the individual may have been lost from the study without a failure being observed. This kind of data point is called a *right-censored* data point, because the researcher can be certain only that the actual failure time is greater than the last recorded survival time. For example, a failure time (in this case, a giving-up time) cannot be assigned to a forager that was still in a study patch when a predator was observed to eat it, but the researcher knows that the time of abandoning foraging (i.e., the failure time) would have been at least as great as the time of predation (i.e., the censoring time). Similarly, a flowering time or death time cannot be assigned to a plant that has lost its marker, because the researcher is uncertain as to its fate at any time later than the prior census, although it is clear that the failure time was later than the prior census.

Researchers often discard right-censored data points either because they assume that these data are not useful or because they believe that methods for handling censored data are unnecessarily complex. Both beliefs are wrong. The former belief leads researchers to discard potentially important data and can contribute to highly biased results. To understand this, consider a simple example from clinical medicine. In a study of the effect of a drug on tumor regrowth, some patients are killed by heart attacks or accidents, with no regrowth at time of death. Discarding these cases would not only be an inefficient waste of data, but more important, it could bias results because those in whom the treatment has been effective are more likely than others to be killed by causes other than cancer.

It is also possible to have *left-censored* data points. This is the case when failures have already observably occurred for some individuals when the study begins (Kalbfleisch and Prentice 1980; Lawless 1982). An example is pupal eclosion times for a Lepidopteran population in which there are open chrysalises present at the beginning of the study. Left-censoring can frequently be avoided by careful planning (e.g., by beginning the study earlier). In many controlled experiments, left-censoring is not even possible because failures cannot occur before the beginning of the experiment (e.g., because seeds cannot germinate before the researcher plants them). Left-censoring is sometimes unavoidable in observational studies. Since this book concerns experimental studies, however, I will not discuss left-censoring further; see Collett (1994) for a useful discussion. In the examples of both left- and right-censoring discussed so far, individuals are not censored as a planned part of the experiment.

Finally, *interval-censoring* occurs when measurements are taken at intervals large enough so that we only know that the event occurred within a particular interval. Obviously, since we always record data at discrete intervals, we might regard most right-censored data as being interval-censored. There are both biological and statistical issues to consider here. Biologically, if I record data daily, I know only that the event occurred between yesterday and today. Is the hourly data meaningful and important? This is not likely for plant mortality, but perhaps so in studies of animal behavior. It seems reasonable to say that we should design studies so that sampling is on a timescale relevant to the questions of interest. Statistically, as measurements occur at shorter intervals, the results of the two kinds of analyses (treating the data as interval-censored and treating them as right-censored) converge (Collett 1994). The best advice is probably this: if the sample interval is long (in a biological sense), regard the data as interval-censored, otherwise regard them as right-censored. If in doubt, analyze the data both ways. Professional packages like SAS and S-Plus have the built-in ability to handle interval censoring.

Experimental designs often include censoring: researchers may often plan to end studies before all individuals have failed. This is obviously necessary when individuals have very long lives. Two types of designs can be distinguished. With Type I censoring, we can plan to end a study at a particular time, in which case the number of failures is a random variable. With Type II censoring, the study is completed after a particular number of failures, in which case the ending time of the study is a random variable (Lawless 1982). In ecological studies, Type I

censoring is more common because studies often end when field seasons, graduate careers, or grants end. In most ecological studies, some of the censoring is
unplanned and random; for example, some experimental individuals die or are
otherwise lost to the study at random times during the course of the study. Consequently, the discussion in this chapter assumes that censoring is Type I and/or
random. For discussion on data analysis with Type II censoring, see Lawless
(1982).

This kind of structure means that there are three important elements that require consideration in event time studies. First, measurements are repeated over
time. Second, most studies include censored data points. Finally, even if no data
are censored, failure times are usually not normally distributed under any standard
transformation.

The nonnormality of failure-time data can be partly a consequence of the necessary experimental structure. Normal distributions are symmetric, with infinitely
long tails on both sides. But by beginning a controlled experiment at a particular
time, the researcher establishes a sharp line of demarcation: no failure can occur
prior to the beginning of a controlled study. In many ecological settings, this
cutoff line is actually further along in time. For example, in studies of flowering
time that begin when the researcher plants seeds, there is usually some minimum
time that must elapse before any plant begins to flower. In this sense, failure-time
data are often intimately related to the particular time at which the experiment is
begun, unlike most other kinds of data. In my studies of the desert annual plant
Eriogonum abertianum, time to flower in the greenhouse was very far from a
normal distribution; plant size at flowering time, plant fecundity, and related traits
fit normal distributions after undergoing a log transformation (Fox 1990a).

There are also ecological and biological reasons why failure-time data may
not be normally distributed. In studies of the survival of marked plants, for example, there is no reason to expect a normal distribution of time to death. In most
populations, deaths are likely to be concentrated by the timing of major environmental events such as frosts, droughts, or herbivore migrations. Failures that occur when some biological threshold is reached—for example, onset of reproduction or deaths due to senescence—are also likely to lead to data sets that are not
normally distributed. Finally, there are theoretical reasons why failure times are
usually not normally distributed. These involve the nature of the risk of failure,
as discussed in the appendix.

13.2.2 How Failure-time Methods Differ from Other Statistical Methods

Failure-time data are frequently of interest to ecologists, but some experimental
designs and statistical analyses are better than others. Using what Muenchow
(1986) called the classical approach, many ecologists have analyzed failure-time
data by counting the number of failures among a fixed number of experimental
units over a fixed interval. They then compare groups for the mean number of
failures, using any of several statistical approaches. In a widely used alternative

approach, ecologists conduct experiments designed to measure the failure times of individuals, but then analyze these data with ANOVA.

Special methods have been designed to deal specifically with failure time data. These failure time methods differ from classical and ANOVA approaches in both experimental design and statistical analysis. Using a failure-time approach, an ecologist measures the time to failure of each uncensored individual. Statistical tests designed for this problem are then used to compare groups over the entire distribution of failure times.

There are several reasons to prefer the failure-time approach. First, the classical approach can compare groups only on a single time scale. This is because the classical approach compares groups for the cumulative number of failures that have occurred by a single point in time: the time for the experiments' end set by the scientist. By contrast, the failure-time approach compares the groups' survivorship curves over all failure times. Second, ANOVA assumes that the groups' failure times are normally distributed with equal variance and compares their means: the shapes of the failure-time distributions are assumed to be identical. By contrast, failure-time approaches allow us to compare the shapes of these distributions. Third, neither the classical approach nor the ANOVA approach can account for censored data. Fourth, under the classical approach, it is not clear what to do with multiple failures of an individual—for example, multiple insect visits to a single flower—because these may not be independent. Finally, ANOVA and t-tests can be seriously biased methods of analyzing failure-time data, because these tests require approximate normality of the data. This bias can lead to spurious results. Worse yet, the direction of the bias depends on the shapes of the distributions of failure times and on the pattern of data censoring. Consequently we cannot say in general whether ANOVA would be biased for or against a particular hypothesis.

Consider a hypothetical case. Figure 13.1A shows clearly that type A plants have a strong tendency to begin flowering before type B plants. Estimates of the mean flowering date depend strongly on the completeness of the data set (figure 13.1B). Most notably, there are times during the season that if we truncated the study, we could conclude that the mean flowering date of type B plants was actually earlier than that of type A plants! Figure 13.1C shows a failure-time approach, based on life tables. Not only is the correct relationship between flowering dates of the two types always preserved; the graph itself provides important information about the timing of events, and we can test for differences at any point..

13.3 Example: Time to Emergence and Flowering

13.3.1 Study Species

As part of a larger study on the ecology and evolution of flowering time and other life history traits in wild radish *Raphanus sativus* (Brassicaceae), I studied the time to emergence and flowering in three populations of wild radish and in

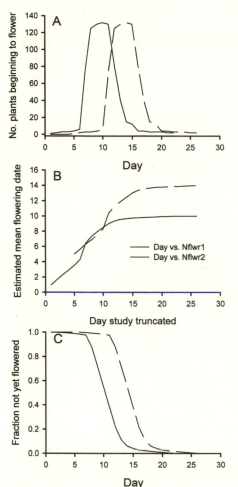

Figure 13.1 A hypothetical example. (A) Type A plants tend to begin flowering before type B plants. (B) Estimates of the mean flowering date depend on when we truncate the study. During part of the season, the estimated mean for type B would be earlier than for type A. (C) A survivorship curve plotting the fraction of plants that have not yet flowered. The relationship between the two types is always preserved correctly, and much more information is provided than a simple comparison of means.

two domestic radish cultivars. Wild radish is a common annual (sometimes biennial) weed in many parts of the northern hemisphere. In North America, *R. sativus* is especially common in coastal areas of California and in the Sacramento Valley. There are two principal sources of these old world natives: inadvertent introductions by European settlers and escapees from cultivated radish varieties. The latter are of the same species as wild radish. There are also hybrids with a closely related introduced weed, *R. raphanistrum*.

In this chapter, I will discuss two relatively narrow questions from this research: (1) What are the distributions of emergence and flowering times in these populations? and (2) Do these differ among populations? I consider these questions by using two different statistical methods: life table analysis and accelerated failure-time models. These methods are complementary, each having distinct advantages: life tables and associated statistical tests are quite useful for exploratory data analysis, but they lack the statistical power of accelerated failure-time models. On the other hand, accelerated failure-time models require more assumptions than life-table methods. I illustrate the use of a third method, proportional hazards models, with a data set from Muenchow (1986). The power of this method is similar to that of accelerated failure-time models, but it depends on somewhat different assumptions.

13.3.2 Experimental Methods

Seeds were collected from randomly selected individuals in populations around Santa Barbara, California. The Campus Point site is on a bluff overlooking the Pacific Ocean, and the Coal Oil Point site is several hundred meters from the ocean. Both sites are consequently subjected to cool fog during much of the year. The third site, Storke Road, is several kilometers inland and is therefore drier and warmer. The domestic cultivars used were Black Spanish, a late-flowering crop cultivar, and Rapid Cycling, a product of artificial selection for short generation time (Crucifer Genetics Cooperative, Madison, Wis.).

In designing failure-time experiments—either in field of laboratory settings—it is important to consider the shapes of failure-time distributions, and the patterns and magnitudes of data censoring. For example, if there is much data censoring, large samples, often involving hundreds of individuals, are necessary to compare treatments. The reason for this is simple: to compare groups, we need at least a minimum number of actual event times, as opposed to censoring times. If all data points are censored, no between-group comparison is possible. Large samples are also necessary if the tails of the distribution are of particular ecological interest (Fox 1990b).

Because the present study was a pilot experiment, I had no a priori estimate of either the number of censored data points to expect or the shapes of the failure-time distributions. Consequently, I used relatively large samples, planting 180 seeds from each of the two domestic cultivars, 140 from both the Storke Road and Campus Point sites, and 130 from the Coal Oil Point site, or a total of 770 seeds.

One seed was planted in each 10-cm pot. Positions on the greenhouse benches were randomized in advance. Because the hypotheses of interest concern differences among populations in the timing of emergence and anthesis, seeds were considered as the experimental units and individual seeds from the same population as replicates.

Seeds were planted in late October. Plants experienced natural photoperiod in Tucson, Arizona, in a greenhouse with untinted windows, throughout the experiment. A sterile 1:1 mix of peat and vermiculite was used as a growing medium. Watering before emergence was always frequent enough to keep the soil surface wet. After emergence plants were fed weekly with a solution of 20:20:20 N–P–K. Temperatures were allowed to vary between approximately 15° and 27°C. Time constraints dictated that the experiment end by mid-March 1992.

For each individual seed, I recorded an emergence time TEMERG and an anthesis time TANTH. Each TEMERG and TANTH could be either an actual emergence or anthesis time, respectively, or a censoring time. I also recorded variables EMRGCENS and ANTHCENS, assigning values of 0 for uncensored data points and 1 for censored data points. Each data record also included a column for population. In general, we could include other variables of interest, such as treatment factors.

The next section discusses general methods for analysis of failure-time data. These methods are then illustrated by applying them to the radish data.

13.4 Statistical Methods

There is no doubt that the anthesis data require some special handling, because many of these data points are censored. But in the emergence data, the small number of censored data points (figure 13.2) might lead us to believe that ANOVA would be a useful way to compare populations. However, there are biological reasons to expect these data to depart from a normal distribution. The germination process began sometime after water was first applied. Since there is no evidence for seed dormancy in this species, we might expect emergence events to be clustered at some point after the start of the experiment and to trail off after

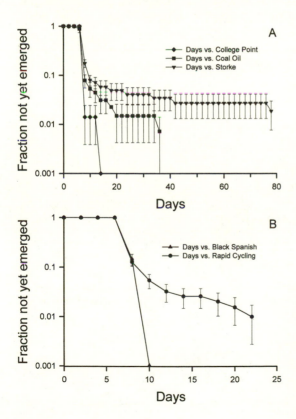

Figure 13.2 Survival curves for radish seedling emergence. Bars are for standard error of the life table estimate. Each curve ends when all seeds had emerged (curves intersecting *x*-axis) or when last seedling emerged (all other curves). (A) Santa Barbara populations. (B) Crop cultivars.

that point. This pattern is suggested by the data in figure 13.2, and normal scores plots of the data showed strong departures from normality. No standard transformations succeed in normalizing these data. Thus, even though there is very little censoring in the emergence data (figure 13.2), comparison of means using ANOVA would be inappropriate.

Fortunately, there are several well-developed methods for analyzing failure-time data that can handle censored data properly and that do not require normally distributed data. These include life table analysis and two different types of regression models, accelerated failure-time models and proportional hazards models. In the following subsections, I describe each of these methods and then apply them in turn.

13.4.1 Life Table Methods: Description

Life tables are a convenient starting point for understanding failure-time statistics. Because life tables historically preceded regression models, they provide a basis for many of the ideas used in regression models. Moreover, life tables are a useful way to begin exploratory data analysis.

Formulas for cohort life table estimates are given in table 13.1. From measurement of the failure rate, it is simple to estimate the proportion of those failing in each interval. There are four other statistics that can be derived from this information. The estimated cumulative survival function, S_i, is the fraction of the cohort

Table 13.1 Definitions of life table statistics

t = time at beginning of ith interval

b_i = width of the ith interval, $t_{i+1} - t_i$

w_i = number lost to follow-up

d_i = number dying in interval

n_i = effective population in interval, $n_{i-1} - w_{i-1} - d_{i-1} - \dfrac{w_i}{2}$

q_i = estimated conditional mortality rate, $\dfrac{d_i}{n_i}$

p_i = estimated conditional proportion surviving, $1 - q_i$

S_i = estimated cumulative survival function, $\displaystyle\prod_{j=0}^{i} p_j$

P_i = estimated probability density function (unconditional mortality rate), the probability of dying in the ith interval per unit width, $\left(\dfrac{S_i q_i}{b_i}\right)$

h_i = estimated hazard function (conditional mortality rate, force of mortality, log killing power), the number of deaths per unit time in the ith interval divided by the number of survivors in the interval, $\dfrac{2q_i}{b_i(1 + p_i)}$

e_i = estimated median life expectancy, $(t_j - t_i) + \dfrac{b_j\left(S_j - \dfrac{S_i}{2}\right)}{S_j - S_{j+1}}$, where the subscript j refers to the interval in which $\dfrac{S_i}{2}$ occurs.

that has not yet failed. The estimated probability density function, P_i, gives the probability that an individual alive at time 0 will fail in the ith interval. Hence, P_i is also called the unconditional mortality rate. The estimated hazard function, h_i, is sometimes called the conditional mortality rate; it estimates the chance of failure in the ith interval given that an individual has survived to the beginning of that interval. Finally, the estimated median life expectancy, e_i, gives the median time to failure of an individual that has survived to the beginning of the ith interval; e_0 gives the median life expectancy of a newborn.

The ecological significance of the cumulative survival function S_i is obvious. Insight into the probability density and hazard functions can be gained by realizing that if measurements are continuous or nearly so

$$P_i = \frac{-d(1 - S_i)}{dt}$$

$$h_i = \frac{-d[\ln(S_i)]}{dt}$$

In other words, the probability density function reflects the rate at which failures accumulate, whereas the hazard function reflects the per capita risk of failure among those remaining.

Median, rather than mean, life expectancies are calculated because the median is often a more useful and robust estimate of central tendency when the data are not symmetrically distributed. Life tables allow for the use of censored data by treating these data points like any other until they reach the censoring interval and then discounting them in the censoring interval.

Formulas for calculation of standard errors of life table estimates are shown in table 13.2. Because life table estimates are population-level statistics, standard errors and sample sizes should always be reported so that readers can judge for themselves how much confidence to place in the estimates. Approximate variances for the life table estimates are given by the terms under the square-root signs in table 13.2.

Life table analyses are always informative in examining failure-time data: their descriptive nature makes them easy to interpret. Consequently, I recommend them

Table 13.2 Standard errors for life table data[a]

$$\mathrm{SE}(S_i) \approx \sqrt{S_i^2 \sum_{j=1}^{i-1} \frac{q_j}{n_j p_j}}$$

$$\mathrm{SE}(P_i) \approx \sqrt{\frac{(S_i q_i)^2}{b_i}\left(\sum_{j=1}^{i-1} \frac{q_j}{n_j p_j}\right) + \frac{p_i}{n_i q_i}}$$

$$\mathrm{SE}(h_i) \approx \sqrt{\frac{h_i^2}{n_i q_i}\left[1 - \left(\frac{h_i b_i}{2}\right)^2\right]}$$

$$\mathrm{SE}(e_i) \approx \sqrt{\frac{S_i^2}{4 n_i\,(P_j)^2}}$$

[a]From Lee (1980).

as a first step in the analysis of most kinds of failure-time data. Hypotheses can also be tested with life table estimates.

There are several ways to compare populations or treatments statistically. Standard errors of the life table estimates can be used for pairwise comparisons. If data are uncensored, standard nonparametric tests such as G-tests or χ^2-tests can be used to test for independence of groups. With censored data, Wilcoxon or log-rank tests [see http://www.oup-usa.org/sc/0195131878/] can be used on the failure-time data to test for heterogeneity among groups. Both of these tests compare observed with expected numbers of deaths in each interval.

The log-rank test is more powerful than the Wilcoxon when the hazard functions of the different samples are proportional to one another and when there is no censoring or random censoring only. The Wilcoxon test is more powerful in many other situations (Lee 1980). In many cases, the tests are likely to give similar results. Lawless (1982) notes that there are circumstances under which neither statistic is likely to be very useful, particularly when distributions are different but the cumulative survivorship or hazard functions cross. Any test is likely to lack power when only a few censuses occur or data are lumped into few intervals and when distributions differ over time but lead to a similar total number of failures (Hutchings et al. 1991). Therefore, if the shapes of the curves are likely to be important, censusing must occur frequently enough to detect these differences. Another way of stating this problem is that ties seriously reduce the power of these tests. Muenchow (1986) observed that, in her study of waiting times to insect visits at flowers, she would have had greater statistical power had she recorded times to the nearest second rather than the nearest minute, because she would have had many fewer ties.

These tests, as well as the life table analyses themselves, can be performed with the SAS LIFETEST procedure [which uses the χ^2-approximation to the log-rank and Wilcoxon tests described in Fox (1993, appendix 1; an updated discussion is at http://www.oup-usa.org/sc/0195131878/)]. If these tests reveal heterogeneity and there are more than two groups, we can use the Wilcoxon or log-rank scores to construct Z-statistics for multiple comparisons among groups [see http://www.oup-usa.org/sc/0195131878/]. SAS code for doing this is available online at http://www.oup-usa.org/sc/0195131878/.

There is one important alternative nonparametric method for analyzing event-time data. The Kaplan–Meier (KM) method (the default in the SAS LIFETEST procedure) differs from the life table method used here in that the cumulative survival function changes only when an event is actually observed. Although this is a maximum-likelihood estimate of the survival function, in practice life table and KM estimates are usually almost the same, and life table estimates allow us to estimate the hazard function. The most important negative side of the life table method is that our choice of time intervals can be arbitrary. Resulting problems can be minimized by trying several intervals and asking whether the answers change.

13.4.2 Life Table Methods: Application

Emergence data for the radish study are shown in figure 13.2. Cumulative survivorship (i.e., the fraction of plants that had not yet flowered) and its standard error were calculated by the SAS LIFETEST procedure, using the code shown on-line at http://www.oup-usa.org/sc/0195131878/. The separation of the survivorship curves implies that the populations may differ in emergence time. This is supported by both the Wilcoxon and log-rank tests [see http://www.oup-usa. org/sc/0195131878/] calculated by the SAS LIFETEST procedure: Wilcoxon $\chi^2 = 439.1$, df = 4, $P = 0.0001$, log-rank $\chi^2 = 335.8$, df = 4, $P = 0.0001$. These tests tell us that the five populations are heterogeneous, but they do not tell us which populations differ from one another. To answer that question, I used the covariance matrix for the Wilcoxon statistic that is automatically generated by the SAS LIFETEST procedure, and calculated Z-statistics for each pairwise comparison [see http://www.oup-usa.org/sc/0195131878/ for a description of the Z-statistic and for SAS code to conduct the multiple comparisons]. As noted in the on-line material for this volume [http://www.oup-usa.org/sc/0195131878/], these multiple comparisons are not conducted automatically by SAS, and performing them requires a small amount of manual data manipulation. These multiple comparisons, conducted at the 0.05 significance level, suggest that all populations differ from one another except possibly Black Spanish and Rapid Cycling (which was a marginally significant comparison) and Coal Oil Point–Campus Point.

Emergence and germination data present a special statistical problem: in general, we do not know whether remaining seeds are capable of germinating, are dormant, have germinated and then died, or were always dead. Inviable seeds should obviously not be considered as part of a study population. In the present case, very few seeds did not emerge, so I analyzed the data by assuming first that these were viable. This means that they were treated as censored data points, with the end of the study as the censoring date. A second analysis assumed that the seeds had always been inviable, so seeds that did not emerge were excluded from the analysis. The results were qualitatively the same; figure 13.2 is based on the first analysis. An alternative approach would be to examine the unemergent seeds for viability using a tetrazolium test (Scott and Jones 1990), and thereby correctly classify each seed. This would be necessary if the two statistical analyses differed qualitatively.

Anthesis data for the radishes are shown in figure 13.3. These cumulative survivorship data and their standard errors were also calculated by the SAS LIFETEST procedure [see http://www.oup-usa.org/sc/0195131878/]. In this case, there is considerable censoring, because many plants had not yet flowered by the time the study had to end, and some deaths did occur. Moreover, the censoring affected some populations much more strongly than others. Many more Rapid Cycling plants than others had flowered by the end of the study. Nevertheless, the survivorship curves again imply that the populations differ. This conclusion is supported by the Wilcoxon and log-rank tests calculated by the LIFETEST procedure: Wilcoxon $\chi^2 = 582.2$, df = 4, $P = 0.0001$, log-rank $\chi^2 = 580.6$, df = 4, $P = 0.0001$. Given that these five populations are heterogeneous, which ones are different from one another? To examine this question, I again used the covariance matrix for the Wilcoxon statistic that is automatically generated by the LIFETEST procedure, and I calculated Z-statistics for each pairwise comparison [see http://www.oup-usa.org/sc/0195131878/ for a description of the statistic and for SAS

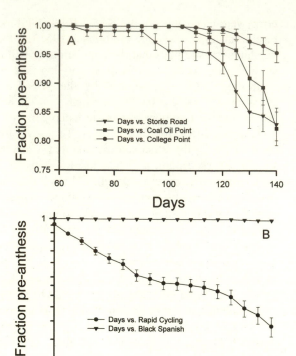

Figure 13.3 Survival curves for radish anthesis. Bars are for standard error of the life table estimate. All curves end at censoring date. (A) Santa Barbara populations. (B) Crop cultivars.

code to conduct the multiple comparisons]. These multiple comparisons, conducted at the 0.05 significance level, suggest that Rapid Cycling differs from all other populations and that Black Spanish differs from Coal Oil Point and Storke Road. There were marginally significant differences between Campus Point on the one hand and Coal Oil Point, Storke Road, and Black Spanish on the other.

These life table analyses suggest that *Raphanus* populations differ in time to emergence and flowering. Moreover they have provided a useful description of the populations' responses. There are many cases in which life table analyses and associated significance tests are fully adequate to examine ecological hypotheses (Fox 1989; Krannitz et al. 1991).

However, two difficulties commonly arise with the use of life tables. First, statistical tests based on life table approaches are sometimes lacking in power. Second, life table methods are difficult to use when there are multiple covariates (Kalbfleisch and Prentice 1980). For example, emergence time seems likely to affect anthesis time, but there is no simple way to account for this effect using life table methods. SAS allows for the calculation of relevant tests (generalizations of the Wilcoxon and log-rank tests), but the algorithms are complex, requiring relatively large amounts of computation, and the results are not easy to interpret in biological terms (see pp. 1046–1048 in version 6 of the SAS/STAT user's guide

(SAS Institute 1989a). The methods discussed in the following section overcome these difficulties, but at the cost of added assumptions.

13.4.3 Regression Models

There are two general types of regression models for censored data: *accelerated failure-time models* and *proportional hazards models*. These models make different assumptions about the affect of covariates such as treatment or initial size.

Accelerated failure-time models assume that treatments and covariates affect failure-time multiplicatively (i.e., the lives of individuals are "accelerated"). An alternative interpretation (Kalbfleisch and Prentice 1980) is that under accelerated failure-time models, covariates make the clock run faster or slower, so that any period of high hazard will shift in time when the covariates shift. Thus the comparisons of *Raphanus* population time to emergence and anthesis are good candidates for accelerated failure-time models. Allison (1995) observes that a classic example of an accelerated failure-time model is the conventional wisdom that 1 dog year is equal to 7 human years.

Proportional hazards models assume that covariates affect the hazard functions of different groups multiplicatively. Under proportional hazards models, the periods of high hazard stay the same but the chance of an individual falling in one of those periods will vary with the individuals' covariates. A good candidate for a proportional hazards model might be a predation experiment in which the predator density or efficiency changes seasonally: individuals with a "bad" set of covariates do not make the predators arrive sooner, but they are more vulnerable to predation when the predators do arrive. Thus, comparisons of *Raphanus* population time to emergence and anthesis are poor candidates for proportional hazards models.

How do we choose between accelerated failure-time and proportional hazard models? Probably the most useful approach is to consider the ecological hypotheses and ask whether treatments are expected to actually change the timing of periods of high hazard (suggesting the use of accelerated failure-time models) or whether treatments merely change the chance of failure (suggesting the use of proportional hazards models). An additional check on the appropriateness of the proportional hazards model is to plot $\log\{-\log[S(t)]\}$ against time. If the proportional hazards model is appropriate, the curves of different groups should be roughly parallel to one another for a given level of a covariate (Kalbfleisch and Prentice 1980). This approach has one limitation: comparisons must be conducted between treatment groups within levels of a covariate. If all of the predictor variables are covariates, no comparison can be made, regardless of whether the proportional hazards model is appropriate. Consequently, it is probably best to rely most heavily on the examination of the ecological hypotheses. Allison (1995) suggests that one might profitably use the proportional hazards model as a default model. This may be a useful guideline for the sociological examples he considers (where there are few a priori considerations to influence a decision), but it is unlikely to be appropriate in most biological situations.

Both kinds of models have been applied in ecological settings; see Muenchow (1986) for proportional hazards models, and Fox (1990a,b) for accelerated failure-time models. These models are closely related; a lucid derivation showing the mathematical relationship between the models is given by Kalbfleisch and Prentice (1980).

Description of accelerated failure-time models.   In an accelerated failure-time model, failure times T are modeled as

$$\ln(T) = X'\beta + \sigma\varepsilon \qquad (13.1)$$

where X is a matrix of covariate values and X' is its transpose, β is a vector of regression parameters, ε is a vector of errors from a specified survival distribution, and σ is a parameter. The covariates in X can be dummy variables corresponding to categorical effects, or continuous variables, or both, or their interactions. For example, X for the radish emergence data is a 5×5 matrix of dummy variables for the five populations of origin, and there are five regression coefficients in β. For the flowering data, X includes an additional variable, emergence date, and there is an additional regression coefficient for emergence date in β.

Consequently, each model consists of two steps: (1) choosing a survival distribution and (2) estimating the parameters for the survival distribution and the regression coefficients β. The most commonly used survival distributions are described in the appendix. All of these are available in the SAS LIFEREG procedure. Some additional distributions are discussed by Lawless (1982) and Kalbfleisch and Prentice (1980).

There are three different ways to choose a survival distribution. First, we could choose a distribution based on a priori ecological or biological grounds. Second, we could take an empirical approach: after the data have been gathered, compare life table estimates of their hazard or survival functions with the hazard or survival functions of various distributions. Several rough empirical methods for doing this are listed in the appendix. We could also use goodness-of-fit tests, but these require modification if any of the data are censored (Lawless 1982). A third method for choosing a distribution is suggested by Kalbfleisch and Prentice (1980). These authors propose using a general form (the generalized F-distribution) that encompasses all of the distributions in the appendix as special cases to determine numerically which distribution best fits the data.

It is often biologically important in its own right that a data set is, say, gamma-rather than Weibull-distributed. For example, the difference among the classic Deevey Types I, II, and III survival functions is precisely that a Type II implies constant risk of mortality (an exponential distribution), whereas Types I and III imply increasing and decreasing risks, respectively. Recent ecological research in which the nature of the failure-time distribution has been important includes work on senescence in birds (McDonald et al. 1996; Ricklefs 1998), the response of plant populations to environmental variation (Bullock et al. 1996), the risk of predation to marine gastropods (Petraitis 1998), and the interval between forest fires (Clark 1989).

To understand the importance of the shape of a survival distribution, we need some model of the failure process. For example, many developmental processes are likely to be gamma-distributed because a sum of exponential distributions (as in waiting times for cell divisions or other developmental steps) yields a gamma distribution. Kalbfleisch and Prentice (1980) discuss examples in other fields. Additionally, knowledge of the survival distribution may sometimes suggest candidates for underlying mechanisms (Lindsey 1995).

Having chosen a survival distribution, the parameters of the distribution and the regression coefficients are determined numerically, using a maximum-likelihood approach. Maximum-likelihood methods in statistics involve two steps. First, we select a statistical model and assume that it is a correct description of the data. The same assumption is made in regression and ANOVA models. Second, we find the model parameters that would make it most probable that we would observe the data. This is done by numerically maximizing a likelihood function, the form of which depends on the model. For more information on construction of likelihood functions for failure-time models, see Kalbfleisch and Prentice (1980) and Lawless (1982). Edwards (1972) provides a comprehensive introduction to likelihood methods, and Shaw (1987) provides an overview of likelihood methods as applied in quantitative genetics.

These methods lead naturally to a set of χ^2-tests for regression coefficients. These test the significance of the covariate's additional contribution to the maximized-likelihood function, beyond the contributions of the other covariates already in the model. Thus, a nonsignificant value of χ^2 means that the model is adequate without that particular covariate, but it does not imply that all covariates with nonsignificant χ^2-values can be deleted. In fact, it is possible for a set of covariates to be jointly significant, even though none of them is individually significant. To examine the possible importance of covariates with nonsignificant χ^2-values, we must delete the nonsignificant terms one at a time, reevaluate the model each time, and examine the changes in the other terms.

When there are multiple levels of a categorical covariate like treatment or population, one level is chosen arbitrarily as a reference level with a regression coefficient of 0. The regression coefficients for the other levels therefore provide comparisons with the reference level. Thus, a significant χ^2-value for a particular level means that only that level is significantly different from the reference level. It may or may not be significantly different from levels other than the reference level; we must perform multiple comparisons to examine this hypothesis. The method for multiple comparisons is analogous to that for life table analyses: we construct Z-statistics from the estimated regression coefficients and the asymptotic covariance matrix generated by SAS [see http://www.oup-usa.org/sc/0195131878/ for a description of the method and for SAS code for its implementation].

These χ^2-tests can also be useful when choosing between two distributions, one of which is a special case of the other. The adequacy of the restricted case can be examined by fitting a model to the more general model, subject to an appropriate constraint. For example, to choose between the Weibull and exponential distributions, we can fit a Weibull model and subject it to the constraint $p = 1$. The result is, of course (see appendix), an exponential model, but SAS automatically calculates a test (called a Lagrange multiplier test) that tests the effect of the constraint on maximizing the likelihood function.

Application of accelerated failure-time models.   For the radish emergence data, it seemed reasonable to choose either the log-logistic or lognormal distribution: Because the species has little or no seed dormancy, we can expect that the rate of germination and subsequent emergence will increase quickly after application of water, reach some maximum rate, and then decline. I chose the log-logistic model because, as noted in the appendix, it has several properties that make it more useful. An a posteriori comparison showed that the log-logistic model provided a better fit to the data than the lognormal.

For the radish anthesis data, I thought it likely that the data might best fit a gamma distribution. The gamma hazard monotonically approaches a constant value, and therefore gamma models may often provide good fits for inevitable developmental processes (see the appendix). Moreover, a gamma distribution is a sum of exponentially distributed variables; if we think of plastochrons or intervals between cell divisions as being reasonably approximated by an exponential distribution, we would again expect a higher level developmental stage like flowering time to be gamma-distributed. Such intuition is often wrong, so I tested the gamma model against Weibull and lognormal models. As shown in the appendix, the Weibull and lognormal distributions are special cases of the three-parameter gamma.

There are several ways to make this test, including a Lagrange multiplier test and a likelihood ratio test. In principle, they are equivalent, but the likelihood ratio test may be more robust for small samples (Collett 1994).

As an example of a Lagrange multiplier test, I constrained the gamma shape parameter to yield the lognormal distribution by assigning a value of 0 to "shape1" (see the appendix) and then setting the option "noshape1" to instruct the LIFEREG procedure not to change this value. SAS code for doing this is available at http://www.oup-usa.org/sc/0195131878/. The Lagrange multiplier test was highly significant ($\chi^2 = 22.68$, df = 1, $P < 0.0001$). This means that constraining the value of "shape1" to 0 had a significant effect on the likelihood function; thus the gamma model does provide a significantly better fit than the lognormal.

I compared the gamma and Weibull distributions with a likelihood ratio test [see http://www.oup-usa.org/sc/0195131878/]. This test is easy to compute by hand, because the test statistic is just twice the logarithm of the ratio of the two maximized likelihood functions. Since I already had estimates of the two maximized likelihood functions, it was simple to calculate

$$R = -2\ln\left[\frac{L(\theta_{gamma})}{L(\theta_{Weibull})}\right] = -2\ln\left(\frac{-232.48}{-276.97}\right) = 0.35 \tag{13.2}$$

which, with 1 degree of freedom, has a large probability of occurring by chance ($P > 0.83$). Thus, the fit of the three-parameter gamma distribution to the data is slightly, but not significantly, better than the fit of the Weibull distribution.

These formal comparisons of models depend on the models being nested. That is, we can use either the Lagrange multiplier method or the likelihood ratio test to compare the gamma with the Weibull or the lognormal distribution, because the latter are special cases of the gamma: they are nested within it for special,

Table 13.3 Analysis of accelerated failure-time model for *Raphanus sativus* emergence time, using 764 noncensored values, 6 right-censored values, and a log-logistic distribution[a]

Variable	df	Estimate	SE	χ^2	P
Intercept	1	1.868	0.014	17442.78	0.0001
Population	4			395.671	0.0001
(Black Spanish)	1	−0.254	0.018	194.03	0.0001
(Coal Oil Point)	1	−0.072	0.020	13.09	0.0003
(Campus Point)	1	−0.084	0.019	19.63	0.0001
(Rapid Cycling)	1	−0.298	0.018	250.84	0.0001
(Storke Road)	0	0	0		
Scale parameter	1	0.100	0.003		

[a]Log-likelihood = 116.3.

fixed values of certain parameters. There is no formal way to test the goodness of fit of nonnested models; in any event, biological criteria are often preferable.

Choosing between the gamma and Weibull distributions based on the present data set requires a biological, rather than a statistical, rationale. The gamma distribution seems a more biologically reasonable description of the time to anthesis of an annual plant, because the gamma hazard tends toward a constant as time becomes large, whereas the Weibull hazard goes to either zero or infinity.

Analyses of the radish failure-time models are shown in tables 13.3 and 13.4, and the SAS code that generated these analyses is available at http://www.oup-usa.org/sc/0195131878/. In both of these examples, the regression coefficients for the Storke Road population are zero. The reason for this is that accelerated failure-time models make the clock run slower or faster for some groups. As men-

Table 13.4 Analysis of accelerated failure-time model for *Raphanus sativus* anthesis time, using 210 noncensored values, 554 right-censored values, and a gamma distribution[a]

Variable	df	Estimate	SE	χ^2	P
Intercept	1	4.975	0.090	3047.02	0.0001
Emergence time	1	0.019	0.010	3.68	0.06
Population	4			495.43	0.0001
(Black Spanish)	1	0.449	0.069	42.90	0.0001
(Coal Oil Point)	1	0.124	0.058	4.579	0.03
(Campus Point)	1	0.324	0.063	26.70	0.0001
(Rapid Cycling)	1	−0.682	0.054	154.35	0.0001
(Storke Road)	0	0	0		
Scale parameter	1	0.394	0.021		
Shape parameter	1	−0.822	0.179		

[a]Log-likelihood = −232.48.

tioned previously, when there are multiple levels of a class variable (in this case, population), one level (the last one) is taken by SAS as a reference level, and all others are compared to this level. The Storke Road population is thus taken to be the reference population by virtue of its order. The coefficients for other populations are therefore compared to Storke Road, and the significance tests for each population test whether it differs from Storke Road.

The analysis of the emergence data (table 13.3) shows that population of origin contributes significantly to the model. The fact that each population's regression coefficient is independently significant shows that each differs from the reference (Storke Road) population. Which other populations differ from one another? To examine this question, I conducted multiple comparisons with Z-statistics [see http://www.oup-usa.org/sc/0195131878/ for a description of the method and for SAS code to implement the multiple comparison]. These multiple comparisons are analogous to the ones used in the life table analyses of section 13.4.2, except that in this case I used regression parameters as the statistics for comparison, rather than Wilcoxon rank scores. As with the life table statistics, the multiple comparisons require a small amount of manual data-handling [see http://www.oup-usa.org/sc/0195131878/].

The life table analyses showed significant heterogeneity among populations, and pairwise comparisons suggested that all pairs differ except Coal Oil Point–Campus Point and possibly Black Spanish–Rapid Cycling. The regression coefficients in table 13.3 and their estimated covariance matrix led to a somewhat different conclusion: the Coal Oil Point–Campus Point comparison still results in a high probability of being from the same population, but the Black Spanish–Rapid Cycling comparison now shows significant differences between these populations. All other comparisons remained qualitatively unchanged. The accelerated failure-time model has thus confirmed the general patterns suggested by the life table analysis, but its greater statistical power has also made it possible to find significant differences among populations that were not revealed by the life table analysis.

To interpret the regression parameters in table 13.3, recall that they correspond to multiplicative effects of covariates on the probability of seedling emergence. The average time to emergence for a Rapid Cycling plant is $e^{1.868-0.298}p_0(t) = 4.807p_0(t)$, whereas for a Storke Road plant it is $e^{1.868}p_0(t) = 6.475p_0(t)$. The more negative the coefficient, then, the earlier the average emergence time. The reference distribution $p_0(t)$ is the log-logistic distribution with the scale parameter shown in table 13.3.

Analysis of the anthesis data (table 13.4) also shows that each population contributes significantly to the model. Late emergence probably tends to delay anthesis. This is to be expected, because the dependent variable is time from planting to anthesis, which must be greater than time from planting to emergence. Including emergence time in the model is thus analogous to including block effects in ANOVA.

The regression coefficients are interpreted in the same manner as they were for the emergence time data. Only the reference probability distributions differ (emergence is distributed as log-logistic, whereas anthesis is taken to be gamma-distributed). However, this does not affect the qualitative interpretation of the regression parameters. For example, analysis of the anthesis data shows that Rapid Cycling plants reach anthesis much earlier and Black Spanish plants much later than all others. This corresponds well with the results shown in figure 13.3.

The significant coefficient for the Coal Oil Point population indicates that it differs significantly from the Storke Road population. By contrast, the pairwise comparisons used following the life table analysis suggest that these populations did not differ. As with the emergence data, multiple comparisons conducted with Z-statistics from this analysis [see http://www.oup-usa.org/sc/0195131878/] point to more among-population differences than were revealed with life table analysis. These comparisons, conducted at the 0.05 significance level, suggest that all populations differ from one another, except possibly the Black Spanish–Campus Point pair, the differences among which were marginally significant. Which analysis is correct? The accelerated failure-time model has more statistical power, and, therefore, it is reasonable to have more confidence in these results than in the life table analysis.

Because the accelerated failure-time model is able to include the effect of emergence time explicitly, we can have more confidence that differences among populations in anthesis time are not simply due to differences in emergence time. Moreover, the accelerated failure-time model—because of its greater statistical power—has revealed several among-population differences that were not identified by life table analysis. Accelerated failure-time models can often provide greater clarity in failure time studies than life table analysis alone. In the following subsection, I illustrate the use of a somewhat different regression model, the proportional hazards model.

Description of proportional hazards models. As mentioned previously, in a proportional hazards model, the effect of covariates is to change the chance of falling in a period with high hazard. The covariates act multiplicatively on the hazard function, rather than on the failure time (as in accelerated failure-time models). The hazard function for the ith group is thus

$$h_i(t) = h_0(t)\exp\left[\sum_{j=1}^{r} \beta_j X_{ij}\right] \tag{13.3}$$

where $h_0(t)$ is a reference hazard function that is changed by the covariates $\mathbf{X}$ and regression coefficient β.

The Cox proportional hazards model, which is very widely used in epidemiology (see Muenchow 1986 for an ecological application), estimates the reference hazard function nonparametrically. The regression coefficients β are then estimated numerically with a maximum-likelihood procedure. The Weibull distribution (see the appendix) can also be used for proportional hazards models comparing two groups, by treating the parameter p as the ratio of two regression coefficients (Kalbfleisch and Prentice 1980).

Because covariates act on the failure time in accelerated failure-time models and on the hazard function in proportional hazards models, the regression coefficients can have opposite meanings. A positive coefficient in a proportional hazards model means that the covariate increases the hazard, thereby decreasing the failure time. A positive coefficient in an accelerated failure-time model means that the covariate increases the failure time. The interpretation of the coefficients can also vary with the particular parameterization used by a statistical package: the best advice is to check the documentation to be sure how to interpret the results and to try a data set with a known outcome.

Applicaton of proportional hazards models.   Muenchow (1986) was interested in testing whether male and female flowers of the dioecious plant *Clematis lingusticifolia* are equally attractive to insects, against the alternative hypothesis that males are more attractive. She recognized that this could be treated as a failure-time problem because differences in attractiveness should lead to differences in time to the first insect visit. By treating this as a failure-time problem, Muenchow was able to examine these hypotheses with a creative experimental design: she watched pairs of flowers on the same plant and recorded waiting times until the first insect visit. For each observation, she also recorded the time of day, air temperature, and a categorization of the flower density within ≈1 m of the target plant. An important part of the design is that, unlike the radish study discussed previously, Muenchow was not following a single cohort through time. Waiting times were recorded to the nearest minute.

This study is one in which the assumptions of the Cox proportional hazards model appear reasonable a priori. If there are differences in attractiveness, they would likely act to increase the hazard (i.e., the chance of a visit) of the attractive gender, relative to the hazard for the less attractive gender. Another way of looking at this is that being a member of the attractive gender should have no effect on the number of insects in the area, but it does affect the chances of a visit once an insect is within some distance of the plant.

Muenchow's estimate of the survival function is shown in figure 13.4. She noted that these data appear to have been drawn from an exponential distribution, but she preferred to use a Cox model for her statistical tests.

An important part of Muenchow's study is the way in which she analyzed the four covariates: gender, flower density category, temperature, and time of day. She reported that the starting time and temperature coefficients were not significantly different from zero; in other words, these factors do not appear to influence waiting time to visitation. Because male plants had more flowers, gender and flower density were correlated. Consequently, Muenchow analyzed the data by stratifying: she examined the effect of flower density within each gender and the effect of gender within each category of flower density.

Both gender and flower density independently had significant effects: within a gender, insects visited dense flower groups faster than other groups, and within a density category, insects visited males faster than females. Muenchow concluded that males were more attractive both because they bore flowers more densely and because they had some unknown attractive character.

SAS has recently developed a PHREG procedure to analyze this model. The PHREG procedure is incorporated in SAS versions 6.07 and later, and it is documented in the SAS/STAT User's Guide for these versions.

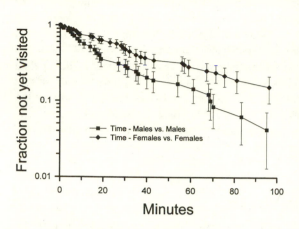

Figure 13.4 Product-limit estimates of the survival function recalculated from Muenchow's (1986) study of waiting times to insect visits at flowers of *Clematis lingusticifolia*. See http://www.oup-usa.org/sc/ 0195131878/for discussion of product-limit estimates. Sample sizes are 47 female (8 censored) and 49 male (2 censored) flowers. Nonparametric tests show the genders to be significantly different: log-rank, $\chi^2 = 5.96$, 1 df, $P = 0.01$; Wilcoxon, $\chi^2 = 5.46$, 1 df, $P = 0.01$.

13.5 Discussion

13.5.1 Assumption of Independence

In the statistical methods discussed here, we assume that the experimental units are independent, that is, we assume that the individuals do not interact. This is certainly a reasonable assumption for the case examined here, time to emergence and flowering of a randomly selected set of seeds grown in individual pots in a greenhouse. However, there are many cases in ecology where this assumption of independence presents problems. Time to flowering in a natural setting often depends on the number and sizes of neighbors of a plant. Consequently, we would have to measure plants that are spaced widely enough that they do not interact. Time to fruit set can depend on the availability of pollen and pollinators and can therefore be strongly frequency- and density-dependent. In this case, it may not always be possible to satisfy the assumption of independence. Life table estimates of survival may still be useful in such situations as descriptions. However, the significance tests discussed here will no longer be valid.

When data cannot be treated as independent, it is usually necessary to treat aggregate measurements of treatment groups (e.g., mean sizes of individuals) as single data points. With uncensored failure-time data, we might take the median failure time of each treatment group as a data point. This approach cannot be used if some data points are censored before 50% of individuals have failed, because then only an estimate of the median and its variance would be available. There are currently no methods for appropriately analyzing failure times in this situation. Rather than comparing failure times, it may be more useful to compare groups for the probability of failing within a specified time.

13.5.2 Data Handling

The assumption that individuals are the experimental units has implications for the treatment of data. Perhaps the most common question I have answered since the first edition of this book is, Why won't SAS handle my data properly? In every case, the data had been aggregated, so that the input data file provided something close to a life table (something the software is better at doing than we are). That is, many ecologists assume that data should be reprocessed so that the software is told that, on day X, n_x individuals died, and the like. A look at the SAS code [see http://www.oup-usa.org/sc/0195131878/] should clarify this. The data set should have an entry for each individual, giving the time to event for each individual.

13.5.3 Problems with Data Censoring

Informative censoring.   One kind of censoring must be avoided at all costs: *informative censoring*. This is defined as a correlation between the chance that an individual is censored and the individual's prognosis, conditioned on its covariates (Cox and Oakes 1984). In other words, if we know that the individual was censored, we can predict the likely time to event, beyond knowing its treatment group, family, or other covariate. A classic example occurs in some agricultural studies (with a close analog in some medical studies). For example, in studying time to fruit ripening, an investigator removes plants that appear sick or stressed from the study because she is interested in "normal" behavior. The problem is that these plants are likely to have a rather different time to event than those that are not censored. Informative censoring biases the results. In some cases, biological information can tell us the direction of the bias. In the present example, it seems likely that the stressed plants would have delayed event times; thus censoring them serves to underestimate the median event time. In most cases, however, the direction of bias is not known, and as a result, informative censoring invalidates most analyses.

There is no statistical test for informative censoring. Allison (1995) suggests (p. 249 ff.) a kind of sensitivity analysis to assess the potential importance of informative censoring. The data are analyzed under two alternative extreme assumptions: censored individuals are at the greatest/lowest risk of an event occurring. By using Allison's SAS code, we can then compare the results to gain insight on how informative censoring might be biasing the results.

Large numbers of censored data points.   If a large portion of the data are censored, analysis can be difficult. Consider the extreme case: all data are censored. Obviously, we would never do this by design, but it certainly is possible to have, for example, all our plants die before any of them flower. In this case, we could not do much to analyze data on flowering time. To understand this, consider a life table analysis: in each interval, the probability of flowering would be estimated as zero, so the cumulative survival function would always be estimated as 1. Thus, we must have a sufficient number of actual events.

Some ecological studies have sufficiently intricate designs so that it is possible for some groups to be 100% censored. For example, a common design is the cross-nested design: there are families nested within populations, with one or more treatments applied to replicates within families. It is not difficult to have all of the individuals in some of these cells be censored.

If nearly all of the data in some cells are censored, there are still problems. Clearly, we cannot estimate means and variances with, for example, one noncensored point in a cell. An additional problem is that some distributions in accelerated failure-time models require numerical integration (especially the gamma distribution), and these estimates can be numerically unstable when there are few uncensored points. There is no solution to this problem except to use larger samples.

Truncating an experiment.   Most ecologists instinctively want to get as much data as practical, so the idea of censoring data voluntarily is not appealing to most of us. However, there are times when censoring makes a lot of sense. Consider a simple germination experiment. The last seed that germinates does so on day 10, but to be sure, an ecologist checks the pots every day for 500 days. Analyzing all 500 days would then mean that the flat tail of the distribution will dominate the data, because we are fitting a curve to the data. If we want to compare the rate of germination among those that germinated, it makes sense to truncate the analysis (if not the data observation) when the tails become flat. In other words, we can impose Type I censoring after the fact. Whether to do so, and when to do so, depends, as always, on what we want to know. Many of us have followed cohorts of marked individuals to the death of the last one, and it may sometimes be informative to do so. But it is worth bearing in mind that survival estimates on the tail of the survival curve must, by their nature, have enormous standard errors.

Censoring at time zero.   What if some individuals are censored at time zero? SAS, and probably most other packages, will not gracefully handle this situation—in fact, it can cause programs to crash. Biologically and statistically, this makes sense: an individual censored at time zero might as well not be in the study at all. If an individual is truly censored at time zero—a marked individual is never again seen—it provides no information.

13.5.4 Hidden Heterogeneity

An important issue arises when there is heterogeneity in the sample population (Keyfitz 1985). For example, if some individuals are inherently more prone to have a failure than others, many of the estimates may be biased. Consider a population with two types of individuals, lucky and unlucky, with lucky individuals always having lower mortality probabilities. Also assume that mortality increases with age for both types. As time proceeds and we watch a cohort, more of the unlucky than the lucky individuals have died. Estimates of the population's mortality rate can actually decrease even though every individual has an increased chance of death with age, because those with the highest risk are no longer strongly represented in the denominator. Thus, hidden heterogeneity can lead to answers that are qualitatively wrong.

There are times when we can directly model this sort of heterogeneity. This leads to so-called frailty models, which include a term describing each individual's inherent "frailty" or "proneness." Obviously, we cannot estimate such quantities when each individual has only a single event (as in mortality). With repeated events, however, this is possible (Lindsey 1993); standard SAS procedures cannot presently analyze such models, but S-Plus can do so quite easily.

In cases where we can observe only a single event (like mortality), we can use the approach employed by several researchers studying senescence in *Drosophila* (e.g., Service et al. 1998). They fit models that assume heterogeneity in mortality rates, and find a maximum-likelihood estimate of the variance in frailties. They are then able to compare the fit of a model that incorporates these frailties with one that assumes no heterogeneity.

Another way to minimize problems that arise from heterogeneity is to include heterogeneity in the experimental design. For example, even if the questions of interest are not principally genetic, if we sample by units like families and then include family in the analysis, we have accounted for much of the heterogeneity resulting from either genetic variation or common environments.

13.6 Conclusion

The kinds of studies discussed in this chapter—time to emergence and flowering in a cohort of experimental plants, and time to insect visits for male and female flowers—illustrate some of the types of questions that can be addressed with failure-time approaches. Research in areas like life history evolution can clearly benefit from using these approaches. However, Muenchow's (1986) insect visitation study also shows how failure-time methods can be used to bring a lot of statistical power to bear on problems that may not initially seem to be failure-time problems.

This said, it is important to note that failure-time methods are not as well developed as ANOVA. We cannot use failure-time methods to estimate variance components. Erratic patterns of censoring and the nonnormality of most distributions of failure times makes such advances seem quite unlikely. Nor is the statistical power of failure-time methods well understood, except for the simple kinds of cases used in much of clinical medicine (e.g., two groups only and no other treatment factors—the survival analog of a simple *t*-test).

On the other hand, there are many types of ecological experiments for which failure-time methods are the best approach, both in terms of experimental design and statistical analysis. As shown in this chapter, the designs are generally quite simple once we acknowledge that censored data points are still data points. Statistical analyses are accessible to ecologists, especially inasmuch as the analyses can generally be related to the study of life tables. In addition to the SAS implementation discussed in this chapter, all of the methods are available in S-Plus, most of them are available in SPSS and BMDP, and many are available in SAS

JMP and in Systat. FORTRAN programs for performing many of these analyses are provided by Kalbfleisch and Prentice (1980) and Lee (1980). These methods should be more widely used by ecologists, and their availability in user-friendly software means that they can be used in undergraduate classes. Several issues related to the presentation of results of failure-time analyses are discussed in Fox (1993).

Acknowledgments   For comments that led to some substantial improvements in the second edition, I thank Tom Ebert, Peter Petraitis, and a large number of biologists who queried me for help and discussed interesting biological problems. This research was supported by NSF grant DEB-9806923.

Appendix: Distributions of Failure Times

Failure-time distributions can be compared in many ways. For ecologists, the most useful are probably comparisons of the hazard functions associated with each distribution, because the assumptions about the failure process are made explicit. Comparisons of the cumulative survivorship function and the probability density function for lifetimes can also be useful, especially because some distributions do not have closed-form expressions for the hazard but do for one or both of these functions.

A number of failure-time distributions have been used in empirical studies of failure times. Although in principle any distribution can be used with an accelerated failure-time model, some are more likely than others to be ecologically relevant. Some of the most important distributions are described in this appendix. Most statistical packages offer a limited choice of distributions. Note that the SAS manual uses considerably more complex, but equivalent, notation. These differences should be taken into account in interpreting SAS output. Other distributions are discussed by Lawless (1982).

Analytical expressions for the hazard, survival, and probability density functions are shown in table 13.5. Kalbfleisch and Prentice (1980) provide readable derivations for these expressions. Representative cases are illustrated for each distribution.

Exponential Distribution

Under the exponential distribution, the hazard function is constant (table 13.5). Fitting data to the exponential thus requires estimation of only the single parameter λ, which must be >0. This generates the cumulative survivorship and probability density functions shown in figure 13.5.

A constant hazard generates a Type II survivorship curve. There are probably few ecological situations in which hazards are truly constant for very long periods of time. However, it may be realistic to treat some hazards as approximately constant.

The exponential distribution is probably of greatest use as a starting point for understanding failure-time distributions. The constant hazard means that the

Table 13.5 Functions of failure-time distributions: h, P, and S are the hazard, probability density, and survival functions, respectively[a].

Distribution	Parameters	$h(t)$	$P(t)$	$S(t)$
Exponential	λ	λ	$\lambda e^{-\lambda t}$	$e^{-\lambda t}$
Weibull	p, λ	$\lambda p(\lambda t)^{p-1}$	$\lambda p(\lambda t)^{p-1} e^{-(\lambda t)^p}$	$e^{-(\lambda t)^p}$
2-parameter Gamma[b]	k, λ	$\dfrac{P(t)}{S(t)}$	$\dfrac{\lambda(\lambda t)^{k-1} e^{-\lambda t}}{\Gamma(k)}$	$1 - \Gamma(k, \lambda t)$
3-parameter Gamma[b]	k, λ, β	$\dfrac{P(t)}{S(t)}$	$\dfrac{\lambda\beta(\lambda t)^{k\beta-1} e^{-(\lambda t)^\beta}}{\Gamma(k)}$	$1 - \Gamma(k, (\lambda t)^\beta)$
Lognormal[c]	p, λ	$\dfrac{P(t)}{S(t)}$	$\dfrac{\sqrt{2\pi}}{pt} e^{-\frac{p^2[\log(\lambda t)]^2}{2}}$	$1 - \Phi[p\log(\lambda t)]$
Loglogistic	p, λ	$\dfrac{\lambda p(\lambda t)^{p-1}}{1 + (\lambda t)^p}$	$\dfrac{\lambda p(\lambda t)^{p-1}}{[1 + (\lambda t)^p]^2}$	$\dfrac{1}{1 + (\lambda t)^p}$

[a]SAS uses these parameterizations, but this is not always obvious from their documentation: in describing the gamma distribution, they fix all but the shape parameter at 0 or 1 so that the other two parameters do not appear in their description.
[b]$\Gamma(x)$ is the gamma function $\int_0^\infty u^{x-1} e^{-u} du$; $\Gamma(k,x)$ is the incomplete gamma function $\dfrac{1}{\Gamma(k)} \int_0^x u^{k-1} e^{-u} du$.
[c]$\Phi(x)$ is the cumulative normal distribution function $\Phi(w) = \dfrac{1}{\sqrt{2\pi}} \int_{-\infty}^w e^{-(u^2/2)} du$.

exponential distribution plays a role in failure-time models that is somewhat similar to that of the normal distribution in linear statistical modeling: the exponential provides a simple null model and a point of departure for more complex failure processes.

A simple empirical check for the exponential distribution is to examine whether a life table estimate for the hazard function is approximately constant. Equivalently, if the exponential distribution adequately describes the data, a plot of $\log[S(t)]$ versus t should be approximately linear through the origin. The SAS LIFETEST procedure produces this plot if you specify Plots = LS in the PROC

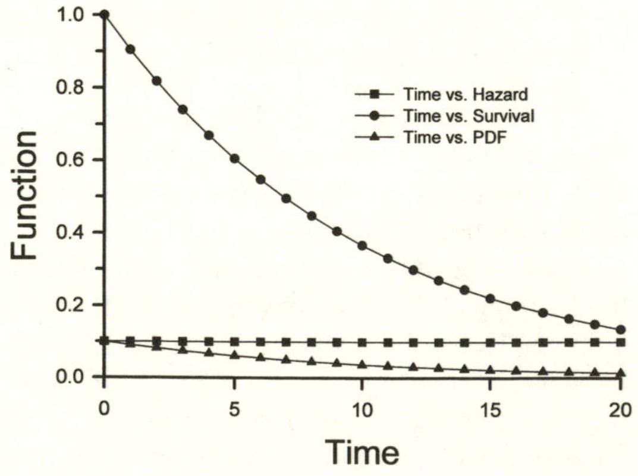

Figure 13.5 Hazard, survival, and probability distribution functions for the exponential distribution for $\lambda = 0.1$.

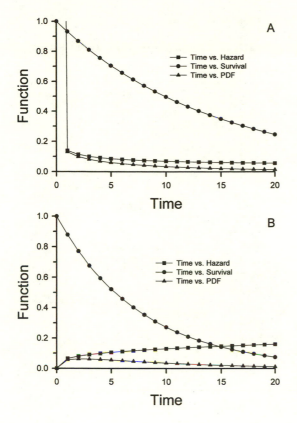

Figure 13.6 Hazard, survival, and probability distribution functions for the Weibull distribution. (A) $\lambda = 0.1$, $p = 0.7$; (B) $\lambda = 0.1$, $p = 1.3$.

statement. Such empirical tests are quite important, because significance tests assuming exponentially distributed data are less robust than tests assuming other distributions (Lee 1980).

Weibull Distribution

The Weibull distribution is a generalization of the exponential. In the Weibull distribution, the hazard function is a power of time (table 13.5). The two parameters p and λ are >0. If $p = 1$, the Weibull distribution reduces to the exponential distribution. Otherwise, the hazard is monotonically increasing for $p > 1$ and decreasing for $p < 1$. Considering its effect on the hazard function, we should not be surprised that the shape of P depends on p, and p is often referred to as a shape parameter. Figure 13.6 shows examples of the hazard, survival, and density functions.

The monotonic trend in the Weibull model means that it can be realistic in many systems. For example, epidemiological models use a Weibull distribution with $p > 1$ for time of onset of fullblown AIDS in HIV-infected patients. Mortality in humans and many other mammals can sometimes be approximated with a Weibull distribution, if only adults are considered. Weibull distributions have

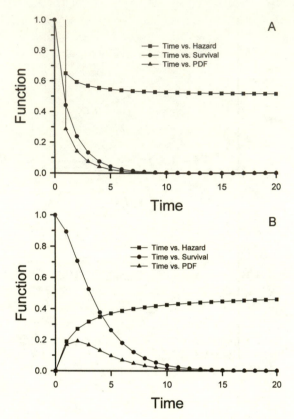

Figure 13.7 Hazard, survival, and probability distribution functions for the two-parameter gamma distribution. (A) $\lambda = 0.5$, $p = 0.7$; (B) $\lambda = 0.5$, $p = 1.9$.

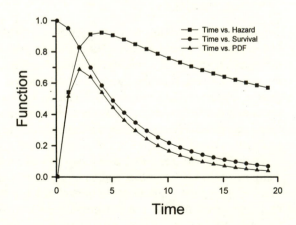

Figure 13.8 Hazard, survival, and probability distribution functions for the lognormal distribution for $\lambda = 0.2$, $p = 1.1$.

been used recently by Petraitis (1998) to model predation rates on mussels, by Ricklefs (1998) to model senescence in birds, and by Bullock et al. (1996) to model aging in plants.

A simple empirical check for the applicability of the Weibull distribution to a data set can be derived from the Weibull survival function: using life table estimates for the survival function, a plot of $\log\{-\log[S(t)]\}$ against $\log(t)$ should be approximately a straight line, with the slope being an estimate for p and the $\log(t)$ intercept an estimate for $-\log(\lambda)$. This plot is produced by the SAS LIFETEST procedure by specifying the Plots = LLS option in the PROC statement.

Gamma Distribution

The gamma distribution is another generalization of the exponential. In contrast to the Weibull distribution, the hazard function in the gamma distribution either increases or decreases monotonically toward a constant as t approaches to infinity. The monotonic nature of the gamma hazard makes it particularly useful for some developmental processes. In addition to examples in this chapter, I have found flowering time (Fox 1990a) and senescent mortality (Fox 1990b) in other plants to be well-described by gamma distributions.

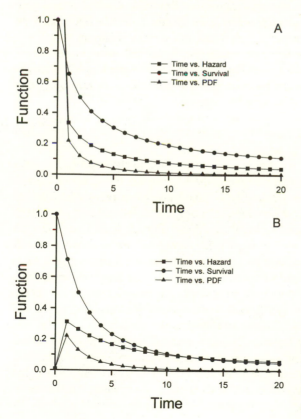

Figure 13.9 Hazard, survival, and probability distribution functions for the log-logistic distribution. (A) $\lambda = 0.5$, $p = 0.9$; (B) $\lambda = 0.5$, $p = 1.3$.

The two parameters k and λ (table 13.5) are both >0; k determines the shape of the distribution and λ its scale. The complicated SAS parameterization allows the shape parameter to be negative. If $k > 1$, the hazard function is 0 at $t = 0$ and increases monotonically toward λ. If $k < 1$, the hazard function is infinite at $t = 0$ and decreases monotonically toward λ with time. If $k = 1$, the gamma reduces to the exponential distribution. Examples are shown in figure 13.7.

The gamma distribution can be made even more flexible by generalizing to a three-parameter model (table 13.5). This three-parameter gamma includes all of the preceding distributions as special cases: the exponential ($\beta = k = 1$), Weibull ($k = 1$), and two-parameter gamma ($\beta = 1$). The lognormal distribution (next section) is the limiting case when k goes to infinity. Clearly, a very wide range of survival data can potentially be fit with this distribution.

Lognormal Distribution

Analytical expressions for the lognormal distribution involve indefinite integrals (table 13.5), but the salient feature of the lognormal is that the hazard function is 0 at time zero, increases to a maximum, and then decreases, asymptotically approaching 0 as t goes to infinity. The latter property means that the lognormal is probably not useful for studies involving long lifetimes (although these are quite rare in ecology) because a hazard of 0 is implausible. The lognormal is a two-parameter distribution in which both p and λ are assumed >0. See figure 13.8 for representative examples of the hazard, survival, and density functions.

As you might guess from the name, under the lognormal distribution the log failure times are normally distributed. This suggests that simple empirical checks for the lognormal distribution can be done by log-transforming the data and using any of the standard methods of testing for normality.

Log-logistic Distribution

The log-logistic distribution is roughly similar in shape to the lognormal, but it is often easier to use the log-logistic, primarily because the functions of interest have closed-form expressions (table 13.5). Especially when there are censored data, the lognormal can involve considerable computation because of the indefinite integrals in the hazard and survival functions. The log-logistic distribution can also be somewhat more flexible than the lognormal, as can be seen by considering the hazard, survivorship, and probability distribution functions (table 13.5, figure 13.9). If $p < 1$, the hazard function is infinite at 0 but decreases monotonically with time. If $p > 1$, it decreases monotonically from λ. When $p > 1$, the hazard is similar to the lognormal hazard: it increases from zero to a maximum at $t = (p - 1)^{1/p}/\lambda$, and then decreases toward zero.

The humped nature of the log-logistic and lognormal hazards may make them especially useful for describing an organism's response to environmental factors. The seedling emergence data analyzed in this chapter is one example. On the other hand, few mortality hazards are likely to begin as very small, reach a maximum, and then decline.

14

The Bootstrap and the Jackknife

Describing the Precision of Ecological Indices

PHILIP M. DIXON

14.1 Introduction

Quantitative ecology uses many indexes and coefficients, including diversity indices (Magurran 1988), similarity indices (Goodall 1978), competition coefficients (Schoener 1983), population growth rates (Lenski and Service 1982), and measures of size hierarchy (Weiner and Solbrig 1984). All of these indices are statistics, calculated from a sample of data from some population and used to make conclusions about the population (chapter 1). To help a reader interpret these conclusions, good statistical practice includes reporting a measure of uncertainty or precision along with a statistic. Although it is easy to calculate the values of many ecological statistics, it is often difficult to estimate their precision. This chapter discusses two techniques, the bootstrap and jackknife, that can be used to estimate the precision of many ecological indices.

14.1.1 Precision, Bias, and Confidence Intervals

To understand how the bootstrap and jackknife work, we must first review some concepts of statistical estimation. Consider how to estimate the mean reproductive output for a species in a defined area. The statistical population is the set of values (number of seeds per plant) for every plant in the area. The mean reproductive output is a parameter; it describes some interesting characteristic of the population. This parameter is known exactly if every plant is studied, but completely enumerating the population is usually impractical.

267

Instead, a random sample of plants are counted. The average reproductive output for the plants in the sample is an estimate of the parameter in which we are interested. How accurate is this estimate? Plants have different reproductive outputs; a different sample of plants will provide a different estimate of average reproductive output. Some estimates will be below the population mean, whereas other estimates will be above (figure 14.1). The set of average reproductive output from all possible samples of plants is the sampling distribution of the sample average; characteristics of this distribution describe the accuracy of a statistic.

The accuracy of a statistic has two components: bias and precision (Snedecor and Cochran 1989). Bias measures whether a statistic is consistently too low or too high. It is defined as the difference between the population value and the average of the sampling distribution. If the sampling distribution is centered around the population value, then the statistic is unbiased. Precision depends on the variability in the sampling distribution and is often measured by the variance or standard error. Bias and precision are separate components of accuracy. A precise statistic may be biased if its sampling distribution is concentrated around some value that is not the population value.

The distribution of estimates from all possible samples is a nice theoretical concept, but an experimenter has only one sample of data and one value of the estimate. If the method used to sample the population has well-defined characteristics (e.g.,

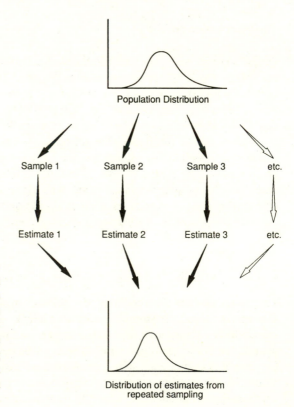

Figure 14.1 Relationship between population distribution and sampling distribution. If the values in the population were known, then the sampling distribution of the estimate could be obtained by repeatedly drawing samples from the population and calculating the statistic from each sample.

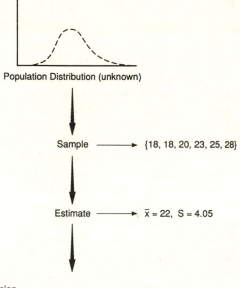

Population Distribution (unknown)

Sample ⟶ {18, 18, 20, 23, 25, 28}

Estimate ⟶ $\bar{x} = 22, \ S = 4.05$

Statements about precision

- Standard error of mean $S_{\bar{x}} = \dfrac{S}{\sqrt{n}} = \dfrac{4.05}{\sqrt{6}} = 1.65$

- 95% confidence interval for μ $(\bar{X} \pm t_{n\text{-}1} \cdot S_{\bar{x}}) = (22 \pm 2.54 \cdot 1.65) = (17.8, 26.2)$

Figure 14.2 Use of data and statistical theory to infer a sampling distribution from one sample of data. The sampling distribution of certain statistics is known theoretically. Statements about the precision can be made from sample information (e.g., sample mean and standard deviation).

simple random sampling), the sampling distributions of sample averages and a few other statistics can be calculated from one sample of data because there are known mathematical relationships between the properties of the sample and the properties of the population (figure 14.2). For example, the sample average from a simple random sample is unbiased and its standard error, $s_{\bar{x}}$, can be estimated by $s_{\bar{x}} = s_x/(n)^{1/2}$, where s_x is the sample standard deviation and n is the sample size.

Confidence intervals can also be calculated if the sampling distribution is known. Under certain assumptions about the observations, the sample mean and sample variance are independent estimates with known distributions, so that a 95% confidence interval is given by $\bar{x} - t_{n-1}s_{\bar{x}}, \ \bar{x} + t_{n-1}s_{\bar{x}}$, where t_{n-1} is the critical value for a two-sided 95% confidence interval from a t-distribution with $n-1$ degrees of freedom. Confidence intervals are commonly misunderstood. Remember, a confidence interval is a random interval with the property that it includes the population mean, which is fixed, with a given frequency. A 95% confidence interval of (1, 2) does not mean that 95% of possible sample values are between 1 or 2, and it does not mean that 95% of possible population means are between 1 and 2. Instead, it means that 95% of the time, the confidence interval will include the population mean.

14.1.2 Precision and Bias of Ecological Indexes

Many useful ecological indexes are more complicated than a sample mean, and a sampling distribution cannot be calculated mathematically. However, it is more important to choose an ecologically useful coefficient rather than a statistically tractable one. The sampling distributions of many ecologically useful coefficients can be estimated using the bootstrap or the jackknife. The jackknife estimates the bias and variance of a statistic. In addition to estimating the bias and variance, the bootstrap also determines a confidence interval.

The jackknife and bootstrap have been used in many ecological applications, such as the following: population growth rates (Meyer et al. 1986; Juliano 1998), population sizes (Buckland and Garthwaite 1991), toxicity assessment (Bailer and Oris 1994), ratios of variables (Buonaccorsi and Liebhold 1988), genetic distances (Mueller 1979), selection gradients (Bennington and McGraw 1995), diversity indexes (Heltshe and Forrester 1985; Heltshe 1988), species richness (Smith and van Belle 1984; Palmer 1991), diet similarity (Smith 1985), home ranges (Rempel et al. 1995), and niche overlap indexes (Mueller and Altenberg 1985; Manly 1990). General introductions to the bootstrap and jackknife can be found in Manly (1997), Crowley (1992), and Stine (1989). More of the statistical theory can be found in Hall (1992), Shao and Tu (1995), and Davison and Hinkley (1997). Useful surveys of applications and problems include Léger et al. (1992), Young (1994), and Chernick (1999).

In this chapter, I will describe the jackknife and bootstrap and illustrate their use with two indices: the Gini coefficient of size hierarchy and the Jaccard index of community similarity [see http://www.oup-usa.org/sc/0195131878/ for the computer code]. I also describe some of the practical issues applying the techniques to answer ecological questions. The focus is on nonparametric methods for independent observations, but some approaches for more complex data are discussed at the end.

14.1.3 Gini Coefficients and Similarity Indices

The Gini coefficient, G, is a measure of inequality in plant size (Weiner and Solbrig 1984). It ranges from 0, when all plants have the same size, to a theoretical limit of 1, when one plant is extremely large and all other plants are extremely small. The coefficient G can be calculated from a set of data:

$$G = \frac{\sum_{i=1}^{n} (2i - n - 1)X_i}{(n - 1)\sum_{i=1}^{n} X_i} \tag{14.1}$$

where n is the number of individual plants and X_i is the size of the ith plant, when plants are sorted from smallest to largest, $X_1 \leq X_2 \leq \ldots \leq X_n$. The bootstrap can be used to estimate the precision of G (Dixon et al. 1987).

A similarity index describes the similarity in species composition, which can be estimated by sampling the species in each community and computing the similarity between a pair of samples. Many similarity indexes have been proposed (see Goodall 1978 for a review); the Jaccard (1901) index, which depends on the similarity in species presence, is computed as

$$J = \frac{a}{a + b + c} \tag{14.2}$$

where a is the number of species found in both communities, b is the number of species found only in the first, and c is the number of species found only in the second. The coefficient is ecologically useful, but its precision is difficult to calculate analytically. The bootstrap and jackknife are methods to describe the accuracy of similarity indexes (Smith et al. 1986).

14.2 The Jackknife

The jackknife procedure (Miller 1974) is a general technique to answer the question, How precise is my estimate? It can estimate bias or standard error for a statistic, but not for a confidence interval (Efron 1982). The basic idea is that the bias and standard error can be estimated by recalculating the statistic on subsets of the data. Although the bootstrap has theoretical advantages, the jackknife requires considerably less computation and includes no random component. The jackknife can also be used in conjunction with the bootstrap (see sections 14.4.3 and 14.5.6).

14.2.1 Ecological Example: Gini Coefficients to Measure Size Hierarchy in *Ailanthus* Seedlings

The tree-of-heaven, *Ailanthus altissima*, is an aggressive introduced tree. Evans (1983) studied whether seedlings grown in a competitive environment have a stronger size hierarchy (higher size inequality) than individually grown plants. Six seeds were randomly chosen from a large collection of seeds and planted in individual pots. Another 100 seeds were planted in a common flat so that they could compete with each other. After 5 months, each survivor was measured (table 14.1). The Gini coefficient for the individually grown plants ($G = 0.112$) was smaller than that for the competitively grown plants ($G = 0.155$), consistent with the hypothesis that competition increases the inequality in the size distribution (Weiner and Solbrig 1984). The sample sizes are small, especially for the individually grown plants, so it is important to estimate the precision of each estimate and calculate a confidence interval for the difference.

14.2.2 Jackknifing the Gini Coefficient

The jackknife estimates of bias and standard error (Miller 1974) are calculated by removing one point at a time from the data set. Consider the Gini coefficient for individually grown plants (table 14.1). The observed value, based on all six plants is $G = 0.112$. If the first point is removed, G_{-1} calculated from the remain-

Table 14.1 Number of leaf nodes for 5-month-old *Ailanthus altissima* grown under two conditions: in individual pots and in a common flat

6 plants grown individually:

18	18	20	23	25	28

75 surviving plants grown together in a common flat:

8							
10							
11	11	11					
12	12	12					
13	13	13	13	13	13	13	
14	14	14	14	14			
15	15	15	15	15	15	15	15
16	16						
17	17	17					
18	18	18	18	18			
19	19	19	19				
20	20	20	20	20	20		
21	21	21					
22	22	22	22	22			
23	23	23	23	23	23	23	
24	24	24					
25	25	25	25				
26							
27	27	27					
30							

ing five data points is 0.110. If the fourth point is removed, G_{-4} calculated from the remaining five data points is 0.124. Each perturbed value is combined with the original statistic to compute a pseudovalue, p_i, for each of the n data points:

$$p_i = G + (n - 1)(G - G_{-i}).$$ (14.3)

The six jackknife samples for individually grown *Ailanthus*, their G_{-i} values, and their pseudovalues are given in table 14.2.

Table 14.2 Jackknife samples with Gini coefficients for individually grown *Ailanthus*

Jackknife sample					Gini coefficient	Pseudovalue
18	20	23	25	28	0.110	0.124
18	20	23	25	28	0.110	0.124
18	18	23	25	28	0.120	0.070
18	18	20	25	28	0.124	0.053
18	18	20	23	28	0.117	0.089
18	18	20	23	25	0.091	0.216
					mean $\bar{p}$:	0.1128

The jackknife estimates of bias (Efron 1982) are

$$\hat{\text{bias}} = G - \bar{p} \tag{14.4}$$

where G is the sample Gini coefficient and $\bar{p}$ is the mean of the jackknife pseudo-values. For the individually grown *Ailanthus*, the jackknife estimate of bias is $0.1121 - 0.1128 = -0.0007$. This bias is very small, but if it were larger, it could be subtracted from the observed value to produce a less biased estimate. The jackknife is especially effective at correcting a statistic for first-order bias, a bias that is linearly proportional to the sample size (Miller 1974).

The jackknife estimate of the standard error is just the standard error of the pseudovalues,

$$s_G = \sqrt{\frac{\Sigma(p_i - \bar{p})^2}{n(n-1)}} \tag{14.5}$$

where n is the sample size and $\bar{p}$ is the mean of the pseudovalues. From the data in table 14.2, the estimated standard error (s.e.) is 0.024. For the 75 plants grown in competition, the s.e. is estimated to be 0.010. The two samples are independent, so the s.e. of the difference is $(s_1^2 + s_2^2)^{1/2}$. This is estimated to be 0.026, which is about half of the observed difference (0.043). These calculations have been demonstrated using a small sample of data, but estimates of bias and standard error from small samples must be treated with caution. As with most statistical techniques, estimates from larger samples are more precise.

In general, the jackknife cannot be extended to calculate confidence intervals or test hypotheses (Efron 1982). Some attempts have been made to construct confidence intervals by assuming a normal distribution (Meyer et al. 1986; Smith et al. 1986). Such confidence intervals would have the form ($G \pm t_k s_G$, where t_k is a critical value from a t-distribution with k degrees of freedom. The problem with this approach is that the appropriate number of degrees of freedom is unknown, in spite of a lot of theoretical research (Efron and LePage 1992). However, using $n - 1$ degrees of freedom, where n is the original sample size, has worked well in some cases (Meyer et al. 1986).

The jackknife has also been used to estimate the precision of various similarity measures, including measures of diet similarity (Smith 1985), community similarity (Smith et al. 1979; Smith et al. 1986; Heltshe 1988), and niche overlap (Mueller and Altenberg 1985). Because these methods require comparing two-samples, the jackknife procedure is slightly different. Details of the two-sample jackknife can be found in Dixon (1993) or in the original articles.

14.3 The Bootstrap Method

The bootstrap has become a popular method for estimating confidence intervals and testing hypotheses about many ecological quantities. Although the bootstrap is applicable to many ecologically problems, it is not appropriate for everything. I will describe the principles of bootstrapping, compare some of the many varieties of the bootstrap procedure, and discuss when the bootstrap is not appropriate.

The bootstrap is a two-step procedure to approximate the unknown sampling distribution. First, the unknown distribution of values in the population is approximated using information from the observed sample (figure 14.3). Then, many bootstrap samples are drawn from this distribution. The unknown sampling distribution is approximated by the distribution of estimates from many bootstrap samples (figure 14.3). The bootstrap distribution is used to calculate a confidence interval, test a hypothesis, and estimate the standard error and bias for a statistic.

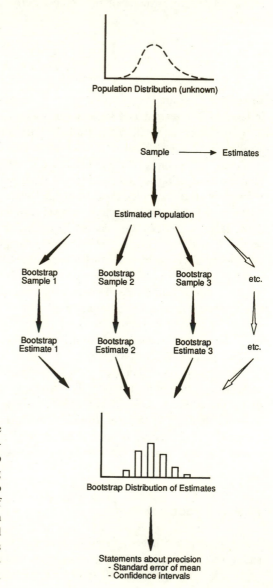

Figure 14.3 Use of data and the bootstrap distribution to infer a sampling distribution. The bootstrap procedure estimates the sampling distribution of a statistic in two steps. The unknown distribution of population values is estimated from the sample data, then the estimated population is repeatedly sampled, as in figure 14.1, to estimate the sampling distribution of the statistic.

Although the concepts are very simple, there are many possible varieties of boot-straps. These differ in three characteristics: (1) how the population is approximated, (2) how bootstrap samples are taken from the population, and (3) how the endpoints of confidence intervals are calculated.

14.3.1 Approximating the Population: Parametric and Nonparametric Bootstraps

The first step in the bootstrap is using information in the observed sample to approximate the unknown distribution of values in the population. A parametric bootstrap approximates the population by a specific distribution (e.g., a lognormal or Poisson distribution) with parameters that are estimated from the sample. For example, the number of leaves on the competitively grown plants fits a Poisson distribution with a mean of 18.4 reasonably well. There are only six individually grown plants, so almost any discrete distribution can fit those data. A Poisson distribution with a mean of 22 is not an unreasonable choice. These parametric distributions are used to describe the unknown populations. Each parametric bootstrap sample of competitively grown plants is a random sample of 75 observations from a Poisson distribution with a mean of 18.4; each parametric bootstrap sample of individually grown plants is a random sample of six observations from a Poisson distribution with a mean of 22.

In a nonparametric bootstrap, the population is approximated by the discrete distribution of observed values (figure 14.3). The estimated population of individual *Ailanthus* sizes is a discrete distribution in which the value 18 occurs with probability 2/6 (because 2 plants had 18 leaves) and the values 20, 23, 25, and 28 each have probability 1/6. Most ecological applications use the nonparametric bootstrap because it requires fewer assumptions about the population. However, the nonparametric bootstrap assumes that the observed sample is representative of the population. With very small samples, the parametric bootstrap is often better, as long as the assumed distribution is not wildly incorrect (Davison and Hinkley 1997).

14.3.2 Drawing a Bootstrap Sample From the Population: Ordinary, Balanced, and Moving Block Bootstraps

Once the population is approximated, samples must be drawn from this population. The simplest way is to draw a simple random sample, with replacement, from the values in the population. If a nonparametric bootstrap is used, the bootstrap sample is a simple random sample of the observed values. The statistic, (e.g., the Gini coefficient) is then calculated from the values in the bootstrap sample. Five bootstrap samples and Gini coefficients are shown in table 14.3, which illustrates an important characteristic of bootstrap samples. Each bootstrap sample omits some observed values and repeats others because the observed data are sampled with replacement. Some of the bootstrap samples have Gini coefficients larger than the observed value, $G = 0.112$, whereas other samples have smaller coefficients.

Table 14.3 Five bootstrap samples for the data from individually grown plants

Bootstrap sample						Gini coefficient
18	18	23	25	28	28	0.117
18	18	18	25	25	25	0.098
18	18	23	23	28	28	0.116
18	18	18	20	23	28	0.107
18	18	20	23	25	28	0.112

This process is repeated for many bootstrap samples. Typically, 50 to 100 bootstrap samples are used to estimate a standard error; 1000 or more bootstrap samples are recommended to calculate a confidence interval (Efron and Tibshirani 1993). The number of bootstrap samples will be discussed further in section 14.5.5.

One potential concern with simple random sampling is that each observation may not occur equally often in the bootstrap samples. In the five samples illustrated in table 14.3, the value 20 occurs a total of two times, but the value 28 occurs a total of six times. Hence, the aggregated bootstrap samples do not represent the population, in which the values 20 and 28 are equally frequent. The balanced bootstrap forces each value to occur equally frequently. One algorithm to draw 100 balanced bootstrap samples is to write down 100 copies of the observed data. For the six individually grown plants, this population has 600 values, 200 of which are 18, 100 are 20, 100 are 23, 100 are 25, and 100 are 28. Randomly permute the 600 values, then take the first 6 as the first bootstrap sample, the next 6 as the second, and so on. The balanced bootstrap can markedly increase the precision of bias calculations, but it is less useful for confidence interval calculations (Davison and Hinkley 1997).

Both the simple random sample and balanced bootstraps assume that there is no temporal or spatial correlation among the values. If there is any correlation, it is eliminated by the randomization used in the ordinary and balanced bootstraps. The moving block bootstrap generates bootstrap samples that retain some of the correlation (Davison and Hinkley 1997). It, and other methods for bootstrapping correlated data, are discussed in section 14.5.4.

14.3.3   Estimating Bias and Standard Error From the Bootstrap Distribution

The bootstrap distribution provides the information necessary to estimate the bias and standard error of a statistic like the Gini coefficient. The bootstrap estimate of bias is simply the difference between the average of the bootstrap distribution and the value from the original sample. For the individually grown *Ailanthus* plants, the sample Gini coefficient is 0.1121 and the average Gini value in the five bootstrap samples of table 14.2 is 0.1099. Hence, the bootstrap estimate of bias is $0.1099 - 0.1121 = -0.0021$. In practice, you should use 50 to 100 bootstrap samples to estimate the bias (Efron 1987). Using 1000 bootstrap samples (data not shown), the biases for the individually and competitively grown *Ailanthus* are estimated to be -0.018 and -0.0020, respectively. In both cases, the bias is small. If the bias were larger, the estimate of bias could be subtracted from the observed value to produce a less biased estimate.

The standard error of the sample Gini coefficient is estimated by the standard deviation of the bootstrap distribution. For the five bootstrap samples shown in table 14.2, the estimated standard error is 0.0079. Again, I use only five bootstrap samples to demonstrate the calculations. Using 1000 bootstrap samples (data not shown), the estimated standard errors for the Gini coefficient of individually and competitively grown plants are 0.022 and 0.0097, respectively. These numbers are not too different from those estimated by the jackknife. This is often the case; theoretical calculations show that the jackknife is a linear approximation to the bootstrap (Efron 1982).

A standard error is a single measure of precision. It can be turned into a confidence interval if we assume a particular distribution, such as the normal. However, confidence intervals can be estimated from the bootstrap distribution without assuming normality.

14.4 Calculating Confidence Intervals From the Bootstrap Distribution

Confidence intervals can be calculated from the bootstrap distribution in at least five different ways: the percentile bootstrap, the basic bootstrap, the studentized bootstrap, the bias-corrected bootstrap, and the accelerated bootstrap (Davison and Hinkley 1997). Different authors have used other names for some of these methods, which increases the confusion. Some of the synonyms are given in the documentation to the SAS JACKBOOT macro (SAS Institute 1995). These different methods, which are discussed subsequently, represent a trade-off between simplicity and generality. No method is best for all problems.

14.4.1 Percentile Bootstrap

The percentile bootstrap is the simplest, and most commonly used, method to construct bootstrap confidence intervals. In this method, the 2.5 and 97.5 percentiles of the bootstrap distribution are used as the limits of a 95% confidence interval. To calculate the 2.5 percentile from N bootstrap replicates, sort the estimates from the bootstrap samples in order from smallest to largest. The pth percentile is the $(N + 1)p/100$th largest value. A histogram of 999 bootstrap Gini coefficients for competitively grown *Ailanthus* is shown in figure 14.4. The 2.5 and 97.5 percentiles of $N = 999$ observations are given by the 25th and 975th largest values. They are 0.133 and 0.171, respectively, so (0.133, 0.171) is the 95% confidence interval using the percentile method.

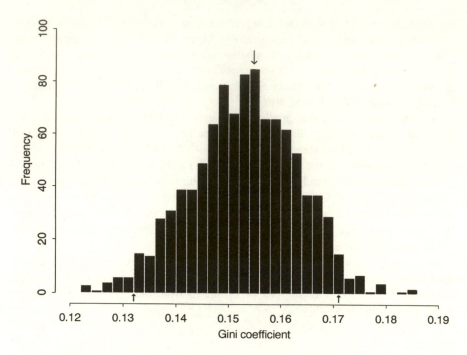

Figure 14.4 Histogram of 999 bootstrap values for competitively grown *Ailanthus*. Arrows below the histogram mark the 25th and 975th largest values. Arrow above the histogram marks the Gini coefficient for the observed data ($G = 0.155$).

Percentile bootstrap confidence intervals can be calculated for any statistic, but they do not always work very well (Schenker 1985). A confidence interval can be evaluated by simulating many data sets from a population with some known parameter, calculating a confidence interval from each data set, and counting how many confidence intervals bracket the true parameter. A 95% confidence interval should include the true parameter in 95% of the simulated data sets. When sample sizes are small (i.e., less than 50), the percentile confidence intervals for the variance (Schenker 1985) and Gini coefficient (Dixon et al. 1987) are too narrow. For example, with samples of 20 points, 90% confidence intervals for the variance include the true variance only 78% of the time, and 95% confidence intervals for a Gini coefficient of a lognormal population include the true value of 0.30 only 85% of the time (Schenker 1985; Dixon et al. 1987).

The percentile bootstrap produces correct confidence intervals when the bootstrap distribution is symmetrical and centered on the observed value (Efron 1982). This is not the case for the Gini coefficient in this example, where 56% of the bootstrap values are smaller than the observed value (figure 14.4). For individually grown *Ailanthus*, 74% of the bootstrap values are smaller than the observed value. In both cases, the observed values are larger than the median of the bootstrap distribution, so the upper and lower confidence bounds are too low.

The plethora of bootstrap confidence intervals comes from different ways of adjusting the percentile bootstrap confidence intervals to improve their coverage. The bias-corrected bootstrap adjusts for bias in the bootstrap distribution and the accelerated bootstrap adjusts for both bias and skewness. The studentized bootstrap is based on the bootstrap distribution of the statistic adjusted by its mean and standard error. The ratio $(G_i - G)/s_i$ is like a Student's t-statistic, hence the name of the method. This approach is especially useful when the variability in the data depends on the mean, as is common in ecological data. The basic bootstrap, a "turned around" version of the percentile bootstrap, is based on the fundamental relationship between hypothesis tests and confidence intervals.

14.4.2 Bias-corrected and Accelerated Bootstraps

The bias-corrected percentile bootstrap adjusts for a bootstrap distribution that is not centered on the observed statistic. The bounds of the bias-corrected intervals are found by determining F, the fraction of bootstrap replicates that are smaller than the observed value and the value, z_0, the probit transform of F. The appropriate percentiles for a 95% confidence interval are calculated as follows:

$$P_1 = \phi(2z_0 - 1.96) \tag{14.6}$$

$$P_u = \phi(2z_0 + 1.96) \tag{14.7}$$

where ϕ is the normal cumulative distribution function (c.d.f.). The normal c.d.f. and probit transformation are available in many statistical packages, or they can be computed from tables of areas of the normal curve (e.g., Rohlf and Sokal 1995, table A). The values ± 1.96 are the critical values for a 95% confidence interval of a statistic with a normal distribution. They can be changed to generate other confidence intervals. The upper and lower bounds of the bias-corrected confidence interval are given by the values in the bootstrap distribution that match the calculated percentiles, P_1 and P_u. When the observed value is the median of the bootstrap distribution, there is no difference between the bias-corrected and percentile confidence intervals.

For the bootstrap distribution shown in figure 14.4, 56.4% of the values are smaller than the observed Gini coefficient of 0.1548, so $F = 0.564$ and $z_0 = 0.166$. The desired percentiles of the bootstrap distribution are $P_1 = \phi(-1.628) = 5.181\%$, and $P_u = \phi(2.29) = 98.91\%$. The number of bootstraps (1000) times the percentiles (5.181% or 98.91%) are not exact integers, so the observations corresponding to the next most extreme integer are used. The bounds of the bias-corrected 95% confidence interval are the 51st and 990th largest values in the bootstrap distribution, which are (0.136, 0.176). Note that the values of the upper and lower bounds are both larger than those for the percentile bootstrap, which were (0.133, 0.171). The bias-correction procedure has shifted the confidence interval upward to adjust for bias.

The accelerated bootstrap makes a second correction that is helpful when bootstrapping statistics like the variance or the Gini coefficient, where a few extreme points will have a large influence on the observed value. This technique is so named because it "accelerates" the bias-correction (Efron 1987). These confidence intervals depend on two constants, z_0 used to correct for bias and the acceleration factor, a, which corrects for skewness. The a coefficient is nonzero when a few points have a large influence on the observed estimate. It can be estimated nonparametrically using the jackknife (see Dixon 1993, p. 302, or Efron and Tibshirani 1993, p. 186, for the details). The percentiles for the upper and lower 95% confidence bounds are calculated as

$$P_l = \Phi\left(z_0 + \frac{z_0 - 1.96}{1 - a(z_0 - 1.96)}\right) \tag{14.8}$$

$$P_u = \Phi\left(z_0 + \frac{z_0 + 1.96}{1 - a(z_0 + 1.96)}\right) \tag{14.9}$$

where ϕ is the normal cumulative distribution function, as with the bias-corrected confidence intervals, and the value 1.96 is the normal critical value for a 95% confidence interval.

Using the number of leaf nodes of competitively grown *Ailanthus* (table 14.1), the constant a was estimated to be 0.0193, using the method described in Dixon (1993). For the bootstrap distribution in figure 14.4, $z_0 = 0.166$, so $P_l = \phi(-1.568) = 5.85\%$, and $P_u = \phi(2.383) = 99.14\%$. Hence, the bounds of the accelerated confidence interval were the 59th and 992nd largest values in the set of 1000 bootstrap replicates: (0.137, 0.176). For these data, both the bias-corrected and accelerated bootstrap procedures adjust the confidence interval upward to account for the skewed and biased sampling distribution.

14.4.3 Studentized Bootstrap

The percentile and bias-corrected bootstraps assume that the sampling distribution has constant variance. In many practical cases, the variance of the sampling distribution depends on the mean. Some ecological examples include average plant sizes, average survival time under exposure to some toxicant, and estimated proportions. The variance of the observations is related to the mean, so the variance of the sampling distribution is related to the mean. The acceleration constant, a, used in the accelerated bootstrap confidence interval is one way to adjust for this unequal variance. A second way is the studentized bootstrap.

When a statistic, g, has a normal sampling distribution, the bounds of a confidence interval are given by

$$(g - ts_g, g + ts_g) \tag{14.10}$$

where s_g is the standard error of g, and $\pm t$ are quantiles of a t-distribution with the appropriate degrees of freedom. The idea behind the studentized bootstrap is to compute a confidence interval with the same form as equation 14.10, but use different critical values and relax the assumption of normality. The studentized bootstrap confidence intervals are given by

$$(g + b_l s_g, \ g + b_h s_g) \tag{14.11}$$

where b_l and b_h are percentiles from the bootstrap distribution of a studentized version of the statistic.

This bootstrap distribution is computed by drawing a bootstrap sample of observations and calculating the statistic, G_i, and its standard error, s_i. The studentized statistic is then

$$b_i = \frac{G_i - G}{s_i}$$

where G is the value observed in the original sample. Note that a standard error is calculated for each bootstrap sample. A large number (e.g., 999) of values of b_i are calculated from a large number of bootstrap samples. For a 95% confidence interval, b_l and b_h are the 2.5 and 97.5 percentiles. If $N = 999$, estimates of b_i are sorted from smallest to largest; these percentiles are the 25th and 975th observations. These percentiles are combined with the observed statistic, g, and observed standard error, se_g, to compute the studentized confidence interval (equation 14.11).

Unlike the other bootstrap methods, the studentized bootstrap requires the standard error of the statistic. For some statistics, standard errors can be computed using appropriate formulas. The Gini coefficient is one of many statistics for which there is no formula for the standard error. A second bootstrap could be used to estimate the standard error (see section 14.3.3), but this requires considerable computation to estimate a standard error for each bootstrap sample. A more practical approach is to use the jackknife (section 14.2.2) to estimate the standard error of the observed statistic and each bootstrap statistic.

The bootstrap distribution of studentized Gini statistics for the 75 competitively grown individuals has a median of 0.26 and is slightly skewed (figure 14.5). The observed Gini coefficient is 0.155 with a jackknife standard error of 0.0102. The 2.5 and 97.5 percentiles are −1.80 and 2.48, respectively, so the 95% studentized bootstrap confidence interval is $(0.155 - 1.80 \times 0.0102, \ 0.155 + 2.48 \times 0.0102) = (0.137, 0.180)$.

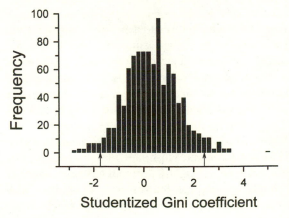

Figure 14.5 Histogram of 999 studentized bootstrap values for competitively grown *Ailanthus*. Arrows below the histogram mark the 25th and 975th largest values.

14.4.4 Basic Bootstrap

The percentile bootstrap is quite easy to interpret but somewhat controversial. The bootstrap distribution approximates the sampling distribution when the unknown parameter, θ, is set equal to the observed value, G. The controversial part is how to calculate an appropriate confidence interval from the bootstrap distribution (Davison and Hinkley 1997). The basic bootstrap determines confidence interval endpoints by exploiting the fundamental connection between hypothesis tests and confidence intervals. One way to find the endpoints of a 95% confidence interval is to use a test of the hypothesis $\theta = x$ to find the values $x = 1$ and $x = u$ for which the hypothesis is rejected at exactly $P = 5\%$. Using the bootstrap to test these hypotheses requires estimating the sampling distribution when $\theta = 1$ and the sampling distribution when $\theta = u$. Under the assumption that the s.e. of the sampling distribution is not related to the mean, the sampling distribution when $\theta = u$ is estimated by shifting the observed sampling distribution by $u - G$. The hypothesis that $\theta = u$ is rejected at $P = 5\%$ when the observed value, G, is exactly the 2.5th percentile of the shifted sampling distribution. Similarly, the hypothesis that $\theta = 1$ is rejected at $p = 5\%$ when the observed value, G, is exactly the 97.5th percentile. Hence, the basic confidence interval bounds are given by $(2G - G_{0.975}, 2G - G_{0.025})$, where G is the observed statistic and $G_{0.025}$ and $G_{0.975}$ are the 2.5th and 97.5th percentiles of the bootstrap distribution.

For the competitively grown individuals, the 2.5th and 97.5th percentiles of the bootstrap distribution are 0.133 and 0.171, respectively, so the basic 95% confidence interval is $(2 \times 0.155 - 0.171, \ 2 \times 0.155 - 0.133) = (0.139, \ 0.177)$. When the sampling distribution is symmetrical about the observed value, G, the basic intervals and percentile intervals are the same. Because the sampling distribution of the Gini coefficient is slightly asymmetric, the endpoints of the basic interval are adjusted slightly upward from those for the percentile interval.

14.5 Practical Details

14.5.1 Choice of Method

The bootstrap and the jackknife are two techniques that answer the same question, How precise is a particular statistic? The bootstrap estimates either a standard error or a confidence interval, but the jackknife is most appropriately used to estimate a standard error. Confidence intervals based on the jackknife standard error assume that the statistic has a normal distribution. If the distribution is skewed or heavy tailed, then jackknife confidence intervals are likely to be inaccurate.

In extremely large samples, theoretical results show that both techniques give the same answers (see Efron 1982 and Efron and Tibshirani 1986 for details). For realistic data, the two techniques usually give slightly different answers. Which is better? The jackknife is often simpler to compute, but that is less important with modern computers. Often, both techniques give similar answers, but if the differences are important, the performance of the two techniques should be evaluated on a case-by-case basis.

14.5.2 Choice of Bootstrap Method

As stated in section 14.3, the choice of bootstrap specifies three characteristics: how to approximate the population, how to draw a bootstrap sample, and how to construct a confidence interval. Some decisions are more important than others. Most ecological applications use the nonparametric percentile bootstrap. It is certainly the simplest method to implement or describe. The choice of parametric or nonparametric bootstrap is not very important when the assumed parametric distribution is reasonable and the sample size is sufficiently large. If the sample size is small, the parametric bootstrap may provide more reliable answers. The choice of ordinary or balanced bootstrap rarely affects confidence intervals (Davison and Hinkley 1997).

The appropriate choice of which method to use to calculate the bootstrap confidence interval can be more difficult, but this also may not make much difference. For the competitively grown plants, all five methods for calculating endpoints give very similar confidence intervals (table 14.4). This is not always the case, especially if the sample size is small (e.g., Efron and Tibshirani 1993, p. 183; Davison and Hinkley 1997, p. 231; Manly 1997, p. 55).

When the differences between intervals are large enough to matter, it may help to consider the characteristics of the problem and each method. The percentile, bias-corrected, and accelerated confidence intervals are restricted to valid parameter values. For example, the Gini coefficient must be between 0 and 1. The endpoints of confidence intervals from these three methods will always fall between 0 and 1. The basic and studentized methods can lead to confidence interval endpoints that are negative or are larger than 1.

Theoretical arguments suggest that the accelerated bootstrap is generally better than the bias-corrected or percentile methods (Efron 1987). However, simulation studies suggest that the practical differences among bootstrap techniques are often small. If there is a difference, the studentized bootstrap has the best coverage (Davison and Hinkley 1997, p. 231 for ratios and P. M. Dixon, unpubl. data, 1999, for Gini coefficients and log-normal means). The adequacy of a method

Table 14.4 Lower and upper endpoints of 95% confidence intervals for the Gini coefficient in competitively grown *Ailanthus*

Bootstrap method	95% confidence interval
Percentile	(0.133, 0.171)
Bias-corrected	(0.136, 0.176)
Accelerated	(0.137, 0.176)
Studentized	(0.137, 0.180)
Basic	(0.139, 0.177)

can be evaluated by repeatedly simulating samples from some population, calculating a confidence interval for each sample, and counting the number of confidence intervals that included the true population parameter. Ideally, a 95% confidence interval includes the true parameter in 95% of the samples. However, studentized bootstrap confidence intervals are usually longer than the other confidence intervals. In some cases, the studentized intervals are ridiculously long, so some people prefer the percentile or accelerated methods (e.g., Efron and Tibshirani 1993).

14.5.3   What Should I Bootstrap?

The key assumption behind the bootstrap is independence. The observations are assumed to be independent samples from some population. The choice of what to bootstrap becomes very important in more complicated problems. For example, consider bootstrapping a regression or correlation coefficient. Bootstrapping a correlation coefficient between two traits on individuals is relatively easy (Efron 1982). If individuals were randomly sampled from some population, then the bootstrap replicates are formed by randomly choosing individuals with replacement from the sample, just as was done for Gini coefficients in my example. Bootstrapping a regression is more complicated because it can be done in two different ways (Efron and Tibshirani 1986, section 5). We can fit a regression, estimate the residuals and predicted values, and generate bootstrap replicates by adding randomly sampled residuals to the predicted values. This procedure estimates the precision of the regression coefficients, assuming that the regression model is correct (Efron and Tibshirani 1986). The other procedure is to randomly select individuals, just like bootstrapping a correlation coefficient. This procedure estimates the precision of the regression coefficients, even if the regression model is not correct (Efron and Tibshirani 1986). This bootstrap can also be used to estimate the precision of model selection procedures like stepwise selection and nonparametric regressions (Efron and Tibshirani 1991). It is also more appropriate when the independent variables are random, not fixed in advance by the investigator.

Bootstrapping is also more complicated when the data include multiple sources of variation (Davison and Hinkley 1997). Although bootstrapping complex survey data has received some attention (Sitter 1992), methods for bootstrapping complex experimental data are not well developed. For example, the zooplankton example had two sources of variation: between samples and between individuals in a sample. Samples could be bootstrapped, or individuals could be bootstrapped. This decision can be made based on what constitutes independent observations. If there is any patchiness in the zooplankton, individuals are unlikely to be independent, so bootstrapping individuals is not appropriate. Bootstrapping samples is more appropriate if the observed samples can be assumed to be random samples from locations in the reservoir. This approach ignores the variability between individuals in a sample.

Consider bootstrapping data with two sources of variation: samples and individuals. One obvious approach is to bootstrap both samples and individuals. First, generate a bootstrap sample by randomly selecting samples with replacement from the collection of samples, then bootstrap individuals (again with replacement) in each one of the selected samples. Theoretical calculations for simple problems (e.g., estimating the mean of all observations) show that this approach tends to overestimate the variance (P. M. Dixon unpubl. data, 1999, using results from Davison and Hinkley 1997, p. 101). Another simple approach is to bootstrap only samples, the highest level in the hierarchy of sources of variation. Theoretical calculations show that this approach underestimates the variance by a consistent factor of $(p-1)/p$, where p is the number of samples. Bootstrap confidence intervals with the correct width can be calculated (Davison and Hinkley 1997, p. 102), but it is not clear how such adjustments would be made in more complex, more realistic problems. Some preliminary work suggests that a generally appropriate approach is to bootstrap at all levels of the hierarchy and use studentized bootstrap to calculate confidence intervals (P. M. Dixon, unpubl. data, 1997). Although the bootstrap estimates are more variable than they "should be," this is corrected by the studentized bootstrap.

14.5.4 Bootstrapping Correlated Data

Another practical issue is how to bootstrap data that are spatially or temporal correlated. The ordinary and balanced bootstraps ignore any correlation and treat all observations as independent. Two approaches can be used to generate bootstrap samples from correlated data. If the correlation structure can be modeled using a time series or spatial correlation model, a model-based bootstrap can generate bootstrap samples that maintain the correlation structure. The process is similar to bootstrapping regression residuals. Fit the time series or spatial model to the data to estimate the correlation parameters and a set of independent residuals, bootstrap the residuals, then use the model to generate a correlated sample (Stoffer and Wall 1991; Davison and Hinkley 1997). This approach assumes that the correct model is used for the correlation structure. If so, this approach works well.

The moving blocks bootstrap is a nonparametric approach to generating bootstrap samples that maintain some of the correlation in the original data. The idea is more clearly explained with time series data. The observed sample is divided into b blocks each with of l sequential observations. A moving blocks bootstrap sample is constructed by randomly sampling the b blocks, with replacement, and concatenating them to form a bootstrap sample of bl observations. The idea is that the correlation among observations is strongest within each block of l observations and that observations in different blocks are (or are almost) independent. The choice of b and l are crucial. If b is too small, there are very few possible patterns of observations in the bootstrap samples. If l is too small, observations in different blocks are no longer independent. These and other practical details implementing the moving blocks bootstrap and moving tiles bootstrap for spatial data are discussed in Davison and Hinkley (1997).

14.5.5 How Many Bootstrap Replicates Should I Use?

Increasing the number of bootstrap replicates increases both the precision of the estimated standard error or confidence interval and the cost of computing it. Efron (1987, section 9) recommends using approximately 100 bootstrap replicates to estimate a variance or standard error and 1000 or more to estimate a confidence interval. Variances can be estimated more precisely because the variance is an average of squared deviations, whereas the endpoints of the confidence interval are individual data points. If the bias-corrected or accelerated techniques are used, more bootstrap replicates are necessary. For 95% confidence intervals, about 50% more (i.e., 1500) bootstrap replicates should be used (Efron 1987, section 9). The extra replicates are necessary because the estimation of the z_0 and a constants introduces extra variability into the endpoints of the confidence intervals.

Although more bootstrap replicates increase the precision of the estimated standard error or confidence interval, there is a limit. Bootstrap estimates have two sources of variability, one due to variability among the bootstrap replicates and one due to sampling variability in obtaining the original data. More bootstrap replicates can not substitute for more observations in the original data.

14.5.6 Diagnostics for Bootstraps

Bootstrap diagnostics are tools to assess whether the bootstrap confidence intervals reflect the entire sample or whether they are heavily dependent on one or a few observations (Efron 1992). One diagnostic method combines the jackknife and the bootstrap: remove points one at a time and calculate a bootstrap confidence interval from each reduced data set. If all the bootstrap confidence intervals are similar, then the bootstrap results are indeed summarizing all the observations. If one or a few bootstrap confidence intervals are very different from the rest, then the bootstrap is heavily dependent on the associated observations. Further details and some other diagnostics are illustrated in Davison and Hinkley (1997, pp. 113–120) and Efron and Tibshirani (1993, pp. 271–280).

14.5.7 Computing

Both the jackknife and the bootstrap are easy to use. All that is required is some way to draw random samples of observations and compute the desired estimates from multiple data sets. These computations can be programmed into many statistical packages (e.g., Dixon 1993), but macros and functions to simplify the process are increasingly available. Some of the more complete implementations include the SAS JACKBOOT macro (SAS Institute 1995), the boot library for S-Plus (Davison and Hinkley 1997), and the bootstrap functions for S-Plus (Efron and Tibshirani 1993).

14.5.8 Testing Hypotheses

The focus in this chapter has been statistical estimation, especially calculating confidence intervals, because estimation is generally more useful than hypothesis testing (Gardner and Altman 1986; Salsburg 1985). However, the bootstrap procedure can be used to test hypotheses by estimating the sampling distribution of a test statistic under some null hypothesis, but, considerable care is required (Hall and Wilson 1991; Tibshirani 1992; Becher 1993). Solow and Sherman (1997) illustrate the difficulty in simulating data from ecologically relevant null hypotheses. A related computer-intensive technique, Monte Carlo randomization (see chapter 16) can also be used to test simple statistical hypotheses.

14.5.9 When Does the Bootstrap Fail?

The bootstrap method does not always work (Chernick 1999). Although it provides standard errors and confidence intervals for many problems that are otherwise intractable, it can fail because of asymptotic properties, inherent inaccuracy, or wild data. In "well-behaved" problems, the bootstrap (in any flavor) has many desirable asymptotic properties (characteristics of the procedure as the sample size gets very large). One of these is that the bootstrap sampling distribution converges to the true sampling distribution sufficiently quickly. For some problems described in Shao and Tu (1995, section 3.6), the bootstrap "fails" because it does not converge on the true sampling distribution "fast enough."

Users of the bootstrap should be more concerned about types of problems where the bootstrap is inherently inaccurate. These are problems where some (but not necessarily all) types of bootstrap will give incorrect answers because of the characteristics of the problem. One important ecological problem that requires the use of caution is the estimation of species richness. Consider sampling N individuals (or N quadrats) and counting the total number of species observed. This is likely to be an underestimate of the true number of species in the population, because some species (especially rare species) were not sampled. However, bootstrap samples of N individuals or N quadrats will never include more species than were observed in the original data. Hence, percentile bootstrap confidence intervals are inherently incorrect. Other cases that require careful application include prediction of maxima or minima (Bickel and Freedman 1981), smoothing and other nonparametric regression methods (Davison and Hinkley 1997), kernel density estimation (Léger et al. 1992), and some multivariate problems (Milan and Whittaker 1995).

Finally, a bootstrap may fail because of "wild" data. All the bootstrap methods use the sample data to reconstruct the properties of the underlying population. If the sample is unusual in some way, then the bootstrap confidence intervals based on that sample will be unusual. The parametric bootstrap methods require only that the parameters estimated from the sample data are close to the true population parameters, but the nonparametric bootstrap methods require that the entire distribution of values in the sample is close to that in the population. One numerical illustration of this problem is given in the documentation for the SAS

JACKBOOT macro (SAS Institute 1995). The only solution is to have a suitably large sample of data. The criteria for a suitably large sample depends on the problem and the characteristics of the population.

Unusual samples and "wild" data are likely to be a serious problem when the population of values has a very skewed distribution. Bootstrap confidence intervals for Gini coefficients from lognormal distributions do not have the appropriate coverage because lognormal distributions with large variance parameters are very skewed. Similar problems can be demonstrated using a simpler example: the mean of a lognormal distribution. It is quite dependent on infrequent very large values. For example, the true mean of a lognormal ($\mu = 0$, $\sigma^2 = 3$) distribution is 4.48. If the largest 1% of values are removed, the mean of the remaining 99% values is only 3.27. Bootstrap confidence intervals for the mean have relatively poor coverage (P. M. Dixon, unpubl. data, 2000) because samples of 20, 50, or even 100 observations are unlikely to include very large values.

Both the jackknife and the bootstrap solve an otherwise extremely difficult problem: estimating the precision of a complicated statistic. However, the accuracy of inferences made by either technique is limited by the amount of information in the original data. Neither technique is a panacea for small sample size or poor study design.

Acknowledgments   Preparation of the first version of this manuscript was supported by contract DE-AC09-76-SROO-819 between the U.S. Department of Energy and the University of Georgia. Alexandra Webber drafted figures 14.1–14.3.

15

Spatial Statistics

Analysis of Field Experiments

JAY M. VER HOEF

NOEL CRESSIE

15.1 Field Experiments

More and more, ecologists are turning to designed field experiments. Earlier in the history of ecology, it was enough to collect field observations to generate reasonable hypotheses (McIntosh 1985). However, as these hypotheses multiplied, they needed to be rigorously tested. A glance at any current ecological journal reveals that many ecologists are designing and analyzing field and laboratory experiments to test such hypotheses.

There are several difficulties associated with conducting field experiments. One is that they are usually expensive. Another is that true replication may be unattainable (e.g., Carpenter 1990). Finally, there is often considerable "noise" in the data, both because the environment is heterogeneous at many scales (e.g., Dutilleul 1993; Legendre 1993), and because field measurements are often crude compared to those achieved under laboratory conditions, resulting in greater measurement error in field studies (Eberhardt and Thomas 1991). Even at smaller scales, it can be difficult to find relatively homogeneous areas. All of these factors contribute unwanted variability to the experiment. Hence, ecologists typically use as many experimental units as they can afford and hope for the best. Too often, natural variability simply swamps many of the treatment effects that we try to detect.

In light of the need for ecological experiments and the expense and difficulty in conducting them, it is imperative that the designs and analyses of field experiments maximize the ability to detect treatment effects. One of the common misuses of statistics is the use of less powerful techniques when more powerful techniques are available. Most ecological field experiments are spatial in nature, yet spatial information is not used in a classical analysis of variance (ANOVA; chapter 4). This chapter reviews the generalized-least-squares–variogram method and its heuristic value. In addition, the methods of spatial maximum-likelihood (ML) estimation and spatial restricted maximum-likelihood (REML) estimation are introduced. Each of the spatial methods use the underlying spatial variation to estimate treatment contrasts (contrasts will be described in greater detail in section 15.3.2) with greater precision than estimates using classical ANOVA.

The example used in this chapter is a designed experiment that examines the effect of time of burn on plant species diversity. The data consist of the numbers of different vascular plant species, or species richness, in a 5×5 grid of 7×7-m contiguous plots (figure 15.1A). The data come from a glade in the Ozark Plateau area of southeastern Missouri. Glades are grassy openings, usually caused by shallow and droughty soils, in a predominantly forested landscape (Kucera and Martin 1957). This particular glade is on a relatively steep slope of 14% (based on the average of 25 measurements from the center of each plot) and has a dolomitic substrate that often weathers into exposed bedrock "benches" in a stairstep fashion. These 25 plots came from the center of the glade and are as homogeneous as possible from this particular glade. There were scattered small trees in the plots, but the main forest vegetation was at least 14 m from the edge of the plots in each direction.

Actually, the number of different plant species per plot was recorded from the plots before any fire treatments, so the underlying natural variability was obtained without any treatments applied (called a uniformity trial in the statistical literature). The overall mean was 24.08 species per plot. Treatment effects were artificially added to the real data to simulate a real experiment for the purposes of demonstration and to evaluate the classical ANOVA and spatial analysis methods. A completely randomized design was employed (chapter 4). Thus, according to table 15.1, we chose the number of species to change by +6 if the plot were randomly assigned a November burn, we chose the number of species to change by −3 if a plot were randomly assigned a May burn, and so on. Therefore, the treatment effect of a November burn is $\tau_4 = +6$, that of a May burn is $\tau_4 = -3$, and so forth. The reason for simulating the experiment, rather than actually conducting it, is to have known treatment effects. In a real experiment, the treatment effects are superimposed on the natural variability among the plots and hence are unknown parameters to be estimated. By analyzing the experiment and estimating the parameters (treatment effects) as if they were unknown, the "closeness" of the treatment effects to the true values can be compared for a classical ANOVA and for the spatial methods introduced in this chapter.

Although the effect of fire on species diversity was chosen to illustrate the spatial analysis of designed experiments, the method is not specific to any particular ecological problem. Spatial analysis may be applied to any field experiment where spatial coordinates are available (in one, two, or three dimensions) and where you would ordinarily use a classical ANOVA. Three-dimensional examples include the assignment of treatments to fish in a lake or to insects in a tree, followed by a spatial analysis.

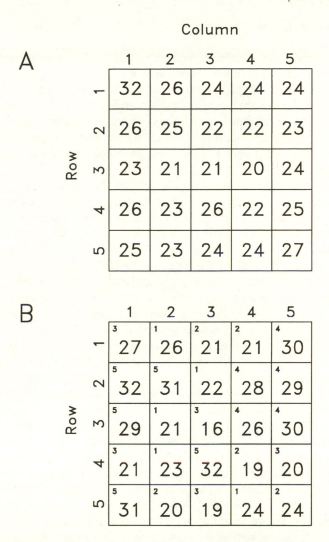

Figure 15.1 (A) Uniformity trial data set of species richness in a 5×5 grid of plots; each plot was 7×7 m. (B) Uniformity trial data set with treatment effects from table 15.1 added randomly. The treatment number is given in the upper left corner of each plot.

Table 15.1 Treatment effects for simulated data

Treatment	Treatment Symbol	Treatment Effect
Control	τ_1	0
May burn	τ_2	−3
August burn	τ_3	−5
November burn	τ_4	+6
February burn	τ_5	+6

15.2 Statistical Issues

A field experiment has spatial components; that is, the experimental units are located in one-, two-, or three-dimensional space. This spatial environment is usually heterogeneous. Typically, locations that are close together tend to have more similar values, or are more positively correlated, than those that are farther apart; this tendency is termed spatial autocorrelation. (It is also possible to have temporal autocorrelation, often manifest as a time series; chapter 9.) A cursory inspection of the data (figure 15.1A) seems to indicate that plots closer together have more similar values.

Well-designed experiments classically rely on three basic concepts: randomization, blocking, and replication (chapter 4). In general, an experimental unit is the unit to which treatments are applied according to some design. In the example from the Ozarks (figure 15.1A), the experimental units are 7×7-m plots, to which fire treatments are applied. Replication is obtained by the repeated application of the set of treatments to all of the experimental units. Replication allows us to estimate treatment effects by averaging over the underlying variability in the experimental units (see discussion in chapter 4). Blocking helps control natural heterogeneity by assigning experimental units to relatively homogeneous groups. Blocks may consist of experimental units that are spatially close or related in some other way. For example, if a field has a shallow center drainage, it might be blocked into those areas that are higher and drier and into those that are lower and wetter. Randomization is the process of assigning treatments, at random according to some design, to experimental units. Randomization helps to provide unbiased estimates of parameters.

Next we discuss a subtle concept. Suppose that the pretreatment responses on the experimental units are themselves random variables. For example, if we were to go back in time, to, say, 10 years before the Ozark data were collected and let time start over again, the number of species in each plot would be slightly different as a result of the random or stochastic processes of nature. If we could generate species numbers for the same set of plots again and again by going back in time repeatedly, then we would obtain a statistical distribution for the number of species in each plot, as well as autocorrelation values among all pairs of plots. In this case, the data from the experimental units are the result of what is termed a spatial stochastic process. The alternative view is that if we went back in time again and again, we would always get exactly the same number of species per plot—nature is inherently deterministic. In the latter case, all of the "randomness" in an experiment comes from assigning treatments at random to experimental units. In other words, all of the probability statements (e.g., P-values, tests of significance, confidence intervals) must come only from the randomness in the *design* of the experiment.

If nature is inherently stochastic, a randomized design has two sources of randomness: the design and the experimental units. These experimental units may be autocorrelated. Even Fisher (1935), the father of modern experimental design based on randomization, noted "After choosing the area we usually have no guidance beyond the widely verifiable fact that patches in close proximity are commonly alike, as judged by the yield of crops, than those which are far apart." Randomization helps neutralize the effect of autocorrelation (Yates 1938; Grondona and Cressie 1991; Zimmerman and Harville 1991). Intuitively, however, there is a cost to randomization. If the experimental units are considered to be random variables with their own natural variability, then randomization introduces more variability. This additional variability allows us to use classical theory, which assumes the random variables are independent.

An alternative approach to classical ANOVA is to *model* the spatial autocorrelation among the experimental units, making possible a more powerful analysis of the experiment. This is a relatively new approach to analyzing designed experiments in a spatial setting, and several recent articles show it is very powerful in a variety of data sets (Baird and Mead 1991; Cullis and Gleeson 1991; Grondona and Cressie 1991; Zimmerman and Harville 1991). In this chapter, we model the spatial autocorrelation with variograms that contain information on the spatial autocorrelation among the experimental units.

15.3 Example

Suppose we conducted an experiment to examine the effect of fire, at different times of the year, on species richness in the Ozark glade (figure 15.1A). Also suppose that there were four treatments where plots were burned at four different times of the year: May, August, November, and February. Species richness was measured in the summer following the February burn. There was also a control, with no burn, so there were a total of five treatments. For illustration purposes, suppose that the true effects of burn time are those given in table 15.1. These treatment effects allow our artificial experiment to be conducted as follows. To the 25 original plots (figure 15.1A), 5 replications of each treatment were assigned randomly (figure 15.1B). The only information available to the scientist is the plot location, the treatment applied to each plot, and the datum at each plot (with treatment effect added), in other words, the information in figure 15.1B. Hence, we will analyze the data of figure 15.1B with a classical ANOVA, the GLS-variogram method, spatial ML estimation, and spatial REML estimation.

15.3.1 Statistical Model and Assumptions

The following statistical model summarizes the experiment:

$$Y_{ijk} = \tau_k + \delta_{ij} \tag{15.1}$$

where Y_{ijk} is a random variable (species richness), i is the ith row ($i = 1, \ldots, 5$, numbered from top to bottom), j is the jth column ($j = 1, \ldots, 5$, numbered from left to right), and k is the kth treatment ($k = 1, \ldots, 5$); τ_k is the kth treatment effect, and δ_{ij} is the random error for the (i, j)th plot, with possible spatial autocorrelation among the δ_{ij}. The statistical distribution of δ_{ij} provides probability statements about estimates of treatment effects or combinations of treatment effects (e.g., τ_1 or $\tau_1 - \tau_2$). For a classical ANOVA, all δ_{ij} are assumed to be independent, normally distributed, have zero expectation, and have constant variance (σ^2).

The randomization distribution yields random errors δ_{ij} that have small autocorrelation, hence allowing us to use classical theory (Grondona and Cressie 1991), as confirmed by several simulations (Besag and Kempton 1986; Baird and Mead 1991; Zimmerman and Harville 1991). However, only one randomization actually occurs. If we ignore all possible randomizations and just concentrate on the one that occurred, then natural variation in the plots implies that the δ_{ij} are autocorrelated (see section 15.3.3). The problem then is how to model the autocorrelation and subsequently use this information to obtain the best parameter estimates and associated probability statements. This problem will receive further attention in section 15.3.4.

15.3.2 Treatment Contrasts

Often, the ultimate goal of an experiment is to estimate a treatment contrast (sometimes called a planned comparison). If the initial hypothesis—no significant difference among treatment means—is rejected, the next step is to ask, How different are the treatment means? (Snedecor and Cochran 1989, p. 224; Hicks 1982, p. 46; Day and Quinn 1989). Treatment contrasts express the difference in treatment effects, or the difference of a combination of treatment effects. For the Ozark example, we might be interested in how a summer burn affects the subsequent summer's species diversity. In terms of the treatments listed in table 15.1, this is expressed as the average of May and August burns minus the control. Using the mathematical notation from table 15.1, this contrast becomes

$$c_1 = \frac{1}{2}(\tau_2 + \tau_3) - \tau_1 \tag{15.2}$$

From table 15.1, the true value is

$$c_1 = \frac{1}{2}(-3 - 5) = -4$$

This means that the actual effect of a summer burn was to decrease by 4 the number of species per plot in the subsequent summer. In the case of c_1, the multiplying coefficients for all treatment effects τ_i, $i = 1, \ldots, 5$, are $\{-1, 0.5, 0.5, 0, 0\}$. It is characteristic of treatment contrasts that the coefficients sum to zero. Several more contrasts, and their true values, are given in table 15.2.

Next, we will see how a classical ANOVA compares to the spatial methods for estimating the five treatment contrasts in table 15.2. Recall that the known treatment effects (table 15.1) were added to the original, untreated data (figure 15.1A) in a completely randomized design (figure 15.1B). Now the data in figure 15.1B will be used to estimate contrasts with classical ANOVA, GLS-variogram, spatial ML estimation, and spatial REML estimation. Because the experiment was conducted artificially, the estimated values can be compared to the known values.

Table 15.2 Contrast estimates for example data (figure 15.1B)[a]

Contrast	True Value	OLS est.	GLS est.	Iter1 GLS est.	ML est.	REML est.
$c_1 = (\tau_2 + \tau_3)/2 - \tau_1$	−4.00	−2.40	−2.80*	−2.82*	−2.89*	−2.95*
$c_2 = (\tau_4 + \tau_5)/2 - \tau_1$	6.00	6.60*	6.69*	6.71*	6.76*	6.81*
$c_3 = (\tau_4 + \tau_5)/2 - (\tau_2 + \tau_3)/2$	10.00	9.00*	9.49*	9.53*	9.65*	9.77*
$c_4 = (\tau_2 - \tau_3)$	2.00	0.40	0.45	0.46	0.49	0.53
$c_5 = (\tau_4 - \tau_5)$	0.00	−2.40	−2.07	−2.04	−1.98	−1.94

[a]The true values are given first. The next column contains the classical ANOVA estimate (ordinary least squares est). The next two columns contain contrast estimates from the GLS-variogram method for two iterations(GLS est. and Iter1 GLS est.). Spatial maximum-likelihood contrast estimates are denoted by ML est., and spatial restricted maximum-likelihood contrast estimates are denoted by REML est. Contrast estimates that test as significantly different from 0, at $P < 0.05$, are marked with an asterisk.

15.3.3. Comparing the Methods

Before giving details on the spatial methods, we provide reasons for their use by comparing them with classical ANOVA. For each method, contrast estimates are shown in table 15.2 and their estimated standard errors in table 15.3. For a classical ANOVA, table 15.2 shows that the null hypothesis of zero would be rejected at $P < 0.05$ for only the second and third contrasts. For example, to test whether the first contrast is significantly different from zero, the t-value is $(-2.4 - 0)/1.29 = -1.86$. Because

$$|-1.86| < t_{\alpha=0.025, df=20} = 2.07$$

the first contrast is not declared to be significantly different from zero for a two-tailed test at significance level $\alpha = 0.05$. Of course, we would not want to reject the hypothesis that contrast c_5 equals zero (but we will, 1 out of 20 times on average, when $\alpha = 0.05$). Therefore, of the four contrasts that had true values not equal to zero, the classical analysis had enough power to detect only two of them.

Table 15.3 Contrast standard errors for example data (figure 15.1B)[a]

Contrast	OLS s.e.	GLS s.e.	Iter1 GLS s.e.	ML s.e.	REML s.e.
$c_1 = (\tau_2 + \tau_3)/2 - \tau_1$	1.29	0.96	0.94	0.83	0.87
$c_2 = (\tau_4 + \tau_5)/2 - \tau_1$	1.29	1.09	1.07	0.97	1.05
$c_3 = (\tau_4 + \tau_5)/2 - (\tau_2 + \tau_3)/2$	1.05	0.88	0.86	0.78	0.84
$c_4 = (\tau_2 - \tau_3)$	1.49	1.15	1.13	1.00	1.07
$c_5 = (\tau_4 - \tau_5)$	1.49	1.60	1.60	1.50	1.68

[a]The first column contains the classical ANOVA estimate of standard errors for the contrasts (OLS s.e.). The next two columns contain estimated contrast standard errors from the GLS-variogram method for two iterations (GLS s.e. and Iter1 GLS s.e.). Spatial maximum-likelihood standard error estimates are denoted by ML s.e., and spatial restricted maximum-likelihood contrast standard errors are denoted by REML s.e.

With the GLS-variogram method, with each iteration, the contrast estimates generally migrate from the classical ANOVA contrast estimates toward the true contrast values, and the standard errors decrease in size. Further iterations only changed the estimates past the second decimal place. The estimated standard errors reflect the fact that the contrast estimates using the GLS-variogram method are better than the classical contrast estimates. For the contrast estimates using the GLS-variogram method, the null hypothesis of zero contrast values can be rejected at $\alpha = 0.05$ for the first three contrasts. For example, to test whether the first contrast c_1 is significantly different from zero, the Z-value is $(-2.8 - 0)/0.96 = -2.92$. Because

$$|-2.92| > Z_{\alpha/2=0.025} = 1.96$$

the first contrast is declared to be significantly different from zero at $\alpha = 0.05$. In fact, the true value is -4. The inference procedure for the GLS-variogram method is only approximate. We defer further discussion on distributions until section 15.3.4. Alternatively, a 95% confidence interval for c_1 is

$$-2.8 \pm (1.96)(0.96) = (-4.68, -0.9)$$

which does not contain the value 0.

For the spatial ML, the estimates generally move even closer to the true values than the GLS-variogram method. The estimated standard errors indicate that the contrast estimates using ML are better than the estimates from ANOVA and the GLS-variogram. For ML, the null hypothesis of zero contrast values can be rejected at $\alpha = 0.05$ for the first three contrasts using the Z-value in the same way as we did with the GLS-variogram method.

Of all the methods, the REML estimates generally move the closest to the true values. However, notice that the standard errors for REML are larger than for ML and GLS-variogram. We discuss the reasons for this subsequently.

In general, the spatial methods have more power than the classical ANOVA to detect contrast c_1. Also, the absolute values of the true contrast values increase by increments of two (table 15.2). Basically, the classical ANOVA had enough power to detect contrasts with absolute values greater than or equal to six, whereas all three spatial methods had enough power to detect contrasts with absolute values greater than or equal to four. No methods had enough power to detect contrast c_4 with a magnitude of two, and no methods committed the Type I error of falsely rejecting the null (true) hypothesis that contrast c_5 is equal to zero.

It is interesting and instructive to compare the standard error estimates of contrasts c_4 and c_5 for both classical ANOVA and the spatial methods. For c_5, the spatial methods give a higher estimated standard error than classical ANOVA (table 15.3). Inspection of figure 15.1B shows why. The randomized assignments for treatments 4 are clustered in the upper right, whereas those for treatments 5 are clustered toward the left. The spatial methods account for the "poor" randomization for contrast c_5; that is, treatment effects may be confounded with local variation. On the other hand, for contrast c_4, treatments 2 and 3 are dispersed and intermixed throughout the plots—a "good" randomization (Hurlbert 1984)—and, consequently, the standard errors estimated from the spatial methods are lower than those from classical ANOVA. Because classical ANOVA relies on the average over all randomizations, it does not recognize the singular features of this particular randomization and hence does not differ in the standard errors for the two contrast estimates. It is also apparent that contrast c_5 is farther from its true value than contrast c_4 is from its true value, which is reflected in the standard errors of the spatial methods but not in the classical ANOVA standard error.

15.3.4 Statistical Methods

Parameter estimation presents several problems. Two groups of parameters can be discussed in the context of the linear model (for more discussion, see Ver Hoef et al. 2001). Equation 15.1 can be written for all of the data as a set of equations in matrix notation,

$$\mathbf{y} = \mathbf{X}\boldsymbol{\beta} + \boldsymbol{\delta} \tag{15.3}$$

where $\mathbf{y}$ is the data vector, $\mathbf{X}$ is the design matrix containing 0's and 1's, $\boldsymbol{\beta}$ is the parameter vector $(\tau_1, \tau_2, \tau_3, \tau_4, \tau_5)'$, and $\boldsymbol{\delta}$ contains the random errors. We assume that the random errors are normally distributed, with $E(\boldsymbol{\delta}) = \mathbf{0}$ and $\mathrm{var}(\boldsymbol{\delta}) = \Sigma_\theta$, where θ is a vector of the covariance parameters that defines Σ_θ.

The two groups of parameters we wish to estimate are $\boldsymbol{\beta}$ and θ. In the case of classical ANOVA, θ contains a single parameter, σ^2, for the variance, and $\Sigma_\theta = \sigma^2 \mathbf{I}$, where $\mathbf{I}$ is the identity matrix. In this case, it turns out that $\boldsymbol{\beta}$ can be estimated without knowledge of σ^2; the estimate is the ordinary least-squares solution $\hat{\boldsymbol{\beta}} = (\mathbf{X}'\mathbf{X})^{-1}\mathbf{X}'\mathbf{y}$. However, $\mathrm{var}\,(\hat{\boldsymbol{\beta}}) = \sigma^2(\mathbf{X}'\mathbf{X})^{-1}$, so we must estimate σ^2 to assess the uncertainty in $\hat{\boldsymbol{\beta}}$. There are two obvious estimators of σ^2. Let the sum of squares error be $\mathrm{SSE} = (\mathbf{y} - \mathbf{X}\hat{\boldsymbol{\beta}})'(\mathbf{y} - \mathbf{X}\hat{\boldsymbol{\beta}})$. Then the maximum-likelihood estimator of σ^2 is SSE/n. However, this estimator is biased. The unbiased estimator is $\mathrm{SSE}/(n - p)$, where p is the number of linearly independent rows in $\mathbf{X}$ (equal to the number of parameters in $\boldsymbol{\beta}$ if the model is not overparameterized); in our example, $p = 5$. It turns out that this is an REML estimator. Of course, as n gets large, the differences become small.

If we wish to make an inference on $\hat{\boldsymbol{\beta}}$, or some function of $\hat{\boldsymbol{\beta}}$ such as a linear contrast, then we must know something about the distribution of $\hat{\boldsymbol{\beta}}$. When σ^2 is known and we divide each element in $\hat{\boldsymbol{\beta}}$ by its standard error, we obtain a standard normal distribution (Z-distribution) that can be used to test hypotheses. However, when σ^2 is unknown (the usual case) and we divide each element in $\hat{\boldsymbol{\beta}}$ by its estimated standard error, we obtain a t-distribution that can be used to test hypotheses. The t-distribution has slightly heavier tails than the Z-distribution to account for the fact that σ^2 is estimated.

Now we generalize this discussion to the spatial setting. Again, we must estimate the two groups of parameters $\boldsymbol{\beta}$ and θ. For the spatial setting, θ typically contains several parameters where the values of Σ_θ depend on the spatial relationships among the data. For spatial models, estimation of $\boldsymbol{\beta}$ depends on θ, and the generalized least-squares estimate is $\hat{\boldsymbol{\beta}} = (\mathbf{X}'\Sigma_\theta^{-1}\mathbf{X})^{-1}\mathbf{X}'\Sigma_\theta^{-1}\mathbf{y}$ and $\mathrm{var}\,(\hat{\boldsymbol{\beta}}) = \sigma^2(\mathbf{X}'\Sigma_\theta^{-1}\mathbf{X})^{-1}$. Again, we can consider maximum-likelihood and REML estimators, but we cannot, in general, write down their explicit forms. Instead, we use numerical solutions such as those found in the procedure MIXED in SAS (Wolfinger et al. 1994). As was the case previously, for small sample sizes, spatial ML estimation is biased for θ (Mardia and Marshall 1984) and spatial REML estimation is less biased. As for classical ANOVA, ML estimation typically underestimates the variability (see also Ver Hoef et al. 2001). That is why the standard errors for the GLS-variogram method and ML estimation are smaller than the standard errors for REML in table 15.3, even though REML estimators are usually closer to the true values (table 15.2).

As for classical ANOVA, if θ is known for the spatial models, then dividing each element in $\hat{\beta}$ by its standard error yields a standard normal distribution (Z-distribution) that can be used to test hypotheses on functions of $\hat{\beta}$, such as linear contrasts. However, when θ is unknown, then dividing each element in $\hat{\beta}$ by its estimated standard error does not necessarily yield a t-distribution like it did with classical ANOVA. There are two main approaches to this problem. One is to use the t-distribution anyway, hoping that the fatter tails in the distribution account for the fact that θ is estimated rather than known. This is the approach SAS uses in the procedure MIXED. The other approach is to further inflate the variance of estimating $\hat{\beta}$ to account for the fact that θ is estimated (e.g., Harville 1985; Prasad and Rao 1990; Cressie 1992; Zimmerman and Cressie 1992; Ghosh and Rao 1994) and use the Z-distribution. However, we are unaware of any software that does this.

15.3.5 The GLS-variogram Method

The GLS-variogram method will be used to illustrate some of the principles of the spatial methods [see http://www.oup-usa.org/sc/0195131878/ for the computer code]. From equation 15.1, we can see that if τ_k is known, we can calculate δ_{ij}: $\delta_{ij} = Y_{ijk} - \tau_k$. The autocorrelation can be estimated by fitting a variogram based on all δ_{ij}. The variogram is a function that contains information on the spatial autocorrelation among the experimental units; more details are given subsequently. On the other hand, if the distribution and autocorrelation for all δ_{ij} are known, then statistical methods that rely on the variogram can be used to estimate τ_k optimally (Cressie 1991, p. 328). So we can envision an iterative approach. In step 1, a classical ANOVA is performed to estimate τ_k. Then, the δ_{ij} are estimated with the residuals from the classical ANOVA, $\hat{\delta}_{ij} = R_{ij}$, where

$$R_{ij} = Y_{ijk} - \hat{\tau}_k \tag{15.4}$$

and $\hat{\tau}_k$ is the classical ANOVA estimate (the average for the kth treatment). Next, a variogram is modeled from the R_{ij}. The variogram is then used to obtain better estimates of τ_k. The whole procedure goes through another iteration by starting with $R'_{ij} = Y_{ijk} - \hat{\tau}'_k$, where $\hat{\tau}'_k$ is the new estimate of τ_k. A detailed description of the GLS-variogram method was given in the previous edition of this book. Computer programs are available on the Website. Here, we briefly describe the method.

The results for the Ozark data are given in table 15.2 and figure 15.2. If the parameters τ_k have been estimated well, the residuals (figure 15.2B) should be close to the original data (figure 15.1A) with the overall mean subtracted from it (figure 15.2A). Figure 15.2B does resemble, reasonably well, the spatial patterning in figure 15.2A. That is, the residuals contain spatial patterning that may be modeled as autocorrelated.

The GLS-variogram method should yield better parameter estimates for the following reason. Suppose that by chance, due to the completely randomized design, several of the values of τ_k were assigned to plots close together; for example, refer to treatment 5 in plots (2,1), (2,2), and (3,1) of figure 15.1B. Then, rather than estimating τ_5 with the simple average of all plots assigned treatment 5, a weighted average is used that gives smaller individual weights to those plots close together because they are likely to have similar values. That is, they have little extra information beyond that of a single plot in the same local region. On the other hand, a plot assigned treatment 5 that is spatially isolated from the other plots assigned treatment 5, for example, plot (4, 3) in figure 15.1B, would get a higher weight because it "represents" a larger region. The optimal weights are obtained through formulas that use the semivariogram.

To use the GLS-variogram approach, the semivariogram of residuals must be estimated. The semivariogram can be used to estimate contrasts as shown by Cressie (1991, p. 328). To estimate the semivariogram from the residuals, it is necessary for the residuals to exhibit approximate *intrinsic stationarity* (Matheron 1963; Journel and Hiujbregts 1978; Cressie 1991). Simply stated, this means that if the spatial data were collected repeatedly, the values at all locations would average toward some constant value and that the variogram between two sites does not depend on their exact locations, only on the relative displacement between them. We will assume further that the variogram actually depends only on the *distance* between any pair of locations, which is called the *isotropy assumption*. These assumptions are impossible to test, because it is impossible to go back in time again and again and generate the experiment each time to check whether each experimental unit has the same mean value or whether the correlation is the same for all pairs of plots that are at some fixed distance from each other. However, any gross spatial trends in the residuals (e.g., high values at one end of the study area gradually shifting to low values at the other end) would cause suspicion that they are not stationary. To check these conditions, use the methods of exploratory data analysis for spatial data (e.g., Cressie 1991, section 2.2; see also chapter 3). The residuals (figure 15.2B) for these data appear satisfactory, except possibly the residual at (1, 1), which seems to be rather large (in absolute value) compared to most others.

For step 2, the empirical semivariogram is used:

$$\hat{\gamma}(h) = \frac{1}{2N(h)} \sum_{M(h)} (R_{ij} - R_{st})^2 \tag{15.5}$$

where R_{ij} is the classical ANOVA residual in the ith row and jth column, $M(h) \equiv \{[(i,j), (s,t)]: [(i-s)^2 + (j-t)^2]^{1/2} = h\}$ (the set of all pairs of data that are at a Euclidean distance of h apart), and $N(h)$ is the number of pairs in $M(h)$. An intuitive way to write equation 15.5 is

$$\hat{\gamma}(h) = (1/2) \text{ average } (R_{ij} - R_{st})^2$$

where R_{ij} and R_{st} are at a distance h from each other; h is often called the lag.

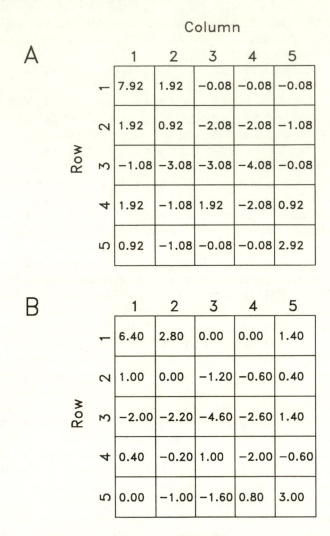

Figure 15.2 (A) Uniformity trial data (figure 15.1A) with the overall mean subtracted from each plot. (B) Residuals from the classical ANOVA.

An example will help illustrate how to use equation 15.5. Let $h = 1$. Then $M(1)$ is the set of all pairs of plots that are 1 unit apart, for example, plots (1, 1) and (1, 2), plots (1, 1) and (2, 1), and so on; there are 40 such pairs in figure 15.2B. Thus, $N(1) = 40$. The empirical semivariogram $\gamma(1)$ is one-half the average over all 40 pairs where each pair is differenced and squared. Figure 15.3A shows that plots close together have more similar residuals than those that are farther apart because $\hat{\gamma}(h)$ is generally smaller for smaller values of h.

Next, a model must be fit to the empirical semivariogram. If the raw empirical values were used, the variance estimate may be negative, which is wrong. There-

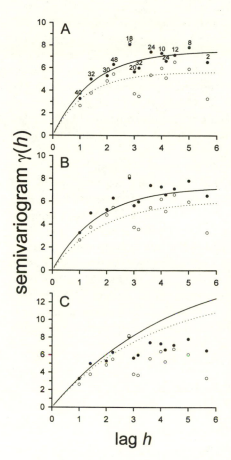

Figure 15.3 Empirical and fitted semivariograms. The solid circles are the empirical semivariogram values (equation 15.5) for the raw data (figure 15.1A) and open circles are the empirical semivariogram values for the residuals of the simulated experiment (figure 15.2B). The numbers above each symbol indicate the number of pairs used to compute each lag of the variogram. The solid line is the fitted semivariogram model (equation 15.6) for the raw data and the dashed line is the fitted semivariogram model for the simulated experiment. (A) GLS-variogram method. (B) Spatial ML estimation. (C) Spatial REML estimation.

fore, in step 3, a model was fit to $\hat{\gamma}(h)$ that satisfies conditional negative-definiteness properties (Cressie 1991, section 2.5). Several models are possible (e.g., Cressie 1991, section 2.3); we chose the exponential model after inspecting a plot of the empirical semivariogram values. Zimmerman and Harville (1991) indicate that the choice of model does not matter greatly as long as it fits the data reasonably well. A spherical model might also have been chosen for these data. But the model "linear with sill" cannot be used for two-dimensional data because it may yield negative variances; however, it may be used for one-dimensional (e.g., transect) data (Webster 1985). Webster (1985) and Cressie (1991, section 2.3) mention several "safe" models that can be used for spatial data in one, two, or three dimensions.

For step 3, the exponential semivariogram model was fit to the empirical semivariogram. The exponential semivariogram is

$$\gamma(h) = \begin{cases} \alpha_0 + \alpha_1(1 - \exp(-h/\alpha_2)) & \text{for } h > 0 \\ 0 & \text{for } h = 0 \end{cases} \qquad (15.6)$$

where α_0, α_1, and $\alpha_2 \geq 0$. Notice that h is always nonnegative (because it is distance) and $\gamma(h) = 0$ when $h = 0$. In step 3, nonlinear regression (weighted least-squares) with weights proportional to $N(h)$ (the number averaged for each h in equation 15.5) was used, as recommended by Cressie (1985), to fit the exponential variogram model equation 15.6 to the empirical variogram equation 15.5. The following estimates were obtained: $\hat{\alpha}_0 = 0$, $\hat{\alpha}_1 = 5.633$, $\hat{\alpha}_2 = 1.203$; the fitted curve is given in figure 15.3A. The empirical and fitted semivariogram (figure 15.3A) on the residuals (figure 15.2B) can be compared to the empirical and fitted semivariogram of the raw data (figure 15.1A).

The semivariogram estimates were used to generate a matrix of semivariogram values between all pairs of experimental units. In the final step, the formulas of Cressie (1991, p. 328) that rely on the semivariogram matrix were used. Contrasts and their standard errors were estimated (tables 15.2 and 15.3). For large samples, a test for a nonzero contrast can be obtained as follows. Construct a confidence interval around the contrast estimate by taking the standard error multiplied by some $Z_{\alpha/2}$, and if the confidence interval does not include zero, the contrast is declared significant at that α-level. It is possible to stop here, but, as was mentioned previously, residuals can again be formed from the current estimates of $\hat{\tau}'_k$ by taking $R'_{ij} = Y_{ijk} - \hat{\tau}'_k$ and then starting the procedure over again at step 2. A stopping rule can be chosen, such as when the contrast estimates stop changing at some decimal place.

15.3.6 Maximum-Likelihood and Restricted Maximum-Likelihood Estimation

Recall the multivariate normal distribution,

$$f(\mathbf{y};\mu,\Sigma) = (2\pi)^{-n/2} |\Sigma|^{-1/2} \exp[-\tfrac{1}{2}(\mathbf{y} - \mu)'\Sigma^{-1}(\mathbf{y} - \mu)] \qquad (15.7)$$

This distribution is very general, allowing a separate mean μ_i for each y_i and a separate covariance among all pairs of data. We can adapt equation 15.7 to a spatial model for designed experiments by replacing μ with $\mathbf{X}\beta$ from equation 15.3 and replacing Σ with Σ_θ, where Σ_θ is obtained from equation 15.6 as follows. If the covariance between two locations separated by distance h is denoted $C(h)$, then the semivariogram is $\gamma(h) = C(0) - C(h)$, where $C(0) = \gamma(\infty)$ for variogram models that have sills (Cressie 1991, p. 67). Thus,

$$C(h) = \begin{cases} \alpha_1 \exp(-h/\alpha_2) & \text{for } h > 0 \\ \alpha_0 + \alpha_1 & \text{for } h = 0 \end{cases} \qquad (15.8)$$

From equation 15.8, Σ_θ is controlled by only three parameters: $\theta = (\alpha_0, \alpha_1, \alpha_2)$. Therefore, we can think of equation 15.7 as a function of the unknown parameters β and θ, called the likelihood $L(\beta,\theta;\mathbf{y})$; once the data have been observed, $L(\beta,\theta;\mathbf{y})$ is maximized for β and θ, yielding the maximum-likelihood estimates. Because the analytical solution is intractable, this is done numerically. The following estimates were obtained: $\hat{\alpha}_0 = 0.001$, $\hat{\alpha}_1 = 6.079$, $\hat{\alpha}_2 = 1.787$. The fitted curve is given in figure 15.3B. The empirical semivariogram on the residuals and raw data can be compared to the fitted semivariogram of the experimental and raw data (figure 15.3B). Spatial ML estimation of θ is biased because we are estimating β simultaneously.

The basic idea behind spatial REML is that, by taking an appropriate linear combination of data, we can create a (restricted) likelihood $L(\theta;y)$ that is free from β. Using this likelihood, $L(\theta;y)$, to estimate θ should decrease bias for θ. The theory of REML is beyond the scope of this chapter, but interested readers are referred to Cressie (1991, p. 93) and references therein. The following spatial REML estimates were obtained from the example data (figure 15.1B): $\hat{\alpha}_0 = 0.002$, $\hat{\alpha}_1 = 13.541$, $\hat{\alpha}_2 = 3.733$. The fitted curve is given in figure 15.3C. The empirical semivariogram on the residuals and raw data can be compared to the fitted semivariogram of the experimental and raw data (figure 15.3C). Spatial ML and REML are available in the procedure MIXED in SAS.

15.3.7 Simulation Experiment

To compare the four estimation methods for the five contrasts in table 15.2, a simulation experiment was conducted. Data were randomly generated 2000 times exactly as was done to create figure 15.1B; that is, the true treatment effects were added randomly to the underlying pattern in figure 15.1A. The results are given in table 15.4.

The mean squared error (MSE) is the first category given in table 15.4. For each contrast, the true value was subtracted from the estimated value, the difference squared, and then the squared differences were averaged over all simulations. The smaller the MSE, the closer the estimated value was to the true value,

Table 15.4 Simulation results

		ANOVA	GLS-variogram	Spatial ML Estimation	Spatial REML Estimation
MSE	c_1	1.810	1.166	1.103	1.037
	c_2	1.822	1.183	1.105	1.040
	c_3	1.139	0.766	0.724	0.687
	c_4	2.364	1.546	1.457	1.380
	c_5	2.433	1.608	1.551	1.461
Coverage	c_1	0.9450	0.9440	0.9300*	0.9505
	c_2	0.9495	0.9415	0.9240*	0.9500
	c_3	0.9575	0.9495	0.9365*	0.9560
	c_4	0.9500	0.9390*	0.9250*	0.9490
	c_5	0.9435	0.9385*	0.9215*	0.9465
Power	c_1	0.8255	0.9750	0.9845	0.9825
	c_2	0.9960	1.0000	1.0000	1.0000
	c_3	1.0000	1.0000	1.0000	1.0000
	c_4	0.2155	0.3370	0.4095	0.3560
	c_5	0.0565	0.0615	0.0785	0.0535

*Coverage outside 95% validity limit.

on average. For all contrasts, it is clear that REML gave the best estimation, and all three spatial methods were significant improvements over ANOVA.

The next category in table 15.4 is coverage, which reports the proportion of times the 95% confidence interval covered the true value. For the spatial methods, confidence intervals were developed using a t-distribution with 19 degrees of freedom, as implemented in SAS. Because we used 2000 independent replications, if the confidence intervals are valid, then coverage should follow a binomial distribution and the intervals should approximately cover the true value within the proportions

$$0.95 \pm 1.96\sqrt{(0.95 \times 0.05)/2000} \approx 0.95 \pm 0.01$$

95% of the time. Therefore, we might declare the confidence intervals invalid if the actual coverage was less than 0.94 or greater than 0.96 for all simulations. All such coverages that are outside the validity limits are indicated with (*). Note that spatial ML estimation only covered the true value about 92–93% of the time. The estimated variance is too small, causing confidence intervals that are too short and causing us to reject a true null hypothesis too often. The GLS-variogram also appears to have confidence intervals that are a bit too short, but not as bad as those using spatial ML estimation. Classical ANOVA and spatial REML estimation appear to produce valid confidence intervals.

The final category in table 15.4 is power. During the simulations, we recorded the percentage of times that we could reject the null hypothesis that each contrast was 0. Notice that for c_1 to c_4, we do want to reject the null hypothesis, but for c_5, which was truly 0, we get an indication of the Type I error rate (with α set at 0.05). From table 15.5, it is clear that spatial ML estimation has the greatest power, followed by spatial REML estimation and GLS-variogram, with ANOVA a distant fourth. However, the reason spatial ML estimation has the greatest power is that, as we saw for coverage, it underestimates the variance of the contrasts. Hence, it is not a fair comparison, because it does not stick to the 0.05 Type I error rate (seen for c_5). Given that spatial ML estimation is invalid, spatial REML

Table 15.5 Comparison of assumptions when performing a classical ANOVA versus the spatial analyses introduced in this chapter

Classical ANOVA	Spatial Methods
Responses from experimental units are fixed in value or they are random variables	Responses from experimental units are random variables
Expectation of random errors = 0	Expectation of random errors = 0[a]
Constant variance for random errors	Constant variance for random errors[a]
Independent random errors	Autocorrelated random errors[a]—variogram and covariance only depend on the spatial displacement between errors
Normally distributed random errors[b]	Normally distributed random errors[ba]

[a]Forms the second-order stationarity assumption.
[b]Not strictly necessary for estimation; only necessary for inference, for example, for confidence intervals and tests of hypotheses.

estimation provides a bit more power than GLS-variogram, even though GLS-variogram seems to slightly underestimate its variability.

These simulations indicate that spatial REML estimation may be the best choice for the spatial analysis of designed experiments. Spatial REML estimation has the smallest MSE, the confidence intervals and tests of hypotheses are valid, and it provides the greatest power among valid tests. Of course, we worked with only one spatial pattern (figure 15.1A) and a single set of true treatment effects (table 15.1), so we do not wish to make sweeping generalizations. However, others have found that spatial REML estimation often outperforms other methods (e.g., Zimmerman and Harville 1991; Ver Hoef et al. 2001), and it is the default method of procedure MIXED in SAS.

15.4 Discussion

This chapter showed how a spatial analysis from a classically designed field experiment can increase the precision of estimating treatment contrasts. It is worthwhile to discuss all of the assumptions for both methods here (table 15.5). There are two main differences. First, as was mentioned in section 15.2, for a spatial analysis, it is assumed that the data are the result of a spatial stochastic process with possible autocorrelation. We believe that this is a realistic assumption for most ecological problems; in fact, ecologists also associate the word "process" to spatial pattern, beginning with Watt (1947). An advantage of classical ANOVA is that the data may be assumed either to be the result of a spatial stochastic process or to be fixed in value. The second main difference is that, for the spatial methods, the random errors need not be assumed independent, only second-order stationary, which allows more precise contrast estimates. We also note that methods are being developed for nonnormal data, such as extending generalized linear models to the spatial setting (e.g., Gotway and Stroup 1997).

The example presented in this chapter consisted of only 25 plots, which kept the illustration simple. This is the minimum number of plots we consider necessary to do an adequate job estimating the variogram and covariances. Spatial REML performed well in our simulations with 25 plots, but small-sample properties and more extensive simulations are lacking. We recommend having more experimental units if possible—as usual, the more the better. Also, to get residuals that reflect the underlying spatial variability, it is important that the initial (i.e., classical) parameter estimates are fairly accurate. Their accuracy depends on the number of replications of each treatment, and we recommend a minimum of about five replications; again, the more the better.

This chapter can be compared to the Mantel method of correcting for spatial effects described in chapter 16. There are two main differences between the spatial methods we present and the Mantel method. First, the Mantel test is nonparametric in comparison to this one with no assumptions about the distribution of variables or the form of their autocorrelation (e.g., the variogram). Here, a covariance or variogram model must be chosen and the random errors must be roughly normally distributed. No formal comparison of the two methods was conducted.

However, we might expect the usual results when comparing nonparametric and parametric methods. The parametric method will be more powerful if the data follow the assumed model, whereas the nonparametric method gives some protection against incorrect model assumptions. The second main difference from the Mantel test is that, here, we are estimating treatment contrasts. That is, we go beyond the question of "Are there differences among treatment effects?" to "What are the specific differences among treatment effects?"

In general, spatial methods that account for autocorrelation appear to be quite robust. Several authors have carried out simulations and tried spatial methods on a variety of data sets, with good results. Zimmerman and Harville (1991) used spatial REML estimation. They examined three different data sets with a total of 11 different blocking configurations and found that spatial REML estimation reduced estimation variance between one-fifth and four-fifths of that of a classical analysis. This translates to much more powerful tests of hypotheses and shorter confidence intervals. Grondona and Cressie (1991) took one large data set, broke it into six subsets, and used a robust semivariogram estimator (Cressie and Hawkins 1980). Their results indicated reduced estimation variance to 75% that of a classical analysis. We point out that it is also possible to use the spatial methods for blocked designs (chapter 4). Both Zimmerman and Harville (1991) and Grondona and Cressie (1991) examined similar spatial methods for blocked designs. In this chapter, we used a completely randomized design to keep the illustration simple.

Using a slightly different methodology of blocking by columns and using a time series type of analysis within blocks, Cullis and Gleeson (1989), in a study of over 1000 variety trials, reduced estimation variance, on average, to 58% that of a classical analysis. Cullis and Gleeson (1991) extend their models to two dimensions. Baird and Mead (1991) simulated data from several models rather than use real data, so they controlled the variability and autocorrelation in the experimental units as well as the variability in the design. They found the spatial methods of Cullis and Gleeson (1989) to be valid over a wide range of simulation models. Additional references for the interested reader are Bartlett (1978), Kempton and Howes (1981), Wilkinson et al. (1983), Green et al. (1985), and Besag and Kempton (1986).

Besides the spatial analysis of an experiment, there is also the notion that, for spatially correlated variables, there are better designs than the classical ones based on randomization. For example, figure 15.1B shows that all applications of treatment 4 were assigned to the upper right corner. This is undesirable, due to the spatial autocorrelation. A better design would spatially distribute the treatments evenly. For example, first-order nearest neighbor balanced block designs (Kiefer and Wynn 1981; Cheng 1983) can be shown to be optimal under certain conditions (Kiefer 1975; Grondona and Cressie 1993), and they contain good interspersion of treatments (Hurlbert 1984). Despite the advantages, the theory is rather difficult, and there is no guarantee that an optimal design will exist for a given number of replications, treatments, and experimental units. The advantage of the classical designs is that they are very easy to construct and well understood. The compromise presented in this chapter is to construct classical designs and analyze them spatially.

In this chapter, we presented model-based spatial methods for analyzing classically designed experiments. That is, the random errors were considered to be the result of a spatially autocorrelated process. The spatial methods did considerably better than a classical ANOVA. In particular, spatial REML estimation provided the best estimation, valid confidence intervals, and increased power. The increased power of the spatial methods can allow ecologists to detect more real treatment effects for a limited amount of time and money.

Acknowledgments   Financial support for this work was provided to J. M. Ver Hoef by Federal Aid in Wildlife Restoration to the Alaska Department of Fish and Game and the Midwest Region of the National Park Service, and to N. Cressie by the National Science Foundation and the Environmental Protection Agency.

16

Mantel Tests

Spatial Structure in Field Experiments

MARIE-JOSÉE FORTIN

JESSICA GUREVITCH

16.1 Ecological Issues

Landscapes are composed of mosaics of patches, with different degrees of spatial autocorrelation within and among them. The phenomenon of spatial autocorrelation, the spatial dependence of the values of a variable, has been widely reported (chapter 15; Cliff and Ord 1981; Upton and Fingleton 1985; Legendre and Fortin 1989). Positive spatial autocorrelation may result from the microenvironment or from dispersal of offspring near the maternal parent, among other causes. Negative spatial autocorrelation may result, for example, from competition for resources. In the case of positive spatial autocorrelation, plants near one another are more similar than are distant plants. In this event, the results of an analysis of variance (ANOVA) will be affected by the spatial pattern because the data violate one of the basic assumptions of parametric inferential methods: the independence of the observations (Cliff and Ord 1981; Legendre et al. 1990).

The use of nonindependent observations affects the estimation of degrees of freedom: because each observation is not independent of the others, it does not contribute a full degree of freedom (Legendre et al. 1990). Furthermore, with positively spatially autocorrelated data, the differences within groups will appear small, which can result in a Type I error such that the differences among groups are declared significant when in fact they are not (Cliff and Ord 1981; Legendre et al. 1990; Sokal et al. 1993). The implications and problems of the use of spatially autocorrelated data for experiments and subsequent statistical analyses have been the focus of several studies (Legendre et al. 1990; Legendre 1993; Sokal et al. 1993).

With field experiments on plants, an additional problem exists: the degree of spatial heterogeneity already present in the field is rarely assessed in advance. This inherent spatial heterogeneity can result in unequal plant responses to the experimental treatments. Even when small-scale pre-experimental sampling or trials can be carried out, there is no guarantee that the nature of the spatial auto-correlation will be the same in response to the actual experimental treatments or when the spatial scale is expanded for the full experiment. Furthermore, when experimental manipulations of resource availability or competitive effects are conducted on spatially autocorrelated plants and the results analyzed with conventional statistical methods, such as ANOVA, it is difficult to disentangle whether the outcome is due to the responses to treatments or to the inherent spatial structure.

Problems caused by spatial and temporal heterogeneity are familiar to field experimentalists (Hurlbert 1984; Mead 1988). Randomized blocks are perhaps the most common experimental design used in ecology to minimize the effects of spatial and temporal heterogeneity (chapter 4). However, even the use of randomized block designs does not guarantee that the block size employed matches the inherent spatial pattern of the plants or their spatial responses to the treatment (van Es et al. 1989; chapter 4). In the worst cases, blocking reduces the power of the analysis without necessarily removing all or even some of the effects of spatial heterogeneity. Other types of experimental designs such as the GLS-variogram method of ANOVA (chapter 15), nearest neighbor analysis (Wilkinson et al. 1983; chapter 15), trend analysis (Tamura et al. 1988), analysis of covariance (chapter 5), and Latin squares (Mead 1988) have been used instead of the randomized block design.

The methods presented in this chapter, the Mantel and partial Mantel tests (Mantel 1967; Sokal 1979; Hubert 1987; Legendre and Fortin 1989; Manly 1997), are based on distance matrices and permutation tests. These methods differ from those just cited primarily in that they are nonparametric and rely on fewer assumptions. Although in this chapter we discuss the use of these methods to account for the presence of underlying spatial autocorrelation of the data, they also can be used to account for other types of autocorrelation. In fact, autocorrelation exists not only in space but genetically, through species dispersal and clonal spread, and temporally, through daily, seasonal, and yearly cycles (chapter 9). In the present study, we will illustrate how these methods can be used first to detect the presence of inherent spatial autocorrelation and then to distinguish the effects of treatments on plant growth (i.e., experimental design factors) from the effects of spatial autocorrelation in a field competition experiment.

16.2 Statistical Solution

Analysis of variance is usually the most powerful statistical method used to test whether there are significant treatment effects. However, in a field experiment, the appropriate statistical method to use depends foremost on whether the data satisfy the specific assumptions of ANOVA such as normality, homogeneity of the variance, and independence of observations. Thus, preliminary tests should be carried out to verify that the data have normally distributed residuals and homogeneity of variances among groups (Sokal and Rohlf 1995). Independence of the observations can be insured by gathering the data using random assignment of experimental units to treatments (Sokal and Rohlf 1995; chapters 4 and 15). However, randomization procedures do not ensure that neighboring units are spatially independent of one another (Fortin et al. 1989; van Es et al. 1989). Day and Quinn (1989) pointed out the importance and statistical implications of deviations from the assumptions of parametric tests such like ANOVA. Although ANOVA has been found robust to deviations from the assumptions of normality and in some cases of homogeneity of variances, it is quite sensitive to nonindependence of the observations (Bradley 1978; Hays 1981; Posten 1984; Kenny and Judd 1986; Milligan et al. 1987). When the data do not satisfy the assumption of normality, nonparametric procedures, such as Mann–Whitney, Kruskal–Wallis, and Friedman tests, or resampling techniques, as suggested by Dixon (chapter 14), should be used; when the data are spatially autocorrelated, Mantel tests may be more appropriate.

When data from a field experiment on plants, for instance, are analyzed by an ANOVA and the treatment effects are not significant, this implies either that the treatments have no influence on the outcome or that the treatment effects are canceled by the spatial responses of the plants or by other unmeasured and uncontrolled factors (Sokal et al. 1993). However, when the differences are significant, there are three possible reasons: (1) the plants show no significant spatial pattern and the treatments really affect plant responses; (2) the degree of spatial autocorrelation of the plants is significant and is inducing spurious significant treatment effects; and (3) both the degree of spatial autocorrelation of the plants and the treatment effects are significant. Mantel and partial Mantel tests enable the investigator to distinguish which of these three cases is occurring.

16.2.1 Mantel Test

To analyze disease patterns through space and time, Mantel (1967) developed a randomization test that takes the spatial and/or the temporal autocorrelation of the data into account by computing the relationship between two distance matrices. The null hypothesis is that the observed relationship between the two distance matrices could have been obtained by any random arrangement in space (time or treatment assignment) of the observations through the study area. The Mantel test has been used by ecologists to evaluate the relationship between ecological data and their spatial structure (Douglas and Endler 1982; Burgman 1987; Legendre and Fortin 1989; Leduc et al. 1992) or to test the goodness of fit of data to a model or hypothesis (Sokal et al. 1987; Legendre and Troussellier 1988; Sokal et al. 1990; Livshits et al. 1991). Previous uses of Mantel and partial Mantel tests have concentrated on detecting existing patterns in sampled populations. We extend this approach to analyze the outcome of designed experiments, given that there may be an underlying spatial pattern in the data.

A Mantel test examines the relationship between two matrices. The entries in the matrices are actually values (such as physical distances) measured between individual "points," which might be sampling or experimental units. In the example we discuss subsequently, the points are individual bunchgrass plants. By carrying out the Mantel test, we are testing the relationship between two different kinds of variables measured on the sampling units, much as we compute and test a conventional correlation coefficient. What actually goes into these matrices? Rather than using the original measurements for the elements of each matrix, we calculate some measure of distance (or similarity) between each point and all of the others. The Mantel test computes a correlation between the two $n \times n$ distance matrices, where one matrix might represent spatial distances, for example, whereas the other represents differences between pairs of plants in some measure of plant status (e.g., mass).

The results of such a test reveals whether small plants are located near other small plants, whereas large ones have large neighbors, as opposed to the null hypothesis of no relationship between spatial location and plant size. In calculating the Mantel statistic, the products of corresponding elements of the distance matrices (A_{ij} and B_{ij}) summed as follows,

$$Z = \sum_{i=1}^{n} \sum_{j=1}^{n} A_{ij}B_{ij} \quad \text{for } i \neq j \tag{16.1}$$

where the variable distance matrix (**A**) might contain some measure that represents the differences in the outcome of the experiment among all n experimental units and the distance matrix (**B**) might contain the actual Euclidean (spatial) distances among the n experimental units. The matrices must be square (i.e., same number of rows as columns): when the square matrices are symmetric, the computation is carried out only on the lower, or upper, diagonal matrices. This is because the distance between points 1 and 2 is exactly the same as the distance between points 2 and 1, so we must include that distance only one time in the matrix. When one of the matrices is asymmetric, the computation has to be carried out on the whole matrix.

A major advantage of using distance matrices is that the values of the matrix elements can be computed using the distance measurements of your choice. This allows us to test the effects of spatial structure or experimental treatments on various types of outcome measurements (qualitative or quantitative), as well for one or more variables at once. The difference in spatial locations can be thus compared with differences in size or in genetic relatedness. The disadvantage of using distances is that by summing the cross products of ecological and geographical distances, the Z-statistic is unbounded and cannot be compared from one study to another. To overcome this, the Z-statistic can be normalized (r) such that it behaves as a product-moment correlation coefficient (similar to Pearson's r, representing a linear relationship), which ranges from −1 to +1. The normalization of each distance matrix is carried out separately using the standard normal transformation, subtracting the mean of that matrix from each element, and then dividing by the standard deviation of the elements in that matrix. This normalized Mantel statistic (r) can be used to compare results from different variables, or studies, by means of confidence limits, as described by Manly (1986b, 1997). When the *r*-statistic is calculated between a variable distance matrix and a geographical distance matrix, the value of *r* corresponds to the average magnitude of spatial autocorrelation of the variable for the entire study area.

Once the Mantel statistic has been calculated, we usually wish to test its statistical significance. Unfortunately it is not possible to use the familiar statistical tests for this purpose. In fact, the Mantel statistic, *Z* or *r*, cannot be tested as an ordinary product-moment correlation because the distances in each matrix are not independent of one another. Therefore, the significance is assessed either by using a permutation test to construct a reference distribution or by using an asymptotic *t*-approximation test (Mantel 1967). In a permutation test, the statistic calculated on the actual data is compared with what happens when the elements of the matrices are shuffled at random. If there is a strong spatial pattern in the data, shuffling the data points will eliminate that pattern.

In practice, only one of the matrices must be reordered to accomplish the task. The reference distribution obtained by permutation is constructed by randomly reassigning the rows and columns of one matrix, each time computing a new Mantel correlation. This randomization procedure creates a population of "experiments," although only one experiment has actually been carried out (Edgington 1985; Manly 1997). Under the null hypothesis of no relationship between the two distance matrices, the observed Mantel statistic is expected to have a value located near the mode of the reference distribution obtained by randomization of the data; that is, the correlation between the two matrices should have neither an extremely low nor an extremely high value. With no relationship between spatial values and sizes, we expect a random reshuffling of the data to result in about as many higher values of the *Z*-statistic as lower ones. On the other hand, if there is a strong relationship (positive or negative) between the two matrices, the observed Mantel statistic (on the actual data) is expected to be more extreme (either higher or lower) than most of the reference distribution values.

For small sample size ($n < 10$), all the possible permutations ($n!$) can be computed (for the exact probability level). However, with a sample size that small, it is difficult to detect significant spatial patterns, and this is true not only for Mantel tests, but for other spatial analysis methods as well (Legendre and Fortin 1989). Legendre and Fortin (1989) recommended sample sizes of at least 30 to detect significant spatial autocorrelation with Moran's *I*-coefficients. Likewise, it is recommended that we use as large a sample size as possible (e.g., 20 and more) to detect significant patterns with Mantel test (P. Legendre, oral, 1999). Therefore, when patches and spatial autocorrelation are suspected to be important, the experimental design should incorporate a relatively small number of treatments so that there are as many replicates as possible (e.g., van Es et al. 1989). With larger sample sizes, the observed Mantel statistic can be tested against a null distribution generated using random sampling (with replacement) of all possible permutations of the data (Smouse et al. 1986). The minimum number of permutations recommended by Manly (1997) is 1000—the higher the number of permutations, the more accurate the significance test.

When n is very large ($n \geq 40$), the Mantel statistic, Z or r, can by transformed into a t-statistic as follows:

$$t = \frac{(Z - \text{Exp}Z)}{\text{S.E.}Z} \tag{16.2}$$

where the expectation of Z is

$$\text{Exp}Z = \frac{\Sigma\Sigma \, \mathbf{A}_{ij} \, \Sigma\Sigma \, \mathbf{B}_{ij}}{n(n-1)} \tag{16.3}$$

and its standard error is

$$\text{S.E.}Z = \sqrt{\Sigma\Sigma \, \mathbf{A}_{ij} \, \mathbf{A}_{kl} \, \text{Exp}(\mathbf{B}_{ij} \, \mathbf{B}_{kl}) - \frac{(\Sigma\Sigma \, \mathbf{A}_{ij})^2 \, (\Sigma\Sigma \, \mathbf{B}_{ij})^2}{n^2(n-1)^2}} \tag{16.4}$$

The significance of the t-statistic is achieved using an asymptotic t-approximation test (Mantel 1967). The larger the number of observations (n), the more reliable is the significance level of the asymptotic approximation test.

16.2.2 Three-matrix (Partial) Mantel Test

The Mantel test allows for a comparison between only two variables or two sets of variables by means of multivariate distance coefficients (Legendre and Legendre 1998). In ecology, this can be quite a limitation, however, when several processes interact with each other. Furthermore, as discussed previously, with field data, space can create spurious relations between two variables that are in fact driven by a spatial gradient or by a third variable that follows the spatial gradient. For example, a positive relationship between plant growth and geographical distances can exist simply because both are related to environmental conditions. But genetic relationships or experimentally imposed treatment factors might also be responsible for differences in plant response. For such cases, we must disentangle the relative contributions of the various factors influencing the outcome measurement.

To address this issue, the Mantel test has been modified in different ways to allow the comparison among three or more distance matrices (Dow and Cheverud 1985; Hubert 1987; Manly 1986b, 1997; Smouse et al. 1986). The third distance matrix, $\mathbf{C}$, can be (1) another variable (species, environmental condition, genetic data) collected independently at the same locations, (2) a design matrix that refers to a structure imposed on the data by an experimental design or by a hypothesis to test, or (3) a matrix of the geographical distances (Sokal et al. 1987; Legendre and Troussellier 1988). Usually, for the geographical distances matrix, Euclidean distances are used, but other distance measures can also be used, such as $1/d^2$, which will take into consideration nonlinear relationships (e.g., patchy structure) between the geographic locations. The third matrix, the one being factored out, can also be a model (contrast) matrix to test the causality between spatially auto-correlated variables (Leduc et al. 1992; Sokal et al. 1993) or a weight (connection) matrix to compute a multivariate correlogram (Oden and Sokal 1986; Legendre and Fortin 1989).

In this study, we use the method originated by Smouse et al. (1986). In this approach, a partial correlation, $r_{AB.C}$, that establishes the degree of relationship between two matrices is calculated, keeping the effects of the third matrix constant. This partial correlation is computed by first constructing a matrix of residuals, $\mathbf{A'}$, of the regression between $\mathbf{A}$ and $\mathbf{C}$, and a matrix of residuals $\mathbf{B'}$ of the regression between $\mathbf{B}$ and $\mathbf{C}$. These are linear regressions based on the standardized matrices. Then the partial Mantel test is computed as in equation 16.1 but using the two residual matrices $\mathbf{A'}$ and $\mathbf{B'}$. The resulting partial Mantel statistic, $r_{AB.C}$, corresponds to a partial product moment varying from -1 to $+1$. Its significance in the two-way Mantel test is assessed by either permutation or by the asymptotic t-approximation test. According to Oden and Sokal (1992), the method of Smouse et al. (1986) has the best statistical properties for analyzing the relationship among three distance matrices when spatial autocorrelation is present.

16.2.3 Limits and Other Methods

Mantel tests may not always work well in detecting spatial autocorrelation where the spatial pattern is complex and not easily modeled with distance matrices. Examination of a variogram or correlogram of the data (chapter 15) is useful where spatial relationships are expected to exist; this may, in such cases, reveal strong spatial patterning even in the absence of significant Mantel test results. One limitation of the use of Mantel tests for field experimental data is that a larger number of data points than are typically available may be necessary for a satisfactory analysis.

Another major problem with the statistics in the Mantel and the partial Mantel tests is that multivariate data are summarized into a single distance, or dissimilarity, value such that the resulting relationship is a global outcome for all of the variables; thus, it is not possible to identify which variable(s) contributed the most to its intensity. Partial Canonical Correspondence Analysis (Partial CCA; ter Braak 1987, 1988; Borcard et al. 1992; Legendre 1993; Palmer 1993) can be used to circumvent this problem when attempting to elucidate the relationship between species' abundances and environmental factors while controlling for the effects of some covariable. Partial CCA is like the partial Mantel test in that it quantifies the relationship between two matrices (i.e., Canonical Correspondence Analysis, CCA), holding the effects of a third matrix constant (partial CCA). In CCA, species ordination axes (i.e., the species matrix) are constrained, using multiple regression, to be linear combinations of the environmental variables (i.e., the environmental matrix) in maximizing species variance. This method is an iterative procedure that estimates the best linear fit between two data sets. Hence, the outcome will be the best possible linear relationship between species and environmental ordination axes. The resulting biplot allows us to identify by which environmental variable(s) each species is the most affected (ter Braak 1994). The outcome of this method is, however, very sensitive to measurement errors in the environmental data and is limited to the environmental variables included in the analysis.

Often the relationship between species and environmental data is induced by other underlying factors such as climate, topography, geographical locations, or historical events and can be controlled for using a third matrix as in the partial Mantel test (ter Braak 1988; Borcard et al. 1992; Legendre 1993; Palmer 1993). Again, the resulting triplot of a partial CCA allows us to disentangle the relative contributions of the factors of a third matrix considered as covariables. Randomization tests are then used to assess the significance of the relationship. As with the partial Mantel test, the use of this third matrix allows us to test specific hypotheses, by coding it as a contrast matrix in ANOVA, or to quantify the relative importance of spatial structure in the data by using the geographic coordinates of the samples (Borcard et al. 1992; Harvey 1996).

Other permutation methods have been developed to address the problem of spatially autocorrelated data in the context of analysis of variance. For example, MRPP (MultiResponse Permutation Procedures) and MRBP (MultiRandomized Block design Procedure) can handle nonlinear data (Biondini et al. 1988; Gurevitch and Collins 1994). Also, as an alternative to the analysis of variance, ANOSIM (ANalysis Of SIMilarities) was developed by Clarke (1993) and is comparable to a Mantel test where the distances are transformed into ranks.

16.3 Example

Ideally, data should be obtained from an experiment where the spatial structure is known, so that an optimal experimental design could have been created and carried out to account for the spatial effects (chapters 4 and 15). However, the real world is not ideal, and it is often difficult or impossible to account accurately for the underlying spatial structure in plant characteristics in nature. Field workers may be forced to compromise, doing their best to carry out the most appropriate experimental design according to the information available. The necessity for compromise being the most common case, the following example will illustrate how the Mantel and partial Mantel tests can be used to test for the presence of and to distinguish the relative contributions of the treatment effect from the effect of spatial structure.

16.3.1 Data

Data from an experiment on plant competition (Gurevitch 1986) will be employed to illustrate these methods. Gurevitch used removal experiments carried out at three sites along a topographic gradient (near Sonoita, Santa Cruz County, Arizona), to test whether the growth of the C_3 grass *Stipa neomexicana* was limited by competition from perennial C_4 grasses. The response of *Stipa* was compared with that of a C_4 grass, *Aristida glauca*. The two species were subjected to several experimental treatments (partial or complete removal of neighbors) and followed over a period of 20 months. We will present the results from the midslope site for only two of the four experimental treatments used by Gurevitch (1986): all neighboring plants within a 0.5-m radius of the target plant (*Stipa* or *Aristida*) were removed (treatment), or neighboring plants were left alone (control). The size (basal area, cm^2) of each target plant was recorded at the beginning (January 1980, hereafter referred to as initial size) and at the end of the experiment (August 1981, hereafter referred to as final size).

A randomized block design was used to assign, at random, two replicates per treatment within each block (figure 16.1). Note that the experimental design was less than perfect, because it would have been better to assign *Stipa* and *Aristida* plants as targets within the same blocks rather than in adjacent ones. We emphasize that the Mantel approach discussed in this chapter can be used with a variety of experimental designs, including completely randomized designs as well as randomized blocks, and with more than two levels per treatment. The experimental unit was an individual mature grass plant, located near the center of a 1.0×1.0-m plot. For each species, there were ten blocks, for a total of $n = 40$ replicates (20 controls and 20 removal treatments for each of the two species).

The following examples were computed using the *Stipa neomexicana* data, as shown in table 16.1. The Mantel (two-matrix) and partial Mantel (three-matrix)

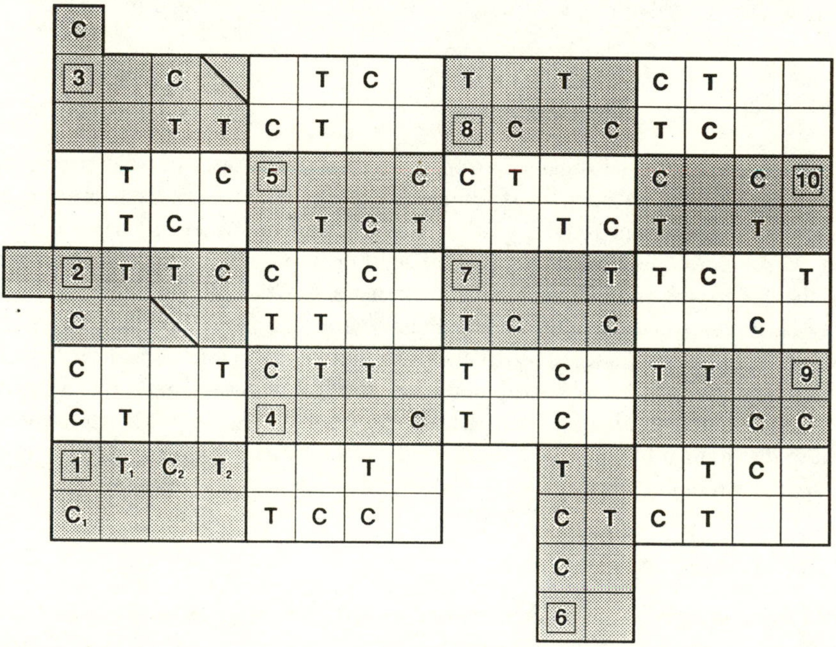

Figure 16.1 Experimental replicates. Each block (solid rectangle) contains eight replicates (1.0×1.0 m), where C indicates controls and T indicates removal treatment replicates used in this study. In the gray blocks, the target species is *Stipa neomexicana*, and in the white ones, it is *Aristida glauca*. The squares with a diagonal were not usable in the experiment. The blocks where *Stipa neomexicana* is the target species are numbered as in table 16.1. In block 1, the replicate numbers are indicated by subscript.

Table 16.1 Initial and final size (basal area, cm^2) of *Stipa neomexicana*[a]

Block	Treatment and replicate	Initial size	Final size	Geogaphical coordinates x	Geogaphical coordinates y	Treatment and replicate	Initial size	Final Size	Geographical coordinates x	Geographical coordinates y
1	C 1	28.34	23.13	3.0	5.0	T 1	79.73	85.88	5.0	7.0
	C 2	30.63	22.97	7.0	7.0	T 2	12.02	32.45	9.0	7.0
2	C 1	75.87	39.80	3.0	13.0	T 1	29.15	58.61	5.0	15.0
	C 2	11.94	7.59	9.0	15.0	T 2	5.18	19.15	7.0	15.0
3	C 1	53.60	49.01	7.0	19.0	T 1	8.95	24.57	7.0	17.0
	C 2	39.25	68.61	3.0	25.0	T 2	32.99	78.11	9.0	17.0
4	C 1	4.32	3.41	11.0	11.0	T 1	23.28	61.38	13.0	11.0
	C 2	5.36	9.53	17.0	9.0	T 2	5.37	18.85	15.0	11.0
5	C 1	22.89	13.62	15.0	17.0	T 1	22.78	33.57	13.0	17.0
	C 2	17.59	0.07	17.0	19.0	T 2	26.14	50.08	17.0	17.0
6	C 1	13.21	30.04	23.0	3.0	T 1	4.90	33.60	23.0	3.0
	C 2	28.84	33.13	23.0	1.0	T 2	9.24	25.92	25.0	1.0
7	C 1	21.32	27.43	21.0	13.0	T 1	46.73	67.80	19.0	13.0
	C 2	15.90	6.48	25.0	13.0	T 2	6.03	26.30	25.0	15.0
8	C 1	4.40	14.42	21.0	21.0	T 1	6.44	50.80	19.0	23.0
	C 2	33.93	16.87	25.0	21.0	T 2	11.64	30.63	23.0	23.0
9	C 1	58.50	15.55	31.0	9.0	T 1	8.95	28.65	27.0	11.0
	C 2	26.30	23.50	33.0	9.0	T 2	27.57	53.85	29.0	11.0
10	C 1	7.66	5.93	27.0	19.0	T 1	14.07	29.99	27.0	17.0
	C 2	8.29	10.05	31.0	19.0	T 2	6.57	9.33	31.0	17.0

[a]T = removal treatment; C = controls. For geographical coordinates, (0, 0) is the bottom left corner of the field pictured in figure 16.1.

statistics, as well as their associated significance values, were computed by the MANTEL program of "The R package: multidimensional analysis, spatial analysis" of Legendre and Vaudor (1991). This package is available for the Macintosh at http://alize.ere.umontreal.ca/~casgrain/R/.

Other software, all for personal computers, can also be used to compute matrix relationships. The Mantel (two-matrix) statistic can be computed with the MXCOMP (for MatriX COMParison) program of the NTSYSpc package. NT-SYSpc is marketed by Exeter Software [http://users.AOL.com/ExeterSftw/]. RT, Randomization Tests (Manly 1997), computes Mantel test (MANT2) as well as Manly's regression approach for a three-matrix (or more) partial Mantel test (MANCOR: Manly 1986b, 1997); it is available at http://ourworld.compuserve.com/homeoages/BManly/. STAT! computes Mantel tests as well as several other types of space–time matrix relationship. STAT! is marketed by BioMedware [http://ic.net/~biomware/]. Two-matrix Mantel tests and MRPP can be computed using PC-Ord marketed by MjM Software [http://www.ptinet.net/~mjm/winspec.htm/]. ANOSIM can be computed with the PRIMER package (Clarke and Warkick 1994). Partial CCA can be computed using CANOCO (ter Braak 1988). CANOCO, available for both Macintosh and personal computers, is marketed by Microcomputer Power [http://ww1.microcomputerpower.com/webpages/mcp/].

16.3.2 Building the Distance Matrices

Given the data from this study, three $n \times n$ distance matrices can be computed: the variable distance matrix, the geography distance matrix, and the design matrix.

1. The *variable* distance matrix, **A**, refers to the differences in the measurements (here, size) of the plants between all possible pairs of replicates where i and j represent different experimental units (here, each plant is an experimental unit). Since we are interested in any treatment effects on size, we compute the distance as the absolute difference between all pairs of replicates as follows:

$$\text{outcome}(i, j) = |\text{size}_i - \text{size}_j| \qquad (16.5)$$

For example, from table 16.1, the difference between the final size of the first control plant, C_1, and the second control plant, C_2, in the first block $(C_1, C_2) = |23.13 - 22.97| = 0.16$. The variable distance matrix can be computed not only for the final size but for the initial size as well. Indeed, the variable distance matrix of the initial size can be used to test whether there was already a significant spatial pattern in plant size at the beginning of the experiment.

According to the hypothesis under study, as well as the type of data available, different distance coefficients can be used to establish the variable (or outcome) distance matrix, **A**. For guidelines to select the appropriate distance coefficient for the type of data (qualitative or quantitative) and for the biological context (community structure, genetic distance), consult Legendre and Legendre (1998) or Gower and Legendre (1986). The actual computation of the distance matrix can be carried out with the program SIMIL of the "R package" (Legendre and Vaudor 1991), which offers 15 distance (dissimilarity) coefficients. The input data for the SIMIL program is an ASCII file containing the raw data where the "rows" are the objects (here, n rows for the n replicates), and the "column" holds the values of a given variable (here, the plant size). The output from SIMIL is the upper triangle of an $n \times n$ distance matrix written in SIMIL binary format. This SIMIL binary file can be used directly with the MANTEL program. An ASCII file of the SIMIL binary file can be obtained by the LOOK program. If the SIMIL program does not have the specific distance coefficient desired, the distance matrix can be computed by another software program. In such cases, the ASCII distance matrix file can be rewritten into the SIMIL binary format using the IMPORT program.

2. The *geography* distance matrix, **B**, contains the physical location distances between each point (here, each plant) and all the others. It can be computed using the Euclidean distance (i.e., the actual physical distance), between the spatial coordinates of all possible pairs of plants as follows:

$$\text{geographic}(i, j) = \sqrt{(x_i - x_j)^2 + (y_i - y_j)^2} \qquad (16.6)$$

So, the geographic distance between the two control replicates of the first block (table 16.1) is

$$\text{geographic}(C_1, C_2) = \sqrt{(3.0 - 7.0)^2 + (5.0 - 7.0)^2} = 4.47$$

that is, the distance between plants C_1 and C_2 in the field is 4.47 meters (see figure 16.1).

When the relationship between plant response and geography is known not to be linear, other types of distance measures can be used. Choices include the inverse of the Euclidean distance, which makes small distances more important than large ones (Estabrook and Gates 1984) and relative position given by a nearest neighbor network (Upton and Fingleton 1985; Harvey et al. 1988).

3. The *design* matrix, **C**, expresses the differences in the treatments to which the plants (experimental units) were exposed. It allows us to test whether plants subjected to different treatments were any different from those with the same treatment. This matrix is coded analogously to a set of contrasts in ANOVA: when two replicates have received the same treatment, their difference is coded 0; when two replicates have received different treatments, their difference is coded 1. A similar type of coding can also be used to test the differences in responses between the two species:

	C treatment					**C** species			
	C_1	C_2	T_1	T_2		*Stipa*	*Stipa*	*Aristida*	*Aristida*
C_1	0	0	1	1	*Stipa*	0	0	1	1
C_2		0	1	1	*Stipa*		0	1	1
T_1			0	0	*Aristida*			0	0
T_2				0	*Aristida*				0

Because the coding of the design matrix is specific to the experimental hypothesis and the data, this matrix should be created as an ASCII file and then rewritten in SIMIL binary format using the IMPORT program. Coding the treatment matrix requires careful thought so that the test really makes the comparison you want. When there are more than two treatment levels, we can code as we did previously: 0 when replicates received the same level of a treatment, and 1 when replicates received different treatment levels, regardless of the level. This coding provides the same weight to all treatments, as does the null hypothesis of an analysis of variance. When directional differences among several treatment levels are to be tested, the coding has to be such that it gives appropriate weights to different pairs of treatment levels.

The interactions among factors, here the interaction between species and removal treatment, can also be tested by appropriately coding the design matrix. For example, within-species coding between the treatment types can be done this way, whereas between-species coding uses opposite signs to test for an interaction between-species and treatment:

		Stipa		*Aristida*	
		Control	Treatment	Control	Treatment
Stipa	Control	0	1	1	−1
	Treatment	1	0	−1	1
Aristida	Control	1	−1	0	1
	Treatment	−1	1	1	0

Here, we used the same the code, 1, for the distance between the plants of different species but having the same treatment, and for plants of the same species but with different treatments. Different weights could have been used to test other hypotheses (Livshits et al. 1991). However, given that the design matrix, **C**, must be a distance matrix, it may be difficult to code complex designs.

16.3.3 Computing the Mantel Statistic

The measure of the correlation between two matrices, such as between the variable and the geography matrices, can be computed using the Mantel test (Mantel 1967; Sokal 1979). The Mantel statistic allows us to evaluate whether there is significant spatial autocorrelation over the entire stand of plants. Thus, we are interested in assessing the correlation between the distance (difference) in size among all pairs of plants, contained in the variable matrix (**A**), and the actual Euclidean distances among the respective pairs of plants contained in the geography (spatial distance) matrix (**B**). If the spatial location of the plants does not affect plant size, the observed Mantel statistic (r) should not be statistically different from zero.

For example, to compute the Mantel statistic between the initial size and the geographical distances of the four replicates of *Stipa* in the first block (replicates C_1, C_2, T_1, and T_2; see table 16.1), we used the following distance matrices:

	A: Initial size					B: Geography					A × B			
	C_1	C_2	T_1	T_2		C_1	C_2	T_1	T_2		C_1	C_2	T_1	T_2
C_1	0.00	2.29	51.39	16.32	C_1	0.00	4.47	2.82	6.32	C_1	0.00	10.24	144.92	103.14
C_2		0.00	49.10	18.61	C_2		0.00	2.00	2.00	C_2		0.00	98.20	37.22
T_1			0.00	67.71	T_1			0.00	4.00	T_1			0.00	270.84
T_2				0.00	T_2				0.00	T_2				0.00

The matrix **A** × **B** is the element-by-element cross product of the values in the two distance matrices **A** and **B**. For example, the first element of **A** (above) multiplied by the first element of **B**, 0×0, equals the first element in **A** × **B** which is zero, whereas the second element in **A** multiplied by the second element in B gives $2.29 \times 4.47 = 10.24$, which is the second element in the **A** × **B** matrix. The

Mantel statistic Z, as mentioned in section 15.2.1, is the summation of the cross-product elements of the matrix $A \times B$ excluding the diagonal $(i = j)$ because the distance of a replicate with itself is zero:

$$Z = 10.24 + 144.92 + 103.94 + 98.20 + 37.22 + 270.84 = 665.07$$

Of course, ordinarily Z would be calculated across both species and all blocks, but this small part of the data set is useful for illustrating the method. Given that the distance between i and j equals that between j and i (in symmetric matrices), the Mantel statistic is evaluated by using only the upper (or lower) diagonal matrices without affecting the relative strength of the relation between the two matrices.

To compute the normalized Mantel statistic, r, each distance matrix is first standardized individually by subtracting the mean of all the elements in the matrix from each observation and then dividing by the standard deviation. The mean distance for this first block of the variable matrix A (initial size) is 34.24 cm^2 and the standard deviation is 25.38 cm^2, whereas for the distance matrix B (geographic distance), the mean is 3.60 m and the standard deviation is 1.68 m. So, for example, to standardize the second element in the first row of A, the calculation is $(2.29 - 34.24)/25.38 = -1.26$. The r is calculated in the same way as the Z-statistic by summing the cross products of the standardized matrices (here abbreviated as "std") and dividing by $(n - 1)$:

$$r = \frac{\sum\limits_{i=1}^{n} \sum\limits_{j=1}^{n} \text{std } A_{ij} \text{ std } B_{ij}}{n - 1} \quad \text{for } i \neq j \quad (16.7)$$

Thus, for the initial size, the normalized Mantel statistic is computed for the first block from the following matrices:

	A: Initial size				B: Geography				A × B			
	C_1	C_2	T_1	T_2	C_1	C_2	T_1	T_2	C_1	C_2	T_1	T_2
C_1	0.00	−1.26	0.68	−0.71	0.00	0.52	−0.46	1.62	0.00	−0.65	−0.31	−0.55
C_2		0.00	0.58	−0.62		0.00	−0.95	−0.95		0.00	−1.15	0.58
T_1			0.00	1.32			0.00	0.24			0.00	0.31
T_2				0.00				0.00				0.00

The normalized Mantel statistic equals

$$r = (-0.65 + (-0.31) + (-0.55) + (-1.15) + 0.58 + 0.31)/$$
$$(6 - 1) = -1.77/5 = -0.354$$

if calculated for the first block only.

We can also test for treatment effects using the same approach. We would expect no significant correlation between the initial size and the treatments, because the treatments were assigned to plants at random and were not begun until after the initial sizes were measured. The example yields the following:

	A: Initial size					**B**: Treatment					**A** × **C**			
	C_1	C_2	T_1	T_2		C_1	C_2	T_1	T_2		C_1	C_2	T_1	T_2
C_1	0.00	2.29	51.39	16.32	C_1	0	0	1	1	C_1	0.00	0.00	51.39	16.32
C_2		0.00	49.10	18.10	C_2		0	1	1	C_2		0.00	49.10	18.10
T_1			0.00	67.71	T_1			0	0	T_1			0.00	0.00
T_2				0.00	T_2				0	T_2				0.00

We calculate $Z = 0.00 + 51.39 + 16.32 + 49.10 + 18.10 + 0.00 = 134.91$, whereas $r = -0.023$. As anticipated, this value is very close to zero.

16.3.4 Assessing the Significance of the Mantel Statistic

As for any statistical test, the significance of the observed value of the Mantel statistic, Z or r, is assessed by comparing it to the reference distribution obtained under the null hypothesis. In practice, the reference distribution can be obtained by permuting the arrangement of the elements of one of the distance matrices randomly at least 1000 times, each time computing the Mantel statistic. The significance of the observed Mantel value is obtained by comparing it with the number of times that the permuted Mantel values are smaller, equal to, or greater than the Mantel statistic for the actual data.

As mentioned previously, the null hypothesis refers to the absence of significant relationship between the two matrices. The alternative hypothesis can be evaluated using either a one- or a two-tailed test. The one-tailed probability of any positive observed Mantel statistic (i.e., right-hand tail probability) is given by first adding the number of equal or greater magnitude than the calculated (actual) value and then dividing by the total number of permutations. Similarly, the one-tailed probability of a negative observed Mantel value (i.e., left-hand tail probability) is given by adding the number of equal and smaller values and then dividing by the number of permutations. In the MANTEL program, the observed value is added to the number of equal values, so to have a reference distribution of 1000, we need ask only for 999 permutations. Although, MANTEL (Legendre and Vaudor 1991) and MXCOMP (NTSYS-pc) programs provide only the number of permutations that are smaller than, equal to, and greater than the observed value, and hence one-tailed probability levels, the MANT2 subroutine provides one-tailed as well as two-tailed probabilities, where the two-tailed probability is given by the number of permutations that, in absolute value, are greater than or equal to the observed statistic.

Using the MANTEL program and thus conducting one-tailed tests, the following relationships were computed for the 10 blocks of *Stipa* across the entire mid-slope site:

	r	Smaller	Equal	Greater	$P(r)$
Initial × Geography	0.1381	977	1	22	0.023
Initial × Treatment	−0.0011	647	1	352	0.648
Final × Geography	0.0389	736	1	263	0.264
Final × Treatment	0.1396	996	1	3	0.004

The results reveal that spatial location (geography) had a significant effect ($r =$ 0.1381, $P < 0.05$) on the initial size of *Stipa* plants (above and table 16.2A). Note that each design factor is tested independently; each line in this table (and in tables 16.2 and 16.3) represents a complete Mantel test of 999 permutations. As expected, competition treatments did not affect the initial size of *Stipa* plants, because treatments were assigned at random after initial size was measured (this table and table 16.2A, $r = -0.0011$, $p \gg 0.05$). By the end of the experiment, the treatment (removal of competitors) had a highly significant effect on *Stipa* plant size (this table and table 16.2B, $r = 0.1396$, $P \leq 0.004$).

In a more complete analysis, we included the entire data set for both *Stipa* and *Aristida*. There were large differences in initial size between the two species (table 16.3A, $r = 0.2208$, $P \leq 0.001$). There were also highly significant differences in final plant size between the two grass species (table 16.3B, $r = 0.2891$, $P \leq 0.001$), a highly significant species × treatment interaction on final size ($r = 0.0496$, $P \leq 0.001$), and a marginally significant effect of blocks at the end of the experiment ($r = -0.0184$, $P < 0.05$).

Table 16.2 ANOVA and Mantel Tests on the Sizes (basal area, cm^2) of *Stipa neomexicana* Plants

	ANOVA		Mantel tests	
Factor	F	P	R	P
A. Initial Size				
Block	1.17	0.3629	0.0147	0.2240
Treatment	0.99	0.3323	−0.0011	0.3650
Treatment.Block			0.0024	0.3120
Treatment.Geography			0.0044	0.2850
Geography			0.1381	0.0230
Geography.Treatment			0.1382	0.0270
B. Final Size				
Block	1.31	0.2920	−0.0002	0.5100
Treatment	10.73	0.0038	0.1396	0.0010
Treatment.Block			0.1409	0.0010
Treatment.Geography			0.1414	0.0010
Geography			0.0389	0.2640
Geography.Treatment			0.0451	0.2300

Table 16.3 ANOVA and Mantel Tests on the Sizes (basal area, cm^2) of *Stipa neomexicana* and *Aristida glauca* plants

	ANOVA		Mantel tests	
Factor	F	P	R	P
A. Initial Size				
Species	43.56	0.0001	0.2208	0.0010
Block	2.26	0.0284	0.0046	0.2700
Treatment	0.02	0.9003	−0.0092	0.0810
Species × Treatment	0.35	0.5541	−0.0065	0.2810
Geography			0.0071	0.3890
Treatment.Geography			0.0067	0.3910
Geography.Treatment			0.0069	0.4250
B. Final Size				
Species	70.94	0.0001	0.2891	0.0010
Block	0.84	0.5422	−0.0184	0.0370
Treatment	29.55	0.0001	0.0876	0.0010
Species × Treatment	10.61	0.0025	0.0496	0.0010
Geography			−0.0387	0.1730
Treatment.Geography			0.0501	0.0010
Geography.Treatment			−0.0371	0.1730

16.3.5 Computing Partial Mantel Tests

The partial Mantel statistic, $r_{AB.C}$ (the correlation between matrix **A** and **B** given **C**), can be computed to test whether there are significant treatment effects (design matrix) on plant size (variable distance matrix) when the effects of the spatial location are kept constant (geography distance matrix). We might also wish to investigate whether plant size (variable distance matrix) is spatially autocorrelated (geography distance matrix) while the treatment effects are kept constant (design matrix). Thus, the partial Mantel test can be carried out to ensure that the effects of the treatments are not canceled by spatial effects. We used the three-way (partial) Mantel test to examine some of the questions that are probably of greatest interest to the experimentalist; for example, what are the effects of the treatments when spatial effects are taken into account? This may be thought of as being analogous to an analysis of covariance, where spatial effects are "held constant" while the effects of the experimental treatment are examined.

When the results were analyzed for *Stipa* alone, taking spatial location into account strengthened the already strong effect of the treatment on final plant size (table 16.2B, Treatment.Geography, $r = 0.1414$, $P < 0.001$, compared with Treatment, $r = 0.1396$), whereas there was no significant effect of the treatments on initial size when geography was held constant (table 16.2A, Treatment.Geography, $r = 0.0044$, NS). Partial Mantel tests could be used to test many relationships of interest: for example, we could test for the effect of the treatments on final plant size, holding initial plant size constant (Treatment.Initial, $r = 0.5518$, $P < 0.001$). For the entire data set including both *Stipa* and *Aristida* (table 16.3), the effects of the experimental treatment on final plant size again remained statistically significant when spatial effects were held constant (table 16.3B, $r = 0.0501$, $P < 0.001$). Spatial location (geography) did not affect plant size, even when treatment effects were taken into account (table 16.3B, $r = -0.0371$, $p > 0.05$).

16.3.6 Mantel Tests versus ANOVA

The results of the previous experiment using the Mantel tests were compared with the results from a standard ANOVA. The results were similar for the two tests (tables 16.2A and 16.3), which is to be expected because strong spatial dependence was not detected in this data set. Therefore, the ANOVA assumption of no spatial autocorrelation (i.e., independence of observations) was not seriously violated in the present study. Nevertheless, the Mantel and partial Mantel tests can bring out complementary information that ANOVA cannot provide, such as the detection of a significant underlying spatial autocorrelation for *Stipa* at the beginning of the experiment (initial size, table 16.2A), which did not hold at the end (final size, table 16.2B). This loss of significant spatial pattern in final sizes offers insight into the spatial responses of *Stipa* and indicates that the treatment effects were strong enough to override the initial spatial pattern. Notice that the Mantel tests were able to detect the spatial pattern regardless of the block size used.

Furthermore, the Mantel approach seems better than ANOVA at detecting block effects and the species × treatment interaction at the end of the experiment as indicated by small P-values, although these relationships were not strong (table 16.3B). The same trend held for the partial Mantel tests (tables 16.2 and 16.3). Recall that even though the spatial effects (geography) were not statistically significant, holding the effects of space constant using the partial Mantel test resulted in a higher r-value (i.e., a stronger relationship) for treatment effects (table 16.2B, $r = 0.1396$ for treatment alone, whereas $r = 0.1414$ for treatment with geography held constant). The stronger treatment effect when spatial effects were held constant was not apparent when *Stipa* and *Aristida* were analyzed together, however (table 16.3B), perhaps because the spatial patterns for the two species differed.

16.4 Conclusion

In this chapter, we have emphasized the importance of detecting and taking into account the underlying spatial pattern in the analysis of field data to obtain a better understanding of the plants' responses. Our results emphasize the value of using larger and simpler experiments to reveal a pattern that may otherwise be obscured (chapter 18; Gurevitch et al. 1992). The trend in ecological experiments is to include many experimental treatments with few replicates, but because the ability to detect spatial pattern increases with sample size, fewer treatments and a larger number of replicates may sometimes be preferable.

Other approaches to the design and analysis of ecological data when spatial autocorrelation is present have also been proposed (e.g., chapter 15; Besag and Kempton 1986; Legendre et al. 1990; Borcard et al. 1992; Legendre 1993). Although we have used Mantel tests to analyze spatially autocorrelated variables, these tests are not restricted to that use and can be implemented to analyze other types of data or used in other contexts (Sokal et al. 1990, 1993). The usefulness and flexibility of the Mantel and partial Mantel tests make them good exploratory tools to detect a posteriori the scale of the spatial pattern or to test the goodness of fit of an alternative hypothesis or model (Legendre and Troussellier 1988; Legendre and Fortin 1989; Sokal et al. 1987).

The extension of the Mantel and partial Mantel tests presented here offers a promising approach in designed experiments where spatial heterogeneity may pose problems for the analysis of the results. Using this approach, it is possible to distinguish the effects of spatial pattern from those of experimentally imposed treatment effects. Therefore, this offers an alternative to classical experimental design and parametric analysis of variance in ecology, especially when the extent and pattern of spatial heterogeneity are not known in advance.

Acknowledgments   Professor Robert Sokal provided invaluable suggestions on how to approach the problem. The work was supported in part by NSF Grant BSR 89-08112 to J.G., which is gratefully acknowledged.

17

Bayesian Statistics

Estimating Plant Demographic Parameters

JAMES S. CLARK

MICHAEL LAVINE

17.1 Introduction

There are times when external information should be brought to bear on an ecological analysis. Experiments are never conducted in a knowledge-free context. The inference we draw from an observation may depend on everything else we know about the process. Bayesian analysis is a method that brings outside evidence into the analysis of experimental and observational data.

With the increasing use of Bayesian methods in ecology, our science has coopted the philosophical controversy that attended the twentieth-century rise of "classical" statistics (Stigler 1986). Limitations of classical hypothesis testing and *P*-values (Berger and Berry 1988) on the one hand, or of Bayesian priors and subjective probability (Dennis 1996) on the other, allow smart people to come down on either side of a polarized debate (Edwards 1996). The debate will undoubtedly continue.

This chapter is not one of the battlegrounds over Thomas Bayes' thinking when he described his famous "billiard" example or its application since (Bayes 1763; Fisher 1959; Stigler 1986). Although we are not always enamored with classical hypothesis testing in general, we often use it. And, although priors can sometimes sound like a bad idea in theory, it is usually harder to abuse them than some people think. Regardless of whether it is always sensible to regard an unknown parameter as having a distribution of values (in a Bayesian sense), this can be the best way to model many ecological problems. For those of us most interested in interval estimation, the fact that both methods usually give similar answers tends to be lost in the fray. The large divergences that can occur with small sample sizes or strong prior opinions are less common than the impression left by some authors. We leave these arguments for others, but refer readers to the lively treatments in the Special Feature of *Ecological Applications* (Dixon and Ellison 1996). Hilborn and Mangel (1998) lay out the utility of Bayesian methods as part of a set of tools for analysis of ecological data. An overview of Bayesian statistics is given in Berger (in press).

Bayesian analysis differs from other topics in this book, so we approach it in a different way. Our gentle introduction is intended for the ecologist who might find either Bayesian or classical approaches useful, depending on the application at hand. So our chapter includes some comparisons, but they are not the insidious examples that rely on strange or unrealistic distributions to generate discord. Although most of this book is designed for the practitioner, providing the bridge from concept to software, Bayesian analysis still requires programming. Thus, although we cannot direct the reader to a broad range of software options, we adopt the general philosophy of this volume by providing a simple and practical introduction to a topic that is generally treated at a more advanced level in graduate statistics courses and beyond.

We cannot go far using Bayesian methods without the routine application of calculus (including numerical methods that require an understanding thereof). Models with multiple parameters get complicated fast, but the conceptual background laid by simpler models generally applies. Rather than attempt a broad survey that would risk losing the reader in technique, we limit this chapter to one sampling distribution (the binomial). This limited scope allows us to introduce a number of concepts (the basic elements of Bayesian methods, conjugacy, comparison with classical methods) that apply generally. Useful introductory texts include Berry (1996), Lee (1989), and Box and Tiao (1973).

17.2 The Basic Elements

Bayesian statistics has two distinguishing characteristics:

1. It combines, in a formal way, data from the experiment at hand with data from any other experiment or information deemed relevant.
2. It summarizes the analysis with a probability distribution that shows how well the various values of the parameter are supported by all of this information.

For the purpose of illustrating concepts, we begin with a simple example. To understand the dynamics of plant populations, ecologists estimate survival from census data. Because annual rates tend to be high (often >95% per year for trees), it can be difficult to obtain data sufficient to make confident estimates (i.e., enough deaths). Information that is external to the study at hand can help to sharpen estimates. This Bayesian example combines census data from a typical field study with external information to evaluate survival of *Acer rubrum* trees in the southern Appalachian Mountains (Wyckoff and Clark 2000).

The probability density in figure 17.1A summarizes the analysis of tree survival. The data include annual censuses of trees; the parameter of interest is the

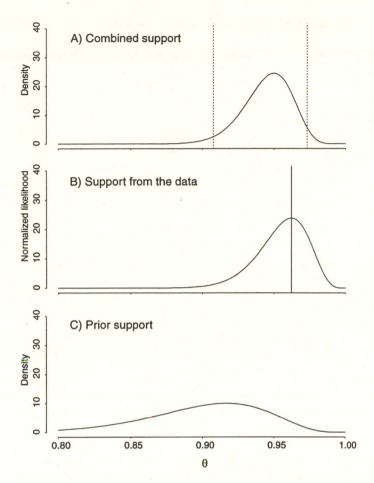

Figure 17.1 The elements of a Bayesian analysis. The posterior density (A) represents the support of values of the survival rate θ in light of data at hand (B) and prior inputs (C).

probability θ that a randomly selected tree survives from one year to the next. Figure 17.1A shows that values of θ around 0.95 are most likely, but all values from about 0.91 to about 0.98 are plausible. The data support the value θ = 0.95 about twice as well as 0.93 or 0.97 and about ten times as well as 0.91 or 0.98.

Where did figure 17.1A come from? The repeated censuses of Wyckoff and Clark (2000) included 127 survivors from a total of 132 individuals. The chance of 127 survivals from 132 total trees is calculated from a binomial distribution as

$$f(\text{data}\,|\,\theta) = \Pr[127 \text{ survivals in } 132 \text{ trees}] = \binom{132}{127} \theta^{127} (1-\theta)^5 \quad (17.1)$$

This function of θ is called a likelihood function and is plotted in figure 17.1B. The likelihood function indicates how well the data support each value of θ. In this example, the data best support the value θ = 127/132, and they lend decreasing support to values on either side. The value best supported by the data has the maximum likelihood and is termed the maximum-likelihood (ML) estimate. The likelihood function in figure 17.1B shows only the support from the data, so it is not yet a Bayesian analysis.

A Bayesian analysis combines the likelihood function of figure 17.1B with other information, which could be data from other experiments or scientific insight. The other information is summarized by a function $f(\theta)$ called the prior density. For example, if we had reason to believe, say, from previous experiments, that θ was most likely to be about 0.9 and very likely to be somewhere between 0.85 and 0.95, then we might use a prior density that looks like figure 17.1C. The name *prior* is used because the other information usually comes to us prior to the information from the experiment we are analyzing. But priority in time is not necessary. Perhaps a better term would be external information (Raftery and Zeh 1993).

The likelihood function and prior density are combined according to Bayes' theorem,

$$f(\theta \mid \text{data}) = \frac{\text{prior} \times \text{likelihood}}{\int \text{prior} \times \text{likelihood } d\theta} = \frac{f(\theta)f(\text{Data} \mid \theta)}{\int f(\theta)f(\text{Data} \mid \theta)d\theta} \qquad (17.2)$$

The left-hand side is called the posterior density and is the combination of likelihood and prior. The theorem says that the posterior density is calculated by multiplying the prior density by the likelihood. (The integral in the denominator is just a normalizing constant.) The posterior represents, at least approximately, how well various values of θ are supported by all the information, including data at hand and the prior information. The prior and likelihood pictured in figures 17.1B and 17.1C, respectively, combine to give the posterior in figure 17.1A.

In summary, a Bayesian analysis takes as inputs both data, by way of a likelihood, and additional information, summarized by the prior, to produce a posterior density. The posterior expresses how the combination of data and prior together support values for the unknown parameter.

17.3 A Few Details

To introduce some of the techniques necessary to arrive at a posterior, we pursue a bit further the example from Wyckoff and Clark (2000), who compared maximum-likelihood and Bayesian approaches to estimate how survival rates change over time.

17.3.1 Arriving at a Posterior

Let *n* be the number of trees counted at a first census, and *k* be those that survive to be counted in the second census. For the *Acer rubrum* example, with *n* = 132 and $k = 127$, intuition tells us that the best estimate of θ is simply $k/n = 0.96$, the fraction that survive. The strong intuitive sense we have about this simple problem helps us grasp the basics of a Bayesian approach. Recall the two ingredients, a likelihood and a prior density for the parameter of interest θ. Rewriting equation 17.1 in terms of generic parameters gives the binomial likelihood,

$$f(k \mid \theta) = \binom{n}{k} \theta^k (1 - \theta)^{n-k} \qquad (17.3)$$

where θ is the probability of survival, and the combinatorial $\binom{n}{k}$ is the binomial coefficient. The prior $f(\theta)$ depends on insights about θ other than those obtained from the data. The flat prior, representing the view that all values of θ are equally probable, is a uniform density on the interval $(0, 1)$:

$$f(\theta) = 1 \qquad 0 < \theta < 1 \qquad (17.4)$$

(dashed line in figure 17.2A). We might use a flat prior if we desire an outcome that is influenced by the data alone and not by external information. To write

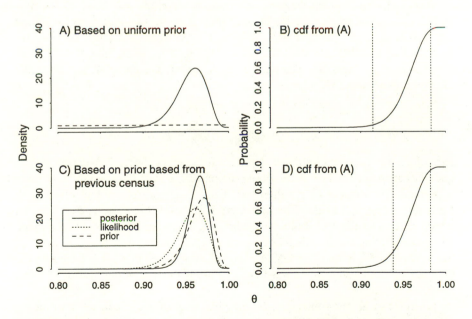

Figure 17.2 Bayesian analyses for uniform (A) and nonuniform (B) priors. In part (A) the posterior coincides with the likelihood. The cumulative plots at right are the cumulative posterior distributions with vertical dashed lines enclosing 95% of the posterior densities.

Bayes' rule (equation 17.2), we note that the combinatorials in the numerator and denominator cancel, leaving

$$f(\theta \mid k) = \frac{\theta^k (1 - \theta)^{n-k}}{\int_0^1 \theta^k (1 - \theta)^{n-k} \, d\theta} = \frac{\theta^k (1 - \theta)^{n-k}}{B(k + 1, n - k + 1)} = (n + 1) \binom{n}{k} \theta^k (1 - \theta)^{n-k} \qquad (17.5)$$

$B(\bullet, \bullet)$ is the beta function. The final step in equation 17.5 makes use of some well-known relationships among beta functions, gamma functions, and factorials that can be found in standard probability texts.

The posterior in equation 17.5 is a beta density, $f(\theta \mid k) = B(k + 1, n - k + 1)$, and expresses the level of certainty assigned to values of θ. The mode of this density is the most probable value of θ and occurs at the critical point, where $df(\theta \mid k)/d\theta = 0$. Differentiation is simplified if we first take logs, $d\ln f(\theta \mid k)/d\theta = \frac{k}{n} - \frac{n-k}{1-\theta}$. Setting this derivative equal to zero shows that the mode of the posterior agrees with our intuitive estimate k/n (figure 17.2A). We can summarize our degree of confidence in θ with quantiles that contain the central $100(1 - \alpha)\%$ of the posterior. The right-hand side (figure 17.2B) is the cumulative distribution for the posterior showing 95% quantiles (dashed lines).

Now consider how our noninformative (flat) prior affects the result. The uniform prior density means that the posterior beta density (equation 17.5) has the same shape as the likelihood function (equation 17.3); the two differ only by a constant and, thus, contain the same information about the parameter θ. The normalized likelihood (divide the likelihood function by the denominator of equation 17.1) coincides with the posterior in figure 17.2A. Because we had no prior insight, the census data governed the result. Before considering how the posterior is influenced by the particular choice of the prior, we compare the Bayesian method with a classical approach.

17.3.2 Comparison with a Classical Approach

How does this Bayesian approach differ from a classical view? A classical (frequentist) approach might involve fitting the parameter θ to data and then deriving a probability statement (a P-value) based on a comparison of that result with some alternative null model. The maximum-likelihood (ML) estimate of θ is that which maximizes the probability of the data set, assuming the model to be correct. By differentiating the likelihood of equation 17.3 with respect to θ, we find the ML estimate of θ to be $\theta_{ML} = k/n$. A classical confidence interval is based on the comparison of this ML estimate with other possible values of the parameter using P-values.

We refer to all such intervals, be they classical or Bayesian, as "confidence intervals"; ecologists do not use the Bayesian jargon "credible interval." To compare confidence intervals, we describe a "likelihood profile" approach. Likelihood profiles are being used increasingly by ecologists; a full description of likelihood profiles can be found in Hilborne and Mangel (1997). The summary of likelihood profiles that follows illustrates the link between classical and Bayesian approaches represented by the likelihood function.

Within a classical context, a probability statement about an estimate requires some alternative hypothesis against which it can be compared. Because there might be many such alternatives, let's consider a broad range. This range is the basis for a classical confidence interval, which we obtain by constructing a likelihood profile. The method involves successively calculating two likelihoods for the same data set, one for each competing hypothesis about θ against the ML estimate, that is, the value obtaining most support from the data. The likelihood ratio (LR) is simply the ratio of the likelihoods of the two models,

$$LR(\theta_0, \theta_{ML}) = \frac{f(k \mid \theta_0)}{f(k \mid \theta_{ML})} = \left(\frac{\theta_0}{\theta_{ML}}\right)^k\left(\frac{1 - \theta_0}{1 - \theta_{ML}}\right)^{n-k}$$

The deviance is the test statistic. It is twice the difference in log likelihoods

$$D(\theta_0) = -2\ln\left(\frac{f(k \mid \theta_0)}{f(k \mid \theta_{ML})}\right) = -2\ln LR$$

and is distributed as χ^2 with 1 degree of freedom (there is one parameter at issue). The deviances increase (figure 17.3A) and associated P-values decrease (figure 17.3B) as the hypothesized value of θ deviates from the ML estimate ($\theta = 0.96$).

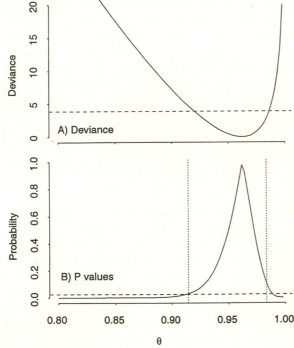

Figure 17.3 Classical confidence intervals for the example in figure 17.2A. The deviance (A) has a horizontal dashed line corresponding to likelihood profile values of $P = 0.05$. The plot of P-values (B) includes the $P = 0.025$ horizontal line and the 95% Bayesian confidence interval (vertical lines) from the example in figure 17.2A.

We might conclude from this example that the data allow us to reject, at the $\alpha = 0.05$ level, the hypothesis that θ lies outside the interval bounded by the horizontal dashed line at $P = 0.025$ in figure 17.3B. We can represent the same interval by the horizontal line through the deviance plot at 3.84 (figure 17.3A): the χ^2-value that yields $P = 0.95$ with 1 degree of freedom. For comparison, we include in figure 17.3B the Bayesian confidence interval obtained with a uniform prior in figure 17.2A (the vertical dashed lines in figure 17.3B).

How different are the interpretations derived from these approaches? The astute reader will note that the confidence intervals from the Bayesian (vertical lines in figure 17.3B) and the classical method (horizontal line in figure 17.3B) nearly coincide. Indeed, with large n, they converge. The lower limits are equivalent, and the upper limits differ slightly. If both methods yield the same confidence intervals, then how important is the distinction? From equation 17.1 (and 17.5), we note that a uniform prior (a reflection of prior ignorance about θ) means that the posterior is simply a normalized likelihood function. Because the likelihood and posterior bear the same shape (we cannot distinguish them in figure 17.2A), they contain the same information about θ. The posterior is completely controlled by the data, without prior bias. And the posterior (normalized likelihood) yields about the same confidence interval as the likelihood profile. This example is general; with large sample size, a noninformative prior produces a confidence interval that converges with the classical one.

Despite similarities, statisticians talk about these two confidence intervals in different ways. The classical confidence interval is taken to cover the fraction of repeated experiments in which the interval would contain the true value of the fixed parameter. If we were to conduct a large number of identical experiments on survival of trees that are subject to the same set of risks, our survival estimate would fall above the dashed lines in figure 17.3A in 95% of those experiments. The Bayesian confidence interval represents our belief that the random parameter spans a certain interval. Here, survival is viewed as random with a density given in figure 17.2A. There are cases where the two approaches can yield importantly different answers (e.g., Cousins 1995). However, from a practical standpoint, it is worth remembering that much of the time the confidence intervals nearly coincide.

17.3.3 An Informed Prior

For problems like tree survival, where estimates suffer from inadequate data, prior (external) knowledge about θ can sharpen our inference. Wyckoff and Clark (2000) incorporated estimates obtained from U.S. Forest Service (USFS) inventories as prior estimates of survivorship. The USFS data might not provide the best estimate for Wyckoff and Clark's (2000) study site, because the data come from a broad region, but they do represent a prior estimate of survival that might be worth combining with field data from their more restricted study area.

USFS data contained $k_0 = 137$ of $n_0 = 142$ surviving *Acer rubrum* trees from the region that includes the study area of Wyckoff and Clark (2000). The prior density, likelihood function, and posterior density are compared in figure 17.2B. For convenience, the prior is taken to be a beta density,

$$f(\theta) = B(k_0, n_0) \qquad (17.6)$$

Note that the parameters of this density are simply the numbers of total and surviving trees. Because the prior from USFS data in equation 17.6 (shown in figure 17.2C) contains far more information about θ than does the uniform one (figure 17.2A), the posterior in figure 17.2C is concentrated about the most probable estimate (still $\theta = 0.96$) to a greater degree than in figure 17.2A. The posterior splits the difference between prior and likelihood, because it incorporates information from both. The greater information that results from the informed prior is reflected in the narrower confidence intervals shown on the right-hand side of figure 17.2.

The beta-binomial example makes obvious the importance of sample size. In our example, the weight of the prior evidence ($n_0 = 142$) and of the new data ($n = 137$) are about the same. The exact solution for the posterior is the beta density with parameters obtained by summing the prior information and data:

$$f(\theta|k) = B(k_0 + k, n_0 + n - k_0 - k) \qquad (17.7)$$

Because the parameters in equation 17.7 are simply the sums of total and surviving trees, it is clear that both contribute similar weight to the posterior. For a small sample size, an informed prior ($n_0 \gg n$) dominates the posterior; the likelihood (i.e., the data) has minimal impact. With increasing sample size ($n \gg n_0$), the likelihood dominates the prior, and the posterior approaches the likelihood. The example using a flat prior (figure 17.2A) is an extreme case, where the weight of the evidence is concentrated in the likelihood. Thus, the impact of the prior is felt most when sample size is low. With an increasing sample size, the posterior tends to normality, with the mean approaching the "true" value of θ_0, and the parameter variance is determined by the curvature of the likelihood surface at θ_0. Thus, with a large sample size, the likelihood alone can be used to estimate the mode and curvature. Provided that the prior assigns nonzero probability to the true value θ_0, the curvature increases with increasing n until the mass of the posterior is concentrated at the point θ_0.

One objection to Bayesian methods is that subjectivity may creep into the analysis through the choice of the prior. In the hope of reducing subjectivity, some practitioners recommend using a flat prior. As we have seen, this approach yields a posterior density of the parameter based on the data and on an initial belief that all values are equally probable. (The foregoing section explains similarities to a classical approach.) Although some strict subjectivist Bayesians might disagree, it is generally good practice to consider several different priors, representing different evaluations of outside information, use them each to compute a posterior, and compare the posteriors. Often the posteriors that result from different priors will be similar (Crome et al. 1996). Wyckoff and Clark (2000) determined how the survival estimates changed when using priors obtained from different data sources. In their analysis, changing the prior had little effect on the posterior, because there were not large discrepancies between the priors obtained from different sources and the likelihood. But, in some cases, they can be quite different, meaning that people can disagree. Wolfson et al. (1996) provide an example of how sample size can be adjusted to ensure a decisive experimental outcome when different parties bring to a problem very different priors. A larger sample size might be needed to demonstrate an outcome that is at odds with strong prior evidence.

17.3.5 Conjugacy

If a probability statement about parameters is the only objective, then a Bayesian analysis can often be done without resorting to the mathematical details behind, say, equation 17.7. Indeed, increasing complexity, such as in the example that follows, demands a computer-intensive approach. Numerical techniques, such Markov Chain Monte Carlo (MCMC) simulation, are well suited to analyzing such models and calculating posterior distributions for the parameters of interest. Gelman et al. (1995) provide an introduction to these methods. The route to the posterior is intractable, and the nonparametric nature of the posterior means that it is not readily transported from one application to the next.

Much ecological investigation is concerned with developing models for understanding and prediction. Knowing the numerical techniques for extracting confidence intervals from high-dimensional posteriors is often not enough. The development of minimal models that permit transparent error propagation and analysis is a goal of ecological research (Hilborne and Mangel 1997; Burnham and Anderson 1998).

A special class of models is analytically tractable when the number of parameters is small and provides a powerful technique for data assimilation. It involves a special relationship between prior and likelihood termed *conjugacy*. A conjugate prior-likelihood pair is one for which application of Bayes' rule results in a posterior having the same form as the prior. Conjugate prior-likelihood pairs can be found for many low-dimensional problems. The beta-binomial is a common example: the prior (equation 17.6) and posterior (equation 17.7) have the same form, and only the parameter values are updated. There are a number of conjugate pairs (we mention the invχ^2-Gaussian conjugate pair subsequently), and their use always simplifies the analysis. Conjugacy is a valuable tool, because it permits an exact result that can be updated repeatedly. For example, a model of forest dynamics can be implemented in fully parametric form. A standard model of this sort using, say, a ML (point) estimate of survival probability does not reflect the uncertainty described by figure 17.2B. The conjugate pair model allows us to draw survival estimates directly from equation 17.7, thus propagating uncertainty in the parameter estimate directly to the model output. Moreover, the next occasion to update the data set requires only a change in the parameters of equation 17.7. Although their calculation requires some math, conjugate pairs provide the most transparent view of the relationship between priors and posteriors, and, when available, they provide a powerful way to assimilate data in ecological models. Of course, in the many cases where a conjugate pair is not available, the analysis must proceed numerically.

17.4 Bayesian Estimation in a Dynamic Model

The methods outlined previously can be extended to more complex ecological problems. Here we provide an example that uses the binomial sampling distribution to estimate the dynamics of seedlings. Ecologists typically study such dynamics by tagging every seedling in a study plot and, through censuses, determining survival probabilities. Such studies are so labor-intensive that few data sets exist (Clark et al. 1999). Moreover, the heavy loss of tags and failure to relocate seedlings necessitate far more complex statistical models than investigators actually use to analyze such data. The following example from Lavine et al. (in press) illustrates how the Bayesian approach can be implemented in a dynamic model to incorporate different types of error and, in the process, extract parameter estimates without the intensive labor required by the standard approach. The model is based on identification of only two classes of seedling age, and it uses local densities at each stage rather than individually tagged seedlings.

Tree seedlings can be conveniently separated into a first-year class and a >first-year class, which is presumably less susceptible to mortality risks. A dynamic model based on this two-stage classification is readily applied to field data, because the two classes of seedlings are distinguished by the presence of bud-scale scars on >first-year seedlings. The data consist of seedling densities of the two classes counted in 1-m^2 quadrats. First-year and >first-year seedlings are termed *New* (*N*) and *Old* (*O*), respectively. Each class has its own survival probability, θ_N and θ_O. The number of new seedlings entering the population in year j, N_j, is determined by input of new seeds. The number of old seedlings, O_j, is the sum of both new and old seedlings from last year that survived to this year. To simplify notation in the equations that follow we define the numbers of both classes that survivor to year j as

$$Y_j = \text{Bin}(N_{j-1}, \theta_N)$$
$$X_j = \text{Bin}(O_{j-1}, \theta_O)$$

The first of these equations says that the number of survivors is a random variable drawn from a binomial distribution with parameters N_{j-1} (the number of potential survivors) and θ_N (the probability that any one individual survives). The total number of old seedlings is the sum of old and new seedlings that survived from year $j - 1$, $O_j = X_j + Y_j$. In other words, the number of old seedlings is the sum of two binomial variates, each with its own survival probability. Data from one of the 1-m^2 quadrates are shown in table 17.1.

Table I Numbers of first-year (new) and >first-year (old) *Acer rubrum* seedlings censused in quadrate 9 from Lavine et al. (in press)

Year	1993	1994	1995	1996	1997
Number of new seedlings, N_j	1	0	2	0	0
Number of old seedlings, O_j	1	1	1	1	0

In 1996 (call this year j), one old seedling was observed. We do not know whether this survivor was old or new in 1995 (year $j - 1$). To estimate the two survival probabilities, we must enumerate the possibilities. As many as 1 or as few as 0 could have been old; the remaining 0 or 1 was new in 1995. From 1 old and 2 new in 1995, the probability (i.e., likelihood) that exactly 1 survived to become old seedlings in 1996 is

$$f(O_{1996} = 1 \mid \theta_O, \theta_N) = \sum_{x=0}^{1} \binom{O_{1995}}{x} \theta_O^x (1 - \theta_O)^{O_{1995}-x}$$

$$\binom{N_{1995}}{O_{1996} - x} \theta_N^{O_{1996}-x} (1 - \theta_N)^{N_{1995}-O_{1996}+x}$$

This likelihood is the sum of two binomial probabilities. The summation from $x = 0$ to 1 adds the two ways we could observe one old seedling. If the old seedling was new last year ($x = 0$ in the summation), we have the probability that the single old individual died (the first binomial) times the probability that one of the two ($O_j - x = 1 - 0 = 1$) N_{j-1} survived. [Lavine et al. (in press) provide simple rules for obtaining the summation limits.] By adding to this value the probabilities that would apply if a single O_{j-1} survived ($x = 1$), we obtain the total probability of obtaining the data. The likelihood function for the whole data set is the product over all Q quadrats and all T years,

$$f(\{O_{i,j}\} \mid \theta_O, \theta_N) = \prod_{i=1}^{Q} \prod_{j=1}^{T} L(O_{i,j} \mid \theta_O, \theta_N) \tag{17.8}$$

$$= \prod_{i=1}^{Q} \prod_{j=1}^{T} \sum_{x=x_{min}}^{x_{max}} \binom{O_{j-1}}{x} \theta_O^x (1 - \theta_O)^{O_{j-1}-x} \binom{N_{j-1}}{O_j - x} \theta_N^{O_j-x} (1 - \theta_N)^{N_{j-1}-O_j+x}$$

In this particular likelihood, we treat each plot in each year as independent. [Lavine et al. (in press) relax this assumption].

Because of the number of parameters involved, calculating a posterior can require some numerical tools. As written, the posterior can thus far be calculated exactly. Combining the likelihood of equation 17.8 with a flat prior results in a posterior density for θ_N and θ_O (figure 17.4A). The posterior shows how well each combination of (θ_N, θ_O) is supported by the evidence. To examine θ_N only, we integrate over θ_O to obtain the marginal density of θ_N:

$$f(\theta_N \mid O) = \int_0^1 f(\theta_N, \theta_O \mid O) d\theta_O$$

This integration is necessary because parameters in complex models can often be correlated (a type of ill conditioning we mention subsequently). Because the two parameters are largely independent (there is little sign of correlation in figure 17.4), the marginal density obtained from this integration might not be too different from a conditional density (at, say, the ML estimate of the other parameter).

In the real world, other sources of error require parameterization. Because not all seedlings will be found in all years, there is random "findability," which can be thought of as the probability that a seedling is counted at all. In particular,

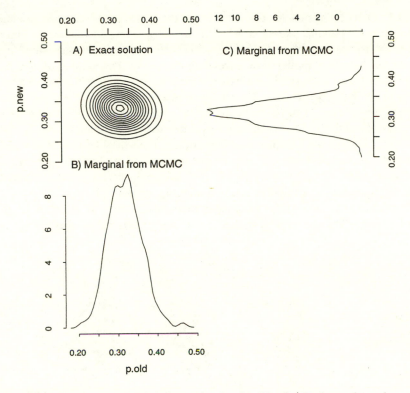

Figure 17.4 (A) Contour plot of the posterior density $f(\theta_N, \theta_O|O)$ shows that values near 0.32 are the best support for both parameters. Marginal posterior densities for the survival parameters θ_N (B) and θ_O (C) obtained by Gibbs sampling from the joint posterior density for the model that includes a findability error (Lavine et al. 2000) are in agreement with those obtained by the simpler model (A).

new seedlings can be small and hard to find. If, for example, there is a Poisson distribution of new seedlings with parameter (mean value) λ and a probability f that a new seedling is actually found, then the problem is too complex to pursue analytically. For the data set considered by Lavine et al. (in press), the marginal posteriors for the survival probabilities are shown in figure 17.4B,C. The marginal distributions are bumpy because they are obtained by numerically (Gibbs) sampling from a joint posterior having these extra parameters to accommodate additional sources of error. Gibbs sampling is a MCMC technique that simulates a posterior and, in the process, accomplishes the integration described previously (without actually integrating anything; see Gelman et al. (1995) for an introduction). The posteriors indicate that survival rates between 0.25 and 0.40 for both new and old seedlings are most likely. This result ran counter to our expectation that survival would be lower for new seedlings, but it may be explained by the fact that some seedlings may still emerge after the July censuses. Note too that the values do not greatly differ from those obtained from the simple model in figure 17.4A.

The biplots for parameter pairs show that there can be substantial correlation between some parameters (figure 17.5). Especially strong correlation occurs between the parameters λ and f. This negative correlation can be understood if we reconsider the verbal model of the foregoing paragraph. The number of new seedlings recorded in a census is the true number of seedlings (with mean value λ) times the probability that a given seedling is observed, f. Because neither quantity is directly observed, both parameters can trade off to yield a particular observed number of seedlings. The correlation evident in the biplot represents a corresponding ridge in the likelihood surface: the fit (i.e., the likelihood) for a high value of f and a low value of λ can be just as good as the fit for low f and high λ. The posterior densities in figure 17.4B,C integrate over such correlations, but it is still important to know that such correlations exist.

Few data sets have sufficient sample size and duration to estimate seedling survival rates. The majority of studies last a year or less, involve sampling from

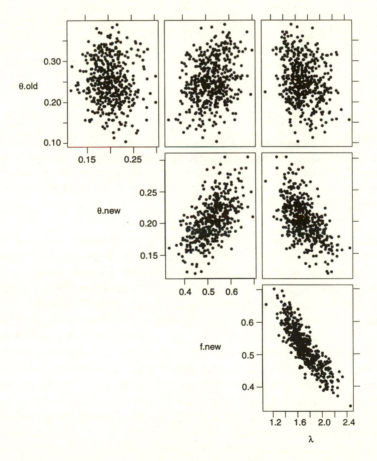

Figure 17.5 Parameter pairs for θ_O (θ.old), θ_N (θ.new), f(f.new), and λ (lambda) showing tendency for correlation among some parameters.

a single stand, and examine effects on a single cohort (Clark et al. 1999b). The lack of adequate data for parameter estimation results from the intense labor required to census tagged seedlings. Comparing data sets obtained from tagged and untagged seedlings, Lavine et al. (in press) showed that this Bayesian approach provides only slightly less information than would the laborious practice of tagging all seedlings. Thus, the method makes it far easier to obtain much larger data sets.

The seedling example demonstrates the common challenge: estimation is based on a dynamic process and a particular observation can be obtained in different ways. By enumerating all of the ways by which a particular observation might arise (equation 17.8), we can accommodate far more complex problems than could be approached using a simple sampling distribution (e.g., equation 17.3). With increasing model complexity, the possibility of ill-conditioning increases, whereby the model asks for more information than the data contain. The parameter trade-offs evident here can be detected by calculating correlation coefficients between pairs of parameters or by examining biplots of the posterior (figure 17.5). Figure 17.4A indicates almost no correlation between θ_O and θ_N, but parameter correlations do arise between some parameters when the model is expanded to include other types of error (Lavine et al. in press). Although ill conditioning arose here in the context of a posterior density, the problem must be considered in any data modeling exercise, not just in Bayesian analysis.

17.5 Some Additional Ecological and Environmental Examples

17.5.1 Incorporating Different Types of Data

Population densities of bowhead whales are difficult to estimate, because the whales move from place to place and they are often underwater. The problems with counting whales motivated Raftery and Zeh (1993) to use a Bayesian analysis that accommodates counts of whales both seen and heard as they migrate past Point Barrow, Alaska. The likelihood function comes from the counts in the 1988 census. The prior takes into account the physical considerations related to locations of observers and sonar arrays, visibility, the physics of sonar location, and the knowledge of bowhead migratory behavior. The posterior that combines this information suggests that the most likely number of bowhead whales is about 7500, but that any size between 6500 and 9000 is reasonably well supported by the data. In listing the advantages of a Bayesian analysis, Raftery and Zeh (1993, pp. 166–168) state the following:

- "It enables us to use a realistic, scientifically relevant model, rather than forcing us to make artificially simple assumptions for the sake of mathematical tractability."
- "It permits us to incorporate the available external, or 'prior' information."
- "It makes elaboration and refinement of the underlying physical assumptions relatively straightforward."

- "It is very hard to develop a non-Bayesian approach that takes account of all the important sources of error."

17.5.2 External Evidence can Change the Inference

Global warming of the oceans could have a large impact on fisheries and coastline management. Change in ocean temperature is difficult to document, because temperature varies with depth, and data sets of the duration needed to document the temperature rise are hard to obtain. An analysis of temperature measurements along the 24.5°N transect at a depth of 1000 m in the Atlantic Ocean suggest that a 0.1°C warming occurred during the time between two voyages completed during the period 1957–1981 (Parilla et al. 1994). Lavine and Lozier (1999) reanalyzed the data using Bayesian methods that allowed them to determine the historic trend in ocean temperatures and to incorporate additional data. The Bayesian approach allowed Lavine and Lozier to consider data from other voyages that were near the 24.5°N transect and thereby to reconstruct the temperature history through time. The temperature history revealed by the Bayesian approach revealed the following:

- 1957 was an unusually cold year in the historical record at 1000 m and, thus, the "trend" resulted in large part to the fortuitous timing of the first voyage,
- Temperatures of isopycnals (surfaces of constant density) are much more constant over time than temperatures at fixed depth.
- The temperature fluctuations are likely due to vertical movement of isopycnals up and down past the fixed depth rather than to a simple increasing trend.

Thus, incorporating the additional evidence brought perspective that changed our interpretation of long-term change in ocean temperatures.

17.5.3 Parametric Empirical Bayes

Parametric empirical Bayes is a term applied to models for data that arise from several sources of variability (see Ver Hoef 1996 for an ecological example). To understand why we say that parametric empirical Bayes is not truly Bayesian, we must discuss mixtures. To demonstrate both the utility of the method and its non-Bayesian nature, we refer to a seed dispersal example of Clark et al. (1999), where both methods were used.

Ecologists have long suspected that seed shadows might have long, "fat" tails, meaning that small numbers of seeds might be dispersed far from the parent plant (e.g., Portnoy and Willson 1993). Recent studies emphasizing how fat-tailed kernels produce patterns of spread that differ qualitatively from traditional models (Kot et al. 1996, Clark 1998, Lewis 1997, Clark et al. 2001) make it important to determine whether fat-tailed dispersal is common. Traditional dispersal kernels, such as Gaussian or exponential, have tails that approach zero rapidly. Unfortunately, kernels with fat tails are difficult to fit to data, and there have not been decisive tests among competing models (i.e., those that assume fat-tailed kernels versus those that do not). Mechanistic models of dispersal are hard to apply to trees, because seeds emanate from broad and diffuse sources (tree crowns) and are released over time, as wind fields and animal dispersers vary. To determine whether seed dispersal data are best described by fat-tailed kernels, Clark et al. (1999a) used more empirical models and the method that is sometimes referred to as parametric empirical Bayes.

Both Bayesian and parametric empirical Bayes involve densities of a parameter $f(\theta)$ and a likelihood function $f(\theta|\text{data})$. In the beta-binomial example, no matter how uncertain we are about a parameter (summarized by the prior $f(\theta)$ in equation 17.4 or 17.6), the data themselves are assumed to have a binomial distribution (the likelihood $f(\text{data}|\theta)$ in equation 17.3). The prior expresses our uncertainty about θ. Uncertainty diminishes as data accumulate. The posterior becomes concentrated at the value of θ that describes the precise binomial distribution that "best" describes the data.

Instead of the binomial of equation 17.3, the dispersal example of Clark et al. (1999a) uses the likelihood for a Gaussian (normal) kernel, $f(\mathbf{r}|\theta) = N(\mathbf{r}|0, \theta^2)$, where θ is a dispersion parameter, and $\mathbf{r}$ is a vector of dispersal distances (i.e., the "data"). A conjugate prior for this Gaussian likelihood is an inverse chi-square density for the parameter θ, $f(\theta) = \text{Inv}\chi^2(u_0, p_0)$, which is shown as a dashed line in figure 17.6B. The parameters u_0 and p_0 determine the spread and shape of the density. The results are insensitive to the precise shape of this prior in this example because it is given low weight, with parameters $u_0 = 25$ and $p = 1$.

As in the beta-binomial example, application of Bayes' rule (ignoring the constant denominator in equation 17.2),

$$f(\theta \mid \mathbf{r}) \propto \text{likelihood} \times \text{prior} = f(\mathbf{r}|\theta)f(\theta;u_0, p_0)$$

yields a posterior having the same form ($\text{Inv}\chi^2$) as the prior,

$$f(\theta \mid \mathbf{r}) = \text{Inv}\chi^2\left(u_0 + \sum_{i=1}^{n} r_i^2, \ p_0 + n\right)$$

(the definition of conjugacy). From prior to posterior, only the parameter values change: the first parameter is increased by adding to it the squared observations, and the second parameter is increased by the sample size n. The combined effect is a posterior that is more "peaked" than the prior, which makes us more certain that we know which Gaussian kernel best describes the data. In figure 17.6B, the posterior lies between the prior and likelihood. It is much closer to the likelihood than the prior, because the prior is "weak." As with the beta-binomial example, our uncertainty about a parameter θ does not affect our assumption about the distribution of data (the likelihood is Gaussian).

Parametric empirical Bayes differs from true Bayes in that the variability in θ is assumed to affect the distribution of the data themselves and not just our understanding of model parameters. Suppose that the Gaussian density (defined by a value for θ) describes dispersal for a given seed released from a particular canopy location at a specific time. Because the canopy, seeds, and transport conditions all vary, there might be a different Gaussian distribution for each seed (represented by a density of values for θ). Unlike the Bayesian case, the density of

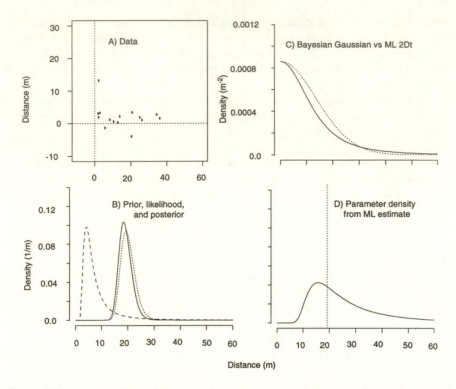

Figure 17.6 A comparison of Bayes and parametric empirical Bayes for seed dispersal data. (A) Dispersal locations of 15 *Fraxinus americana* seeds released from coordinate (0, 0). (B) Bayesian analysis. Symbolism follows figure 17.2. (C) Comparison of the ML dispersal kernel using the mixture model that assumes an Invχ^2 density of dispersal parameter values (solid line) and the Gaussian kernel corresponds to the most probable Bayesian estimate of θ. (D) The Invχ^2 density of θ for ML estimates of u and p corresponding to the 2Dt dispersal kernel in figure 17.7C.

θ-values represents variability in the data, not our range of belief about θ-values. We incorporate this variability into the likelihood itself, because it is part of the process that produces the data. Both r and θ are random variables, and the likelihood that includes both of their effects on the observations is a marginal density (mixture) obtained by integrating over the variability in θ:

$$f(r \,|\, u, \, p) = \int_{0}^{\infty} f(r \,|\, \theta) \, f(\theta; u, \, p) d\theta$$

This expression says that the probability (likelihood) of observing a given dispersal distance r, when r depends on a random variable θ, is their joint probability (their product), accumulated (integrated) over all possible values that θ might assume. Upon integration, the likelihood is Student's t with parameters u and p. The next step generally involves fitting the Student's t-parameters (u, p) directly to data (e.g, using maximum likelihood).

The analysis of Clark et al. (1999a) demonstrated the tendency of dispersal kernels to have long, fat tails (small values of the shape parameter p) (figure 17.6C), which implied that the density of θ values also had a long tail (figure 17.6D). The mixture model provided a test among competing models, which can be represented by different values of the parameter p. Results suggest that the dispersion parameter of the Gaussian density can be viewed as sometimes having large values that correspond to times of high winds or dispersal by animal vectors (the tail of figure 17.6D). This variability can produce a fat-tailed kernel (solid line in figure 17.6C), which, in turn, suggests the plausibility of rapid population spread (Clark 1998).

In summary, although the term *parametric empirical Bayes* sounds Bayesian, it does not involve priors and posteriors. With large sample size, the Bayesian posterior converges to a point mass centered at a single value of θ. This posterior, in turn, implies a Gaussian dispersal kernel, regardless of whether the data better support the fatter-tailed two-dimensional t-distribution 2Dt (which, in fact, they did in the analysis by Clark et al. 1999a). Parametric empirical Bayes assumes the data are distributed as 2Dt, because the variability in θ affects the actual process. The 2Dt does not converge to a Gaussian kernel with increased sample size, because the data are better described by the 2Dt. In other words, the posterior density of θ in figure 17.6B would become increasing peaked with additional data, whereas we can expect the density of θ in figure 17.6D to retain its spread.

17.5.4 Classical Significance versus Bayesian Support

Crome et al. (1996) compared a classical intervention analysis and a Bayesian approach to assess the impacts of logging on recapture rates of birds and small mammals in eastern Queensland. The likelihood of the data set is the product of two lognormal distributions, the joint probability of observing a set of differences between the logged and unlogged sites before and after logging. This is the Before-After-Control-Impact-Pairs (BACIP) model of Stewart-Oaten et al. (1986, 1992; see chapter 9). The Bayesian analysis included three priors, representing expectations about logging impact that ranged from a 25% reduction to a 25% increase in capture rates. Although the mean values differed by these percentages, the priors possessed broad overlap and thus did not represent large differences in perspective.

The classical analysis showed so few significant results as to be unhelpful— because the notion of no logging effect in this context is silly. It would be difficult to convince any ecologist that birds and mammals failed to notice that the trees had vanished. A nonsignificant result does not alter this view; rather, it points to the need for larger sample sizes to obtain a "significant" result.

By contrast, Bayesian confidence intervals help sharpen our understanding of the logging impact. Due to broadly overlapping priors, the posteriors obtained for a given species did not show large differences. The authors refer to these similarities as examples of "concensus" among those bearing different prior views. Posteriors suggest the degrees to which different species responded to the intervention and how those responses differed among habitats.

17.6 Getting the Analysis Done

Unlike most classical statistical methods, the availability of software for Bayesian analysis is limited. Most practitioners program their own models. This is most easily done using a high-level language, such as S-Plus, or specialized Bayesian software, such as BUGS [http://www.mrc-bsu.cam.ac.uk/bugs/Welcome.html], for graphical models, or BATS, for time series [http://www.stat.duke.edu/~mw/bats.html]. More complicated problems require programming in low-level languages, such as C++ or Fortran.

17.7 Conclusion

A Bayesian approach allows us to incorporate external information in the interpretation of experimental or observational data. This approach can place data in perspective of prior insights, it can provide for probability statements in situations that do submit to simple, classical frameworks, and it can minimize sample sizes and study durations necessary to arrive at experimental outcomes. This chapter scratches the surface of a broad and complex topic—one that cannot be ignored, regardless of philosophical leanings, because all ecologists must judge for themselves the increasing number of Bayesian studies in the ecological literature.

Acknowledgments For their helpful comments on the manuscript, we thank B. Beckage, J. Gurevitch, J. HilleRisLambers, J. Lynch, J. McLachlan, A. Pringle, C. Saunders, S. Scheiner, and two anonymous reviewers.

18

Meta-analysis

Combining the Results of Independent Experiments

JESSICA GUREVITCH

LARRY V. HEDGES

18.1 Ecological Issues

In this chapter, we examine methods to statistically combine the results of separate studies to reach general conclusions. The statistical synthesis of the results of independent experiments is known as *meta-analysis*. These relatively new statistical techniques have been introduced fairly recently to the field of ecology (Gurevitch et al. 1992; Arnqvist and Wooster 1995), although they have had considerable influence on research synthesis in medicine (e.g., Sacks et al. 1987; Chalmers et al. 1989) and in the social sciences (e.g., Glass et al. 1981; Hyde and Linn 1986). The handbook edited by Cooper and Hedges (1994) provides a general reference on statistical, methodological, and other aspects of the subject.

At the heart of meta-analysis is the concept that the progress of science depends on the ability to reach general conclusions from a body of research. Scheiner (chapter 1) raised the basic question, What, after all, are the purposes of carrying out experiments in ecology? An experiment is designed to test a particular hypothesis or set of hypotheses. The results formally provide a test of these hypotheses only for specific individual organisms in one place at one time. The information provided by the experiment is limited to those particular circumstances. Experiments can be exactly replicated in different laboratories in some sciences, confirming the general applicability of the results. This is not possible in ecology, but that does not make it any less pressing to evaluate the generality of the findings of an experiment.

In ecology, basic and applied research differ in that in applied work, information about particular local systems that leads to better management may be useful in itself. In basic ecological research, the results of testing hypotheses with individual experiments are of no interest if they cannot be generalized. What can the outcome of any one experiment tell us about nature? To what extent can we extrapolate from the results? Do the results seem to support or refute the conclusions of previous studies? Under what conditions do other studies come to the same or to contradictory conclusions? There are widely divergent views among ecologists on how to answer such questions.

The need to generalize is implicit in the way experimental results are interpreted in research articles, in textbooks, and in communicating the findings of ecological research to the general public. Textbooks commonly generalize from the results of single experiments to general truths about ecological processes and interactions or to entire species or even trophic levels. Some ecologists are hesitant to extrapolate results beyond individual experiments or across studies. Raudenbush (1991, p. 33), however, points out that "summarizing may be regarded as the essential subject matter of statistical science," and that, whether the summary is based on the results of a single study or many studies and is qualitative or numerical, summarizing evidence is a fundamental task in all research. Raudenbush raises interesting issues about the tension between completeness and parsimony in generalizing results from data, suggesting that if parsimony were not important, the best summary would simply be the raw data. Most importantly, we risk missing the patterns and truths that may be contained in the results of a group of studies when we concentrate only on the individual details in each one.

Meta-analysis allows us to reach general conclusions about a domain of research differently than the manner with which most ecologists are familiar. In a conventional review, we compare the results of different studies verbally in a research review, yet it is often desirable to compare and synthesize results quantitatively. When a number of independent studies have been carried out on a particular question, it might be important to assess, for example, the overall impact of doubling atmospheric CO_2 levels on plant growth or to evaluate whether the survivorship of juveniles is affected on average by competitors. Usually we would like to know the magnitude and direction of the effect (Does the experimental treatment have a positive or negative effect? Is this effect significantly greater than zero? Is the effect small or large?), and how variable that effect is among studies (Do the results of the studies seem to agree, or are there statistically significant differences among them?). In many cases, differences among classes of studies in the magnitude of the effect are of the most interest. Are the effects of predation different at different trophic levels? Do C_3 plants respond differently to CO_2 enrichment than C_4 plants, as predicted by theory? Is competition among plants greater in productive habitats than in unproductive habitats? In the field of ecology, these kinds of questions have been addressed using a number of different approaches.

18.2 Statistical Issues

At first, it might seem obvious that we could just count up the number of statistically significant results in the various studies to get some idea of the importance of an effect (i.e., how large it is), how frequently it occurs, and the magnitude of the effect among different kinds of studies. Quantitative ecological reviews have commonly employed this method in recent years. Unfortunately, this "vote-counting" approach is subject to serious flaws, because the significance level of a study is a function not only of the magnitude of the effect, but also of its sample size. Studies with small sample sizes are less likely to have statistically significant results than large studies, even when both have effects of exactly the same magnitude. In fact, because small studies are less likely to produce significant results (i.e., they have low power), it has been shown that vote counting is strongly biased toward finding no effect (e.g., see Hedges and Olkin 1985).

The problems associated with vote counts are particularly acute when effects and sample sizes are modest—almost always the case in ecological experiments. Counting up the number of significant results is not a reliable indication of whether an effect is real (that is, different than zero), how important the effect is, how frequently or under what circumstances it exists, or whether the studies are in agreement in their assessment of the magnitude of the effect (e.g., Hunter et al. 1982; Hedges and Olkin 1985). Narrative reviews may also suffer from a subtle form of vote counting, in that they are also likely to base their conclusions about the existence and frequency of ecological phenomena on the statistical significance of the outcomes or on authors' summaries of their own results based on probability levels, without considering sample size and statistical power.

Typically, a meta-analysis begins by representing the outcome of each experiment by a quantitative index of the effect size. This effect size is chosen to reflect (1) differences between experimental and control groups or (2) the degree of relationship between the independent and dependent variables in a way that is independent of sample size and of the scale of measurement used in the experiment. Meta-analytic techniques most commonly serve to estimate the average magnitude of the effect across all studies, to test whether that effect is significantly different from zero, and to examine potentially causative differences in the effect among studies. Because effect size is not dependent on sample size (in contrast to the dependence of significance level on sample size), meta-analysis procedures are not subject to the problems of vote counting.

Meta-analysis offers a potentially invaluable tool for ecological research. Although ecological experiments are essentially never replicated in any strict sense, it has become increasingly critical to be able to understand what an entire body of experimental data addressing conceptually similar questions tells us. How important and how consistent is a particular effect across a wide variety of systems? Does the effect differ substantially in aquatic and terrestrial habitats, or in disturbed and undisturbed environments, or in large versus small patches? Most ecologists can probably think of questions like this for systems with which they are familiar, where a substantial body of data already exists or could be collected. Furthermore, it is likely to become increasingly urgent for ecologists to be able to generalize from the results of studies conducted in different systems as we are increasingly called upon to address the consequences of global change. Data collected in diverse locations with similar research objectives will have to be summarized to understand emerging patterns and to detect outliers. Quantitative generalizations can help us track both stability and change; outliers may be populations, species, or ecosystems in imminent danger or those that are unusually resilient.

Some of the specificity and richness of detail of individual studies is necessarily overlooked in a meta-analysis for the sake of being able to generalize across studies. Depending on the questions being asked, this may or may not be appropriate. For example, in one of the studies included in the example of a meta-analysis herein (Gurevitch 1986), the effects of competition on two grass species was examined at three positions along an environmental gradient. Although the contrast between these responses was an important part of the primary study, it was ignored in the meta-analysis. Because the meta-analysis seeks to estimate the effect of competition on the growth of terrestrial plants and all of the measurements offer information on this, the particular details about the influence of the environmental gradient are sacrificed. It is always true that when summarizing the behavior of a population, the individual characteristics of members of that population are overlooked, whether the population consists of a group of individual studies or the population is a group of individual animals whose responses are summarized using conventional statistics.

18.3 Statistical Solution

18.3.1 Necessary and Usual Ingredients in Meta-analysis

Carrying out a meta-analysis requires, like any statistical analysis, both data and statistical models to analyze the data. The data in a meta-analysis generally takes the form of standardized metrics of effect size and their associated sampling variances. A number of different metrics of effect size are typically used in meta-analyses. Effect size metrics that have been used in ecological and evolutionary meta-analyses include the standardized mean difference, d (detailed subsequently), Pearson's correlation coefficient, r (analyzed as the z-transform), and lr, the log response ratio (the natural log of the ratio between the mean of the experimental and control groups; Hedges et al. 1999). Other measures of effect size also exist, such as the odds ratio for categorical outcomes (Cooper and Hedges 1994; Rosenberg et al. 1999; chapter 11). The example presented in this chapter uses d, but the analysis is essentially the same for other standard effect size measures. It is important that the statistical properties of the metric of effect size used be well understood. Using well-established measures also makes the results much more readily understandable and useful to readers.

Once we obtain a measure of the effect size from each experiment, the analysis can be carried out. Both continuous and categorical models have been developed in meta-analysis, and both approaches generally depend on weighting by the inverse of the sampling variances of each data point (i.e., each effect size). It is *not* generally appropriate to use familiar methods such as ANOVA and regression for a meta-analysis; doing so risks serious inaccuracies and bias. Effect sizes have different properties than do measurements made on individuals, and it is highly likely that this will result in serious violations of the assumptions of these familiar parametric tests. Gurevitch and Hedges (1999) explain these issues in detail. Fortunately, alternatives exist that are relatively easy to use and that take the properties of meta-analytic data into account properly. In the present chapter, we will only discuss the analysis of categorical (ANOVA-like) models, but continuous models (based on weighted regression approaches) can also be employed in meta-analysis (e.g., Hedges and Olkin 1985, 2000; Rosenberg et al. 1999).

Both categorical and continuous approaches to meta-analysis can rely upon fixed, random, or mixed models. We explain those differences and outline a method for each of these. In addition, statistical tests of significance can be carried out using parametric or randomization approaches (Adams et al. 1997; Rosenberg et al. 1999). We present only a parametric approach here, but readers are encouraged to investigate the use of randomization techniques as well. New methods for the analysis of meta-analytic data are still being developed. For example, a recent development is a method for the meta-analysis of simple factorial experiments, where the same two factors are manipulated in all studies (Gurevitch et al. 2000).

18.3.2 Statistical Models for Meta-analysis: Fixed Effect and Mixed Models

One common metric of *effect size* in meta-analysis is d, the difference between the means of two groups of individuals (typically, an experimental group and a control group), divided by their pooled standard deviation to standardize the effect among studies. The effect size then is the difference in standard deviation units between the experimental and control groups. This number is multiplied by a correction factor, J, to correct for small-sample bias (see subsequent discussions; Hedges and Olkin 1985). Cohen (1969) provides a conventional interpretation of the magnitude of effect sizes: 0.2 is a "small" effect, 0.5 is "medium" in magnitude, 0.8 is "large," and presumably any effect greater than 1.0 standard deviations difference between experimental and control groups would be "very large." The sampling variance of the effect size from each experiment provides a measure of its sampling uncertainty, which is used in subsequent calculations, and can also be used to construct a confidence interval around the effect size. The data necessary from each study to calculate d and its variance are the means of the two groups (e.g., experimental and controls), the standard deviations about these means, and the number of individuals in each group. Thus, it is not necessary to have access to the original raw data to perform a meta-analysis; the basic information and simple statistics which *should* be reported in published articles are sufficient.

Unfortunately, articles published in even the most selective ecological journals may be lacking this basic information on the outcome of the experiments they report; for example, sample size is often omitted. It is also often difficult to understand how numbers in published articles were obtained. Authors should include means, standard deviations (or other measures of variation from which standard deviations can be derived), and sample sizes (describing the units, i.e., $n = 30$ snails, or $n = 5$ cages, with 6 snails per cage; mean weights per cage were used in the analysis) when publishing the results of experiments, and reviewers should look for this information when evaluating articles for publication. We cannot overemphasize that it is insufficient to report probability levels alone as the outcome of an experiment, because they are insufficient to evaluate the results.

Most meta-analyses have been based on *fixed effects* models. That is, it is assumed that a class of studies with similar characteristics share a common true effect size. (Another assumption made in the models discussed here is that the data in the experimental and control groups are normally distributed.) Differences among studies in the actual effect size measured are assumed to be due to sampling error. This assumption is probably rarely justified in ecology, and in many instances may not be very reasonable in other fields either. In a *mixed model*, it is assumed that the studies within a class share a common mean effect but that there is also random variation among studies in a class, in addition to sampling variation. Another way of picturing a mixed model is to think of the effect size of a particular study as being composed of various components: part of the effect is "fixed," or characteristic of all studies in a class (all herbivores share certain characteristic responses, for example), part of the effect is due to the particular characteristics of that one study in which it differs at random from other studies, and part of the effect is due to "error," or sampling variation. Mixed models in meta-analysis are analogous to mixed models in ANOVA (for a discussion of fixed and random effects in primary studies, see chapter 4). Mixed models are often preferable to fixed effects models in ecological data synthesis because the assumptions of the mixed model are more likely to be satisfied. Strict random effects models (in which all variation among studies is random variation) have been previously used in other fields in meta-analysis (Hedges 1983), but they are generally not of interest for ecological research. Both fixed effect and mixed effect models in meta-analysis are demonstrated in this chapter.

18.3.3 Conducting the Meta-analysis: Gathering the Data

The process of gathering references and making decisions regarding the studies to include in a meta-analysis involves complex issues to which we cannot hope to do justice in a short chapter. In many ways it is no different than in any other review of research literature, although the quantitative nature of the review brings certain problems into sharper focus. Many substantive publications offer specific suggestions on gathering and handling the data for a meta-analysis (e.g., Cooper 1989; Cooper and Hedges 1994; Light and Pillemer 1984). If the meta-analysis does not include every article published on a topic, the criteria for inclusion should clearly be reasonable and scientifically defensible. Publication bias exists if articles that demonstrate no effect are less likely to be published than those that come up with statistically significant effects. In that case, a meta-analysis—or any other review—of the published literature will overestimate the effect under consideration. If publication bias exists, it can clearly result in inaccurate conclusions, and should be carefully considered in carrying out a meta-analysis or any other type of review. Problems associated with publication bias have been discussed at length among meta-analysts, and various approaches have been proposed to estimate its magnitude and counter its influence (e.g., see Hedges and Olkin 1985; Cooper and Hedges 1994; Rosenberg et al. 1999).

The extent to which we can extrapolate the conclusions drawn from a research review, whether quantitative or narrative, depends in part on the quality and nature of the data that are collected. As every ecologist who has gathered data is aware, carefully and properly setting up the experiments or observations and collecting the measurements are of the greatest importance in getting good results. Good analysis cannot lead to sound conclusions from bad data. This is true regardless of whether we are carrying out a single study, a conventional research review, or a meta-analytic research synthesis. If the collection of studies is biased or incomplete, the conclusions drawn from a meta-analysis (or any other analysis) of that collection will be suspect or at least limited. Some controversy exists regarding whether all studies on a topic should be included in the analysis, or whether "low-quality" studies (e.g., medical studies in which the treatments were not assigned to subjects randomly) should be excluded. One option is to include all studies, noting which ones are of low quality by some a priori criteria, and to test whether the results of these studies differ from those of higher quality, rejecting them if they do.

Ecological data are often published in the form of graphs rather than tables. The best way to access such data is by digitizing the graph; this is now a simple matter using a scanner with digitizing software such as TechDig (Jones 1999). Some suggestions on how to obtain sample sizes and standard deviations from ecological articles when they are less than obvious have been offered elsewhere (Gurevitch et al. 1992). Once collected, descriptive information and data can be recorded on data sheets and entered onto a spreadsheet. Careful checking and rechecking of the data sheets and computer entries is critical. Although it is not difficult to carry out a small, fixed effects meta-analysis directly on a spreadsheet, as the data set becomes larger, it becomes almost impossible not to incorporate errors (as we did, sorry to say, in the first edition of this book). Mixed model analysis is much more complicated on a spreadsheet, and resampling tests require extensive programming. For this reason, one of us (JG) worked with two colleagues to develop a software package for meta-analysis that was appropriate for ecological and other data (Rosenberg et al. 1999). This software, MetaWin 2.0 (Rosenberg et al. 1999), will carry out all of the analyses described here in addition to other approaches; the results are more accurate and reliable than doing the analyses on a spreadsheet.

Careful thought must be given as to how each experiment should be recorded. The "control" identified by the author may not be what the meta-analyst identifies as a control for the purposes of research synthesis. In the following example, the "control group" was made up of those organisms with natural densities of competitors, whereas the "experimental group" had experimentally manipulated densities of competitors, regardless of what the authors called the groups (this more uniform organization of the data allows us to address additional questions; see Gurevitch et al. 1992). Because competitor densities can be reduced or increased, the expected sign of the response to the manipulation will vary accordingly, if competition has been demonstrated. To make it possible to combine the estimates of the effect of competition, the sign of the response was reversed if competitor density was increased (reduction is much more common). The sign of the response in a meta-analysis can and often be confusing and must be carefully considered and explained. Here, for example, a positive response indicates that organisms responded positively to the removal of competitors, not that they responded positively to competition. Another issue is specifying the units for which the effects are being evaluated and identifying the appropriate measures to use from each study. Although in many cases it is the effect on individual organisms that is summarized by the effect size (as measured by characteristics of the organisms such as growth or seed production), in other cases the "individual" is a plot or other unit (such as when survivorship or density is the measure of outcome). All of these issues must be considered and specified explicitly in a meta-analysis.

Another statistical issue in meta-analysis is the lack of independence among the effect size measures being combined. There are various sources of nonindependence. One simple source of nonindependence is that more than one kind of response (e.g., growth and mass) may be measured on the same organisms. These measurements are not independent and should not be included in the same meta-analysis, but it might be reasonable to conduct separate syntheses for each common measure of response in a group of studies. Not only may several different kinds of measurements be made on each organism, but the same measurements may be repeated over time (chapter 8). In the latter case, it is not correct to use more than one measurement per individual as if they were independent observations on different individuals, and so we may choose to take the final measurement made at the end of each experiment (e.g., Gurevitch et al. 1992). In complex experiments in which various manipulations are compared to a single control (e.g., competition of a target species with several different species of competitors), there may be no way to avoid including the same control group a number of times as the basis for calculating more than one effect size (the effect of each of the competing species) even though this is not ideal. In ecological experiments, more than one species may often be measured. Such cases of nonindependence are less clear; being too conservative will cause the loss of too much valuable data. If the proportion of comparisons in which there is nonindependence (e.g., use of a control group mean for more than one comparison) is small or if the degree of nonindependence is relatively small (i.e., correlations among responses of different species are small), there will be little effect on the conclusions drawn from the meta-analysis. However, a great deal of nonindependence will inflate the significance levels of statistical tests and underestimate confidence intervals. Currently, no clear consensus exists as to the best ways to handle nonindependence in meta-analytic data (nor in primary data, for that matter), and the degree to which the conclusions are robust or sensitive to these problems is not well understood.

The results of an experiment may be tested and reported using many different statistical approaches. We would usually not include results reported using different kinds of statistics (e.g., means, slopes, correlations) in a single meta-analysis. The most common forms of data reported in ecological studies are means and some measure of variance based on data that are assumed to be normally distributed; we use these data here to calculate effect sizes. Most meta-analysis approaches, including the fixed effects and mixed effects models presented here, assume normally distributed data. Resampling methods relax some of these assumptions (Adams et al. 1997; Rosenberg et al. 1999).

Although the scale of the measurements is standardized in calculating effect sizes (so that the responses of large organisms become comparable with those of small ones), the decision about whether to combine results is really a scientific rather than a statistical one. For example, it might not be very meaningful to combine experiments using very different sorts of measurements (such as biomass and survivorship) in a single meta-analysis. It would probably also make little sense to combine experiments conducted under starkly contrasting conditions, such as field and lab experiments, in a single meta-analysis. It is assumed that the measurements can be equated linearly among the studies, and this may not be a biologically reasonable assumption for very different sorts of responses. The decision about what to combine is a substantive one that depends on the generality of the questions the summary is designed to answer. If a very general summary of many different kinds of evidence is the goal, then results measured in many different ways may be combined. Most verbal reviews and vote counts in ecology have taken this approach, combining all measures of outcome in a single summary of the effects. However, combining only those studies that measure similar quantities (e.g., biomass, growth, or survivorship) allows us to ask more specific questions. There may also be biological reasons to expect that different kinds of responses may reveal different information about the phenomena being investigated.

18.3.4 An Example

Data were collected as part of a larger study on the effects of competition as studied in field experiments. The larger study included all measures of outcome (survivorship, reproductive output, density, and so on) in response to the manipulation of competitors for organisms in a wide range of trophic levels and systems. An analysis of organisms' responses measured as biomass has been published (Gurevitch et al. 1992) and includes a detailed description of how articles were selected for inclusion in the study and how the data were collected. For the example presented here, we chose a small part of the larger data set for convenience in demonstrating the methods involved. The data that we examine are for the responses of primary producers to competition, where the responses were measured as recruitment (an increase in the number of individuals) or growth (an increase in the size of individuals). The data set is original and has not been edited other than arbitrarily selecting organisms of only one trophic level. Much more can be done with the data in a full meta-analysis than is attempted here, where we address only a small number of issues so that we can focus on how to actually carry out the analyses.

The studies were categorized a priori into three classes (terrestrial, lentic, and marine) that we wished to compare. The basic information taken from the studies can be found in columns E–J in table 18.1.

First, we calculate the unbiased standardized mean difference, d_{ij}, for each study,

$$d_{ij} = \frac{\bar{X}_{ij}^E - \bar{X}_{ij}^C}{s_{ij}} J \tag{18.1}$$

where d is calculated for the jth study in the ith class, and

$\bar{X}_{ij}^C$ = mean of the control group
$\bar{X}_{ij}^E$ = mean of the experimental group
N_{ij}^C = total number of individuals in the control group
N_{ij}^E = total number of individuals in the experimental group
s_{ij} = pooled standard deviation of the control and experimental groups

Thus,

$$s_{ij} = \sqrt{\frac{(N_{ij}^E - 1)(s_{ij}^E)^2 + (N_{ij}^C - 1)(s_{ij}^C)^2}{N_{ij}^E + N_{ij}^C - 2}} \tag{18.2}$$

where

s_{ij}^C = standard deviation of the individuals in the control group
s_{ij}^E = the standard deviation of the individuals in the experimental group

(Note that we use the terms *study*, *experiment*, and *comparison* interchangeably to indicate a comparison between a single experimental group and its control; several such comparisons may be used from a single published article.) The term J corrects for bias due to small sample size:

$$J = 1 - \frac{3}{4(N_{ij}^E + N_{ij}^C - 2) - 1} \tag{18.3}$$

(Hedges and Olkin 1985). As the sample size increases, J approaches 1. In the majority of these experiments, the experimental group represents a *decrease* in the density of the hypothesized competitors, and a positive effect size indicates a positive response to this decrease—that is, a positive value for d indicates competition. In those cases in which the experimental group represents an *increase* in competitor density, the sign of d is reversed in the table (for a complete discussion, see Morrow 1990). Finally, the variance in the effect for the jth study in the ith class is approximated by

$$v_{ij} = \frac{N_{ij}^E + N_{ij}^C}{N_{ij}^E N_{ij}^C} + \frac{d_{ij}^2}{2(N_{ij}^E + N_{ij}^C)} \tag{18.4}$$

(Hedges and Olkin 1985). This term may be used for comparing effect sizes among studies directly, for calculating a confidence interval around the effect sizes for each study to evaluate its magnitude (e.g., is the effect significantly greater than zero?), and for combining effect sizes across studies (see next section). The values calculated from the example can be found in table 18.1.

Another common measure of effect size for ecological studies is the log response ratio, *lr*, where

$$lr = \ln\left(\frac{\bar{X}_{ij}^{E}}{\bar{X}_{ij}^{C}}\right) = \ln(\bar{X}_{ij}^{E}) - \ln(\bar{X}_{ij}^{C}) \qquad (18.5)$$

(Hedges et al. 1999). If $\bar{X}_{ij}^{E}$ and $\bar{X}_{ij}^{C}$ and below are normally distributed and both are greater than zero, then lr is approximately normally distributed with variance

$$\frac{(s_{ij}^{E})^2}{N_{ij}^{E} \, (\bar{X}_{ij}^{E})^2} + \frac{(s_{ij}^{C})^2}{N_{ij}^{C} \, (\bar{X}_{ij}^{C})^2} \qquad (18.6)$$

The mean effect size of lr in a meta-analysis can be calculated in the same way as when using d as the measure of effect size, using the reciprocal of the variance of each lr as the weight, w_{ij}. Likewise, confidence intervals, homogeneity tests, and so on are calculated in the same fashion, substituting lr and its sampling variance for d and its variance.

18.3.5 Combining Effect Sizes in the Fixed Effects Model

The cumulated (i.e., combined) mean effect size across studies for the fixed effects model within the ith class, d_{i+}, is a weighted average of the effect size estimates for the studies in that class. The weights w_{ij} used for combining effect sizes are the reciprocals of the sampling variances, $w_{ij} = 1/v_{ij}$. The weighted estimate of the true effect size δ_i is:

$$d_{i+} = \frac{\sum_{j=1}^{k_i} w_{ij} \, d_{ij}}{\sum_{j=1}^{k_i} w_{ij}} \qquad (18.7)$$

where k_i equals the number of comparisons in class i (Hedges and Olkin 1985). The cumulated effect size is a weighted average in which larger studies are counted more heavily than smaller studies, on the assumption that larger sample sizes will yield more precise results (Hedges and Olkin 1985). The variance of d_{i+}, $s^2(d_i+)$ is

$$s^2(d_{i+}) = \frac{1}{\sum_{j=1}^{k_i} w_{ij}} \qquad (18.8)$$

The lower and upper limits for the 95% confidence interval for d_{i+}, d^L and d^U, respectively, are:

$$d^L = d_{i+} - [Z_{\alpha/2} s(d_{i+})] \qquad (18.9a)$$

$$d^U = d_{i+} + [Z_{\alpha/2} s(d_{i+})] \qquad (18.9b)$$

where Z is the two-tailed critical value of the standard normal distribution. (The tests presented throughout this chapter for comparing effect sizes are all two-tailed, but calculation of one-tailed tests of particular hypotheses would be a straightforward extension of this approach.) Taking the sums needed in equation

Table 18.1 Data and calculations for the fixed effects model

A	B	C	D	E	F	G	H	I	J	K	L	M	N	O	P	Q
Publication	Species	Category	Source	N_c	N_e	X_c	X_e	S_c	S_e	J	d	v_{ij}	w_{ij}	wd	wd^2	k
Fowler 1986	*Bouteloua rigidiseta*	terrestrial	Table 2	7	7	78.14	79.71	40.650	40.650	0.936	0.036	0.286	3.499	0.127	0.005	1
Fowler 1986	*Aristida longiseta*	terrestrial	Table 2	7	7	18.86	26	9.170	9.170	0.936	0.729	0.305	3.282	2.392	1.744	1
Platt & Weis 1985	*Mirabilis hirsuta*	terrestrial	Table 1	6	6	-1.8	-2.1	0.490	0.490	0.923	0.565	0.347	2.885	1.631	0.922	1
Platt & Weis 1985	*Verbena stricta*	terrestrial	Table 1	5	5	-2.2	-2.8	0.224	0.447	0.903	1.533	0.517	1.932	2.962	4.540	1
Platt & Weis 1985	*Solidaga rigada*	terrestrial	Table 1	7	7	-2.1	-3	0.265	0.529	0.936	2.014	0.431	2.322	4.677	9.421	1
Platt & Weis 1985	*Asclepias syriaca*	terrestrial	Table 1	6	6	-2.3	-4.2	0.490	1.225	0.923	1.880	0.481	2.081	3.912	7.356	1
Gross & Werner 1982	*Verbascum thapsus*	terrestrial	Table 8	3	3	85.3	285.7	115.008	153.806	0.800	1.181	0.783	1.277	1.508	1.780	1
Gross & Werner 1982	*Oenothera biennis*	terrestrial	Table 8	3	3	0	3	0.000	2.425	0.800	1.400	0.830	1.205	1.687	2.361	1
Gross & Werner 1982	*Verbascum thapsus*	terrestrial	Table 8	3	3	0	2.00	0.000	2.078	0.800	1.089	0.765	1.306	1.422	1.548	1
Gross & Werner 1982	*Oenothera biennis*	terrestrial	Table 8	3	3	0	1.67	0.000	1.732	0.800	1.091	0.766	1.306	1.424	1.554	1
Pons & van der Toorn 1988	*Plantago major*	terrestrial	Table 4	5	5	17	17	7.603	5.367	0.903	0.000	0.400	2.500	0.000	0.000	1
Pons & van der Toorn 1988	*Plantago lanceolata*	terrestrial	Table 4	5	5	47	37	10.286	9.391	0.903	-0.917	0.442	2.262	-2.075	1.903	1
Burton & Mueller-Dombois 1984	*Metrosideros polymorpha*	terrestrial	Table 2	4	4	87	272	37.712	183.532	0.870	1.214	0.592	1.689	2.051	2.490	1
Gurevitch 1986	*Stipa neomexicana*	terrestrial	Fig. 4 (log)	18	20	-0.113	0.294	0.255	0.215	0.979	1.694	0.143	6.977	11.821	20.029	1
Gurevitch 1986	*Stipa neomexicana*	terrestrial	Fig. 4 (log)	20	20	-0.163	0.412	0.588	0.218	0.980	1.273	0.120	8.316	10.586	13.476	1
Gurevitch 1986	*Stipa neomexicana*	terrestrial	Fig. 4 (log)	18	20	0.140	0.632	0.380	0.359	0.979	1.303	0.128	7.818	10.190	13.283	1
Gurevitch 1986	*Aristida glauca*	terrestrial	Fig. 4 (log)	20	20	-0.184	0.259	0.326	0.238	0.980	1.519	0.129	7.762	11.789	17.904	1
Gurevitch 1986	*Aristida glauca*	terrestrial	Fig. 4 (log)	20	20	-0.075	0.354	0.487	0.182	0.980	1.144	0.116	8.595	9.829	11.240	1
Gurevitch 1986	*Aristida glauca*	terrestrial	Fig. 4 (log)	20	20	0.147	0.541	0.340	0.299	0.980	1.206	0.118	8.461	10.206	12.310	1
McCreary et al. 1983	*Eleocharis acicularis*	lentic	Fig. 1A&B	4	4	281.11	-201.03	158.038	27.520	0.870	3.696	1.354	0.739	2.730	10.091	1
McCreary et al. 1983	*Juncus pelocarpus*	lentic	Fig. 1A&B	4	4	187.31	-155.32	80.163	41.252	0.870	4.674	1.865	0.536	2.506	11.711	1
Stimson 1985	*Acropora* spp.	marine	Table 7	7	7	11.8	16.0	3.080	3.370	0.936	1.218	0.339	2.953	3.596	4.380	1
Stimson 1985	*Pocillopora verrucosa*	marine	Table 7	3	10	0.4	9.5	1.470	7.230	0.930	1.288	0.497	2.011	2.592	3.339	1

Reference	Species	Habitat	Source													
Reed & Foster 1984	*Pterygophora californica*	marine	Table 1	20	20	0	14.1	7.603	7.603	0.980	1.818	0.141	7.077	12.864	23.384	1
Reed & Foster 1984	*Macrocystis pyrifera*	marine	Table 1	20	20	0	7.1	3.130	3.130	0.980	2.223	0.162	6.182	13.742	30.547	1
Reed & Foster 1984	*Desmarestia ligulata*	marine	Table 1	20	20	0	1.4	1.789	1.789	0.980	0.767	0.107	9.315	7.145	5.481	1
Reed & Foster 1984	*Desmarestia ligulata*	marine	Table 3	10	10	82.2	94	29.093	9.171	0.958	0.524	0.207	4.834	2.533	1.327	1
Reed & Foster 1984	*Desmarestia kurilensis*	marine	Table 3	10	10	8.3	10.5	14.546	11.068	0.958	0.163	0.201	4.983	0.812	0.132	1
Reed & Foster 1984	*Nereocystis luetkeana*	marine	Table 3	10	10	0	20	42.691	42.691	0.958	0.449	0.205	4.877	2.188	0.982	1
Johnson & Mann 1988	*Laminaria longicruris*	marine	Table 2	2	2	3.63	18.5	3.352	4.257	0.571	2.218	1.615	0.619	1.373	3.046	1
Johnson & Mann 1988	*Laminaria longicruris*	marine	Table 2	2	2	0	0.25	0.000	0.354	0.571	0.571	1.041	0.961	0.549	0.314	1
Johnson & Mann 1988	*Laminaria longicruris*	marine	Table 2	2	2	3.63	2.25	3.352	0.707	0.571	−0.326	1.013	0.987	−0.321	0.105	1
Turner 1985	*Rhodemela larix*	marine	Table 4	4	4	0	34.8	0.000	58.200	0.870	0.735	0.534	1.873	1.378	1.013	1
Turner 1985	*Cryptosiphonia woodii*	marine	Table 4	4	4	0	25.3	0.000	35.800	0.870	0.869	0.547	1.827	1.588	1.380	1
Turner 1985	*Phaeostrophion irregulare*	marine	Table 4	4	4	5.4	23.6	10.880	47.000	0.870	0.463	0.513	1.948	0.902	0.417	1
Turner 1985	*Odonthalia floccosa*	marine	Table 4	4	4	1.8	10.5	5.200	24.200	0.870	0.432	0.512	1.954	0.845	0.365	1
Turner 1985	*Microcladia borealis*	marine	Table 4	4	4	0	10.3	0.000	17.400	0.870	0.728	0.533	1.876	1.365	0.994	1
Turner 1985	*Fucus distichus*	marine	Table 4	4	4	0	8.7	0.000	17.000	0.870	0.629	0.525	1.906	1.199	0.755	1
Turner 1985	*Iridaea heterocarpa*	marine	Table 4	4	4	0	5.7	0.000	14.000	0.870	0.501	0.516	1.939	0.971	0.486	1
Turner 1985	*Bossiella plumosa*	marine	Table 4	4	4	10.8	5.4	15.800	8.800	0.870	−0.367	0.508	1.967	−0.722	0.265	1
Dungan 1986	*Ralfsia pacifica*	marine	Table 2	4	4	21.25	37.25	9.540	22.020	0.870	0.820	0.542	1.845	1.513	1.240	1
Dungan 1986	*Ralfsia pacifica*	marine	Table 2	4	4	40.25	20.25	8.780	9.000	0.870	−1.956	0.739	1.353	−2.646	5.177	1
Dungan 1986	*Ralfsia pacifica*	marine	Fig. 3B	5	5	15.8445	11.9533	10.787	6.240	0.903	−0.399	0.408	2.451	−0.978	0.390	1

	$d+$	$s^2(d+)$	Q Sums:				
terrestrial	1.167	0.014	23.301	73.783	86.138	123.864	19
lentic	4.107	0.784	0.300	1.275	5.236	21.802	2
marine	0.798	0.015	43.612	65.739	52.488	85.520	22

18.7 from the bottom of columns N and O in table 18.1, the values for the cumulated effect sizes for the three classes in the example are

$$d_{i+} \text{ (terrestrial producers)} = 86.138/73.783 = 1.17$$
$$d_{i+} \text{ (lentic producers)} = 5.236/1.275 = 4.11$$
$$d_{i+} \text{ (marine producers)} = 52.488/65.739 = 0.80$$

and from equations 18.8, 18.9a, and 18.9b, using the sums at the bottom of column N in table 18.1, their variances and 95% confidence intervals are, respectively,

$$\text{terrestrial} = 1/73.783 = 0.014 \qquad 1.68 \pm 1.96\sqrt{0.014} = 0.940 \text{ to } 1.396$$
$$\text{lentic} = 1/1.275 = 0.784 \qquad 4.107 \pm 1.96\sqrt{0.784} = 2.371 \text{ to } 5.843$$
$$\text{marine} = 1/65.739 = 0.015 \qquad 0.798 \pm 1.96\sqrt{0.015} = 0.557 \text{ to } 1.040$$

We conclude from these results that when examined across all studies, terrestrial, lentic, and marine plants all exhibited statistically significant effects of competition on average. The values are significantly greater than zero (at $P < 0.05$) because the 95% confidence intervals for the effect sizes do not overlap zero. The effects of the removal of competitors (the d_{i+}) are considered to be large for marine plants, and very large for terrestrial and lentic plants (Cohen 1969).

The grand mean effect size across all classes, d_{++}, is

$$d_{++} = \frac{\sum_{i=1}^{m} \sum_{j=1}^{k_i} w_{ij} d_{ij}}{\sum_{i=1}^{m} \sum_{j=1}^{k_i} w_{ij}} \qquad (18.10)$$

where m is the total number of classes (here, $m = 3$: terrestrial, lentic, and marine). The variance of the grand mean,

$$s^2(d_{++}) = \frac{1}{\sum_{i=1}^{m} \sum_{j=1}^{k_i} w_{ij}} \qquad (18.11)$$

can also be calculated. In the example,

$$d_{++} = 143.862/140.797 = 1.02$$
$$s^2(d_{++}) = 1/140.797 = 0.007$$

Thus, the mean effect of competition across all studies was large. The previous calculations could have been carried out by substituting lr (or any other effect size metric) and its variance, if that was judged to be more appropriate for the data being analyzed.

The null hypothesis that all effect sizes are equal, versus the alternative hypothesis that at least one of the true effect sizes in a series of comparisons differs from the rest, can be tested with the homogeneity statistic Q which has approximately a χ^2-distribution with degrees of freedom equal to 1 less than the total number of studies. The greater the value of Q, the greater the heterogeneity in effect sizes among comparisons. The total homogeneity, Q_T, can be partitioned into within-class homogeneity, Q_W, and between-class homogeneity, Q_B, much as one can partition variance in an ANOVA,

$$Q_B + Q_W = Q_T$$

Within-class homogeneity, Q_W,

$$Q_W = \sum_{i=1}^{m} \sum_{j=1}^{k_i} w_{ij}(d_{ij} - d_{i+})^2 \tag{18.12}$$

is a measure of the variation among studies within classes (across all of the classes), whereas the between-class homogeneity, Q_B, is a measure of the variation between classes in mean effect size,

$$Q_B = \sum_{i=1}^{m} \sum_{j=1}^{k_i} w_{ij}(d_{i+} - d_{++})^2 \tag{18.13}$$

which is distributed as a χ^2-statistic with degrees of freedom equal to the number of classes minus 1 (Hedges and Olkin 1985). A computational formula for the Q_T statistic is

$$Q^T = \sum_{i=1}^{m} \sum_{j=1}^{k_i} w_{ij}d_{ij}^2 - \frac{\left(\sum_{i=1}^{m} \sum_{j=1}^{k_i} w_{ij}d_{ij}\right)^2}{\sum_{i=1}^{m} \sum_{j=1}^{k_i} w_{ij}} \tag{18.14}$$

The within-class homogeneity statistic across all studies, Q_W, is calculated as the sum of the within-class statistics $Q_{W1}, Q_{W2}, \ldots, Q_{Wm}$ for the m classes. Each within-class statistic Q_{Wi} can be calculated using the computational formula

$$Q_{Wi} = \sum_{j=1}^{k_i} w_{ij}d_{ij}^2 - \frac{\left(\sum_{j=1}^{k_i} w_{ij}d_{ij}\right)^2}{\sum_{j=1}^{k_i} w_{ij}} = \sum_{i=1}^{m} w_{i+}d_{i+}^2 - \frac{\left(\sum_{i=1}^{m} w_{i+}d_{i+}\right)^2}{\sum_{i=1}^{m} w_{i+}} \tag{18.15}$$

with $(k-1)$ df, Q_B is obtained by subtracting Q_W from Q_T, and w_{i+} is the sum of the fixed effects model weights for each class.

The values for the fixed effect Q_{Wi}-statistics within each class in our example are found using equation 18.15 and the sums at the bottom of table 18.1:

terrestrial: $Q_{W1} = 123.864 - (84.618)^2/73.783 = 23.301$, 18 df, $0.5 > P > 0.1$
lentic: $Q_{W2} = 21.802 - (5.236)^2/1.275 = 0.300$, 1 df, $0.9 > P > 0.5$
marine: $Q_{W3} = 85.520 - (52.488)^2/65.739 = 43.612$, 21 df, $P < 0.005$

The statistical significance of these statistics is evaluated using a standard χ^2-table. We interpret these results to mean that the studies in the first two classes are homogeneous: within these classes, the effect sizes differ by no more than would be expected due to random sampling variation. (Although it is possible that true differences between the two lentic studies could not be detected due to lack of power, we have both observed many instances in which a small number of studies resulted in significant values for the Q-statistic.) Studies in the third class (marine producers) apparently exhibited more variation than can be attributed to sampling error. The variation among studies within classes, even for marine plants, is actually quite small for ecological studies, particularly in light of the rather stringent assumption here that the studies within a class share a common true effect size. Next we calculate

$$Q_T = 231.185 - (143.862)^2/140.797 = 84.192, \text{ 42 df}, P < 0.001$$
$$Q_W = 23.301 + 0.297 + 43.614 = 67.212, \text{ 40 df}, P < 0.005$$
$$Q_B = 84.192 - 67.212 = 16.980 \text{ 2 df}, P < 0.001$$

across the three classes of primary producers. There is a highly significant difference between the three classes ($Q_B = 16.98$, 2 df, $P < 0.001$). Therefore, we conclude that in these studies there is a statistically significant difference among the responses of terrestrial, lentic, and marine producers to competition, when measured in terms of growth and recruitment. The confidence intervals for terrestrial and marine producers overlapped, but the confidence interval for lentic producers did not overlap either of the other classes, suggesting that lentic producers experienced greater competitive effects than terrestrial or marine producers, which did not differ from one another. In addition to an informal evaluation based on confidence intervals, formal procedures for constructing contrasts for mean effect sizes have been developed (section E in chapter 7 of Hedges and Olkin 1985).

18.3.6 The Mixed Model Analysis

The fixed effect variance of d_{ij}, or v_{ij}, is actually a conditional variance, because it depends on the assumption that there is one true effect size, δ, shared by all of the studies in the same class. The unconditional variance of d_{ij}, v_{ij}^* (which can also be thought of as the mixed model variance), assumes that there is random variation among studies in the effect of interest, which therefore do not share a common true effect size. (The asterisk here indicates the mixed model version of a term.) To get v_{ij}^*, we must add a term for the pooled within-class variance component, $\hat{\sigma}_{pooled}^2$, to the usual v_{ij}.

To carry out the mixed model analysis (table 18.2), several additional terms must be calculated. First, we calculate a constant c_i for each class,

$$c_i = \sum_{j=1}^{k_i} w_{ij} - \frac{\displaystyle\sum_{j=1}^{k_i} w_{ij}^2}{\displaystyle\sum_{j=1}^{k_i} w_{ij}} \tag{18.16}$$

where i is the class and k_i is the number of experiments in class i (note that this formula has been corrected from the original version published by Gurevitch and Hedges 1993). Then we compute the estimate of $\hat{\sigma}_{pooled}^2$ via

Table 18.2 Calculations for the mixed model

A Publication	B d	C v_{ij}	D w_{ij}	E w_{ij}^2	F v^*	G w^*	H w^*d
Fowler 1986	0.036	0.286	3.499	12.246	0.514	1.946	0.070
Fowler 1986	0.729	0.305	3.282	10.772	0.533	1.876	1.368
Platt & Weis 1985	0.565	0.347	2.885	8.322	0.575	1.740	0.983
Platt & Weis 1985	1.533	0.517	1.932	3.734	0.746	1.341	2.056
Platt & Weis 1985	2.014	0.431	2.322	5.393	0.659	1.518	3.057
Platt & Weis 1985	1.880	0.481	2.081	4.329	0.709	1.411	2.653
Gross & Werner 1982	1.181	0.783	1.277	1.632	1.011	0.989	1.168
Gross & Werner 1982	1.400	0.830	1.205	1.452	1.058	0.945	1.323
Gross & Werner 1982	1.089	0.765	1.306	1.706	0.993	1.010	1.099
Gross & Werner 1982	1.091	0.766	1.306	1.706	0.994	1.009	1.111
Pons & van der Toorn 1988	0.000	0.400	2.500	6.250	0.628	1.592	0.000
Pons & van der Toorn 1988	−0.917	0.442	2.262	5.117	0.670	1.492	−1.368
Burton & Mueller-Dombois 1984	1.214	0.592	1.689	2.852	0.820	1.219	1.480
Gurevitch 1986	1.694	0.143	6.977	48.679	0.372	2.691	4.560
Gurevitch 1986	1.273	0.120	8.316	69.148	0.348	2.870	3.653
Gurevitch 1986	1.303	0.128	7.818	61.120	0.356	2.808	3.660
Gurevitch 1986	1.519	0.129	7.762	60.249	0.357	2.801	4.254
Gurevitch 1986	1.144	0.116	8.595	73.875	0.345	2.902	3.319
Gurevitch 1986	1.206	0.118	8.461	71.593	0.346	2.887	3.482
McCreary et al. 1983	3.696	1.354	0.739	0.546	1.582	0.632	2.336
McCreary et al. 1983	4.674	1.865	0.536	0.287	2.093	0.478	2.233
Stimson 1985	1.218	0.339	2.953	8.717	0.567	1.764	2.148
Stimson 1985	1.288	0.497	2.011	4.045	0.725	1.379	1.776
Reed & Foster 1984	1.818	0.141	7.077	50.083	0.370	2.706	4.919
Reed & Foster 1984	2.223	0.162	6.182	38.213	0.390	2.564	5.700
Reed & Foster 1984	0.767	0.107	9.315	86.767	0.336	2.980	2.286
Reed & Foster 1984	0.524	0.207	4.834	23.369	0.435	2.298	1.204
Reed & Foster 1984	0.163	0.201	4.983	24.835	0.429	2.332	0.380
Reed & Foster 1984	0.449	0.205	4.877	23.788	0.433	2.308	1.036
Johnson & Mann 1988	2.218	1.615	0.619	0.383	1.843	0.543	1.203
Johnson & Mann 1988	0.571	1.041	0.961	0.923	1.269	0.788	0.450
Johnson & Mann 1988	−0.326	1.013	0.987	0.974	1.241	0.805	−0.262
Turner 1985	0.735	0.534	1.873	3.510	0.762	1.312	0.965
Turner 1985	0.869	0.547	1.827	3.340	0.775	1.290	1.121
Turner 1985	0.463	0.513	1.948	3.794	0.742	1.348	0.624
Turner 1985	0.432	0.512	1.954	3.820	0.740	1.352	0.584
Turner 1985	0.728	0.533	1.876	3.518	0.761	1.313	0.956
Turner 1985	0.629	0.525	1.906	3.632	0.753	1.328	0.836
Turner 1985	0.501	0.516	1.939	3.761	0.744	1.344	0.673
Turner 1985	−0.367	0.508	1.967	3.869	0.737	1.357	−0.498
Dungan 1986	0.820	0.542	1.845	3.404	0.770	1.298	1.064
Dungan 1986	−1.956	0.739	1.353	1.830	0.967	1.034	−2.022
Dungan 1986	−0.399	0.408	2.451	6.009	0.636	1.572	−0.627

		w_{ij}	w_{ij}^2	c_i	w_{i+}^*	$w_{i+}^*d_{i+}^*$	$wi+^*di+^*2$
	terrestrial	73.783	447.185	67.722	33.859	36.642	39.654
	lentic	1.275	0.833	0.621	1.110	4.569	18.809
	marine	65.739	302.582	61.136	35.016	24.517	17.166
	grand sums	140.797	750.600				

$$\hat{\sigma}^2_{pooled} = \frac{Q_W - \sum_{i=1}^{m} (k_i - 1)}{\sum_{i=1}^{m} c_i} \qquad (18.7)$$

where m is the number of classes and Q_W is the within-class homogeneity from the fixed effects analysis. From our example, taking the data from the sums at the bottom of table 18.2, for the terrestrial class,

$$c_i = 73.783 - 447.185/73.783 = 67.722$$

and $c_i = 0.621$ for lentic and 61.136 for marine producers. Continuing with the example,

$$\hat{\sigma}^2_{pooled} = (69.550 - 40)/129.480 = 0.228$$

Now we can find v^*_{ij} for each study,

$$v^*_{ij} = v_{ij} + \hat{\sigma}^2_{pooled}$$

From the example, in the first data line of table 18.2, $v^*_{ij} = 0.514$ (taking v_{ij} in the column C + 0.228). The weights used to carry out the meta-analysis are the reciprocals of the random effects variance estimates (just as the weights in the fixed effects model are the reciprocals of the fixed effects variance estimates),

$$w^*_{ij} = 1/v^*_{ij}$$

and are found in the column G of table 18.2. These are multiplied by the effect sizes (d_{ij} from the column B), as in column F of table 18.2.

The cumulated effect sizes for each class in the mixed model, d^*_{i+}, and their variances, $s^2(d^*_{i+})$, are calculated as in the fixed effects model (equations 18.7 and 18.8). From the example (table 18.2), the values for the mixed model cumulated effect sizes (d^*_{i+}) for the three classes are

terrestrial = 36.642/33.859 = 1.08
lentic = 4.569/1.110 = 4.12
marine = 24.517/35.016 = 0.70

and their variances and 95% confidence intervals are, respectively,

terrestrial = 1/33.859 = 0.030; (0.754, 1.427)
lentic = 1/1.110 = 0.901; (2.256, 5.977)
marine = 1/35.016 = 0.029; (0.370, 1.031)

It is reassuring that the values are very similar to those for the fixed effects model. As expected, the confidence intervals are somewhat larger because an additional variance component is included. The conclusions regarding the effects of competition are essentially the same for the mixed effects model as for the fixed effects model, except that in this case we have not made the conventional assumption that all studies in a class share a common true effect size.

An estimate of the within-class variance component $\hat{\sigma}^2_{i+}$ can be calculated for each class:

$$\hat{\sigma}_{i+}^2 = \frac{Q_{wi} - (k_i - 1)}{c_i}$$ (18.18)

The grand cumulated effect size, d_{++}^*, and its variance, $s^2(d_{++}^*)$, are calculated as

$$d_{++}^* = \frac{\sum_{i=1}^{m} w_{i+}^* d_{i+}^*}{\sum_{i=1}^{m} w_{i+}^*}$$ (18.19)

and

$$s^2(d_{++}^*) = \frac{1}{\sum_{i=1}^{m} \sum_{j=1}^{k_i} w_{ij}^*}$$ (18.20)

(Note that this is slightly different than in the fixed effects model.) From the example,

$d_{++}^* = (36.642 + 4.569 + 24.517)/(33.859 + 1.110 + 35.016) = 0.939$
$s^2(d_{++}^*) = 1/(33.859 + 1.110 + 35.016) = 0.014$

The effect of competition across all studies was large. Finally, the homogeneity among classes can be tested using

$$Q_B^* = \sum_{i=1}^{m} w_{i+}^* d_{i+}^{*2} - \frac{\left(\sum_{i=1}^{m} w_{i+}^* d_{i+}^*\right)^2}{\sum_{i=1}^{m} w_{i+}^*}$$ (18.21)

where w_{i+}^* is the sum of the mixed model weights (the w^*-values) for each class. Note that this equation is also calculated somewhat differently than in the fixed effects model. From the example,

$$Q_B^* = (39.654 + 18.809 + 17.166) - \frac{(36.642 + 4.569 + 24.517)^2}{(33.859 + 1.110 + 35.016)}$$

$$= 13.899$$

with 2 degrees of freedom, which is significant at $P < 0.001$. The conclusions again are the same as for the fixed effects model, that is, the effects of competition are not the same for terrestrial, lentic, and marine primary producers. In the mixed effects model, we do not calculate Q_W, because we are no longer making the assumption that all studies in a class share a common true effect. The test of the homogeneity of the effect sizes within a class, using Q_W from the fixed effects model, can, however, be interpreted in the mixed effects model, as a test that the within-class variance component σ^2 is larger than zero. This test would rarely be useful because it typically leads to rejection of the hypothesis that $\sigma^2 = 0$, and so is not likely to be informative.

18.4 Resampling in Meta-analysis

The tests presented previously are parametric tests. Resampling methods offer an alternative to parametric techniques. Resampling statistics are computer-intensive techniques that allow us to evaluate the significance of a given test value (Manly 1997; Crowley 1992). Resampling tests are often useful when the original data do not conform to the distributional assumptions of parametric tests (Manly 1997). Ecologists are becoming familiar with resampling methods in primary analyses, and these methods have been extended to meta-analytic data as well (Adams et al. 1997).

Resampling is performed by calculating a statistic from the original data and evaluating it by permuting the original data in some way, recalculating the test value of interest, and then repeating this procedure many times. The test values from all of the iterations are then used to generate a distribution of test values, and the original test value is compared to this generated distribution to determine the statistical significance of the original data (e.g., chapters 7 and 14). One type of resampling, randomization tests, are most frequently used to determine the significance level of a given test statistic. For each iteration, the original data are randomly reassigned to the treatment classes. A test statistic is then calculated using the randomly shuffled data. This represents one possible outcome based on the data. By performing many iterations, a frequency distribution of possible outcomes (i.e., test statistics) is generated. The actual test statistic is then compared to this frequency distribution of randomly generated statistics, and the proportion of randomly generated statistics more extreme than the actual statistic is taken to be the significance level for that data set. Randomization tests have been used to calculate the significance levels for the homogeneity statistic, Q_B. This can be done by randomly reassigning the effect sizes among studies, recalculating Q_B each time, to determine whether the actual differences among categories are greater than if the categories were based on chance assignment of the effect sizes.

The second common use of resampling methods is to generate confidence intervals around a given statistic using bootstrapping. Bootstrapping works by choosing (with replacement) N studies (from a sample size of N) and then calculating the desired statistic. For example, if there were 20 studies in total, 20 studies would be chosen for each bootstrap iteration. However, because bootstrapping is sampling with replacement, some of the studies from the original sample would be chosen more than once, whereas others would not be chosen at all. This procedure is repeated many times to generate a distribution of possible values. The lowest and highest 2.5% values are then chosen to represent the lower and upper 95% bootstrap confidence limits.

Bootstrapping can be used to calculate confidence intervals around the mean effect size in meta-analysis. Confidence intervals generated in this way are called *percentile bootstrap confidence intervals*, because they are calculated by merely choosing certain percentile values, in this case the upper and lower 2.5% (chapter 14). These confidence intervals assume that the distribution of bootstrap values is centered around the original value. When this is the case, the percentile bootstrap is known to produce the correct confidence intervals (Efron 1982; chapter

14). However, when more than 50% of the bootstrap replicates are above or below the original value, the bootstrap confidence interval must be corrected for this bias. This is done by first finding the fraction (F) of bootstrap replicates smaller than the observed value. The probit transform of F is then found. Finally, the upper and lower *bias-corrected bootstrap confidence limits* are calculated as $\pm\Phi(2z_0 - 1.96)$, where Φ is the normal cumulative distribution function (chapter 14).

Those well-versed in computer programming can write software to analyze their data using randomization and bootstrapping techniques. Alternatively, the program MetaWin 2.0 (Rosenberg et al. 1999) will calculate both the uncorrected percentile bootstrap confidence intervals and the bias-corrected bootstrap confidence intervals, as well as conduct randomization tests on homogeneity statistics.

18.5 What Can be Done When Published Primary Data Lack Standard Deviations or Sample Sizes?

Conventionally, meta-analysts omitted these data because the common tests depended on having these values to conduct the basic analyses: determining weighted means, confidence limits around those means, and homogeneity tests for differences among kinds of studies. Ecological data is poorly reported so commonly, however, that many people wish to have some way of using studies where means are reported but other basic statistics are lacking. The recent developments reported here present one possibility for carrying out meta-analyses using poorly reported studies. That is, we can do an unweighted analysis (i.e., the weights of all studies are equal to 1.0) using *lr*, calculating bootstrapped confidence limits and using randomization tests for the homogeneity statistics. This approach requires only that we have means, but not standard deviations or sample sizes. It is less accurate and less powerful than the approaches presented previously, but it may be the only choice in some cases (see Gurevitch and Hedges 1999). We recommend that the meta-analysis be carried out using those studies that are reported adequately (i.e., studies that include means, sample sizes, and variances) first, and the results of studies with poorer reporting procedures compared with those with better reporting.

18.6 Interpretation and Conclusions

In the analysis we conducted, there was a large effect of competition on growth and recruitment of primary producers, with lentic plants experiencing very large effects and terrestrial and marine organisms lesser, but still large, effects. The results here contrast to some degree with the results of a meta-analysis of the effects of competition on biomass (Gurevitch et al. 1992). In that study, primary producers shared a common, small to medium effect of competition ($d_+ = 0.34$), which did not differ among terrestrial, freshwater, or marine organisms. It seems reasonable that a change in size and number, as in the present study, might be expected to show a greater response to an experimental manipulation than the total size of the organisms, as in the previous study. The number of experimental comparisons in the three systems was similar in the previous and present studies for primary producers. It would be interesting to investigate lentic plants further, because the large effects of competition in freshwater are based on only two experimental comparisons by a single author in the present meta-analysis. However, we can be fairly confident about the conclusions regarding the generally substantial effects of competition on primary producers found here.

These results are difficult to compare with those of narrative and vote-count reviews of competition in field experiments, for the reasons discussed at the beginning of the chapter. Nevertheless, the results agree in broad terms with those of Connell (1983), Schoener (1983), and Goldberg and Barton (1992), in concluding that competition among primary producers was common. Previous reviewers have been unable to accurately assess the magnitude of competitive effects across studies, or to compare the intensity of competition among categories of studies.

It is reassuring to note the general agreement between the results of the more conventional fixed effects model and the mixed effects model presented here. As we would expect, the major difference between the two is the larger confidence intervals in the mixed effects model, which reflect the additional source of error included in that model. The mixed effects model assumption of random variation between studies within a class is in many cases a much more reasonable one than the fixed effects model assumption of all studies within a class sharing a single true effect size. For that reason, the more widespread use of the mixed effects model is to be strongly encouraged.

18.7 An Overview of the Potential Role of Meta-analysis in Ecology

Underlying the motivation for both the original and revised edition of this book has been the explosion of interest in experimental ecology, particularly in the field, within the past 20 years or so. The preceeding chapters all attempt to assist experimentalists engaged in designing and analyzing their experiments. The sophistication of the statistical tools available to ecologists for analyzing the results of experiments has increased exponentially in recent years, and this book is an effort to make some of those tools more widely available. Before the introduction of meta-analysis, the tools that existed to make the best use of the wealth of ecological data provided by the outcome of these efforts were crude, inaccurate, and inadequate to the task. Meta-analysis establishes a new standard in the tools available to synthesize data gathered in different studies. However, meta-analysis itself, particularly as applied to ecological research, is still in its infancy. Although numerous developments have been made in the years since the first edition of this book was published, there is no question that great potential still exists to develop and improve the statistical models used to conduct meta-analyses in ecology and in other fields.

In the first edition of this book, we predicted that meta-analysis would have a substantial impact on the field of ecology over the next few years. This has been borne out by publications that have increased exponentially over the seven years since that prediction was made, as well as by workshops, special features, review articles on the subject, and a statistical software package designed primarily with ecological meta-analysis as its focus. There are several reasons for this increasing interest in meta-analysis in this field. First, ecologists are now becoming aware of these techniques, which are not difficult to use. In fact, the statistics presented here are perhaps the easiest ones to compute in this entire volume. More substantively, the number of experimental studies in various subdisciplines of ecology has probably reached a critical mass: there is abundant material to summarize. Most critically, the ever-increasing impact of humans on the natural world has made the need to make sense of this growing body of data increasingly urgent. Meta-analysis presents compelling opportunities for revealing the broad patterns contained in the accumulated body of experimental ecological research.

Acknowledgments   This work was supported by NSF Grant BSR 9113065 to J.G. Todd Postol, Janet Morrison, Alison Wallace, Tom Meagher, and Sam Scheiner offered valuable suggestions on the manuscript. The generous help of D. Slice, who wrote a digitizing program that enabled us to obtain data from graphs, was very much appreciated. Alison Wallace and Joe Walsh provided invaluable assistance with the literature survey and data sheets. The methods for the collection and interpretation of the data were developed in part by Laura Morrow, whose efforts and insight in sweating the details substantially improved the quality of the work.

References

Abrams, P. A. 1990. The effects of adaptive behavior on the type-2 functional response. Ecology 71:877–885.

Abrams, P. A., B. A. Menge, G. G. Mittelbach, D. Spiller, and P. Yodzis. 1996. The role of indirect effects in food webs. Pages 371–395 in G. A. Polis and K. O. Winemiller, editors, Food webs: Integration of patterns and dynamics. Chapman and Hall, New York.

Adams, D. C., J. Gurevitch, and M. S. Rosenberg. 1997. Resampling tests for meta-analysis of ecological data. Ecology 78:1277–1283.

Agresti, A. 1990. Categorical Data Analysis. Wiley, New York.

Aldrich, J. H., and F. D. Nelson. 1984. Linear probability, logit, and probit models. Sage University Paper Series on Quantitative Applications in the Social Sciences 07-045. Sage, Thousand Oaks, California.

Allison, P. D. 1995. Survival Analysis Using the SAS System. SAS Institute, Inc., Cary, North Carolina.

Andrews, R. M. 1982. Patterns of growth in reptiles. Pages 273–320 in C. Gans and F. H. Pough, editors, Biology of the Reptilia. Academic Press, New York.

Antonovics, J., and N. L. Fowler. 1985. Analysis of frequency and density effects on growth in mixtures of *Salvia splendens* and *Linum grandiflorum* using hexagonal fan designs. Journal of Ecology 73:219–234.

Arnold, S. J. 1972. Species densities of predators and their prey. American Naturalist 106:220–236.

Arnqvist, G., and D. Wooster. 1995. Meta-analysis: synthesizing research findings in ecology and evolution. Trends in Ecology and Evolution 10:236–240.

Ayala, F. J., M. E. Gilpin, and J. G. Ehrenfeld. 1973. Competition between species: Theoretical models and experimental tests. Theoretical Population Biology 4:331–356.

Bailer, A. J., and J. T. Oris. 1994. Assessing toxicity of pollutants in aquatic systems. Pages 25–40 in N. Lange, L. Ryan, L. Billard, D. Brillinger, L. Conquest, and J. Greenhouse, editors, Case Studies in Biometry. Wiley, New York.

Baird, D., and R. Mead. 1991. The empirical efficiency and validity of two neighbour models. Biometrics 47:1473–1487.

Baker, H. G. 1983. An outline of the history of anthecology, or pollination biology. Pages 7–28 in L. Real, editor, Pollination Biology. Academic Press, Orlando, Florida.

Bard, Y. 1974. Nonlinear Parameter Estimation. Academic Press, New York.

Barker, H. R., and B. M. Barker, editors. 1984. Multivariate Analysis of Variance (MANOVA). University of Alabama Press, Tuscaloosa, Alabama.

Barlow, R. E., D. J. Bartholomew, J. M. Bremer, and H. D. Brunk. 1972. Statistical Inference Under Order Restrictions. Wiley, New York.

Barnett, V. 1976. The ordering of multivariate data. Journal of the Royal Statistical Society, Series A 139:318–354.

Barnett, V. 1999. Comparative Statistical Inference, 3rd ed. John Wiley & Sons, London.

Bartlett, M. S. 1935. Contingency table interactions. Journal of the Royal Statistical Society (Supplement) 2:248–252.

Bartlett, M. S. 1978. Nearest neighbor models in the analysis of field experiments with large blocks. Journal of the Royal Statistical Society, Series B 40:147–174.

Bautista, L. M., J. C. Alonso, and J. A. Alonso. 1992. A 20-year study of wintering common crane fluctuations using time series analysis. Journal of Wildlife Management 56:563–572.

Bayes, T. 1763. An essay towards solving a problem in the doctrine of chances. Philosophical Transactions of the Royal Society 53:370–418.

Beaupre, S. J. 1995. Comparative ecology of mottled rock rattlesnake, Crotalus lepidus, in Big Bend National Park. Herpetologica 51:45–56.

Beaupre, S. J., D. Duvall, and J. O'Leile. 1998. Ontogenetic variation in growth and sexual size dimorphism in a central Arizona population of the western diamondback rattlesnake, Crotalus atrox. Copeia 1998:40–47.

Becher, H. 1993. Bootstrap hypothesis testing procedures. Biometrics 49:1268–1272.

Benditt, J. 1992. Women in science, 1st annual survey. Science 255:1363–1388.

Bennett, B. M. 1968. Rank-order tests of linear hypotheses. Journal of the Statistical Society B 30:483–489.

Bennington, C. C., and J. B. McGraw. 1995. Natural selection and ecotypic differentiation in Impatiens pallida. Ecological Monographs 65:303–324.

Bentler, P. M. 1985. Theory and Implementation of EQS: A Structural Equations Program. BMDP Statistical Software, Los Angeles, California.

Berger, J. O. 1985. Statistical Decision Theory and Bayesian Analysis, 2nd ed. Springer-Verlag, New York.

Berger, J. O. (in press). Bayesian analysis: A look at today and thoughts of tomorrow. Journal of the American Statistical Association.

Berger, J. O., and D. A. Berry. 1988. Statistical analysis and the illusion of objectivity. American Scientist 76:159–165.

Berry, D. A. 1996. Statistics: A Bayesian Perspective. Wadsworth, Belmont, California.

Berven, K. A. 1982. The genetic basis of altitudinal variation in the wood frog Rana sylvatica. I. An experimental analysis of life history traits. Evolution 36:962–983.

Besag, J. E., and R. A. Kempton. 1986. Statistical analysis of field experiments using neighboring plots. Biometrics 42:231–251.

Bickel, P. J., and D. A. Freedman. 1981. Some asymptotic theory for the bootstrap. Annals of Statistics 9:1196–1217.

Biondini, M. E., P. W. Mielke Jr., and E. F. Redente. 1988. Permutation techniques based on Euclidean analysis spaces: A new and powerful statistical method for ecological research. Coenoses 3:155–174.

Bishop, J. G., and D. W. Schemske. 1998. Variation in flowering phenology and its consequences for lupines colonizing Mount St. Helens. Ecology 79:534–546.

Blackburn, T. M., J. H. Lawton, and J. N. Perry. 1992. A method of estimating the slope of upper bounds of plots of body size and abundance in natural animal assemblages. Oikos 65:107–112.

Boecklen, W. J., and G. J. Niemi. 1994. Multivariate association of graph-theoretic variables and physicochemical properties. SAR and QSAR in Environmental Research 2: 79–87.

Bollen, K. A. 1989. Structural Equations with Latent Variables. Wiley, New York.

Bookstein, F. L. 1986. The elements of latent variable models: A cautionary lecture. Pages 203–230 in M. E. Lamb, A. L. Brown, and B. Rogoff, editors, Advances in Developmental Psychology. Lawrence Erlbaum, Hillsdale, New Jersey.

Borcard, D., P. Legendre, and P. Drapeau. 1992. Partialling out the spatial component of ecological variation. Ecology 73:1045–1055.

Borenstein, M. 1994. Planning for precision in survival studies. Journal of Clinical Epidemiology 47:1277–1285.

Bormann, F. H., and G. E. Likens. 1979. Pattern and Process in a Forested Ecosystem. Springer-Verlag, New York.

Box, G. E. P. 1954. Some theorems on quadratic forms applied in the study of analysis of variance problems: I. Effect of inequality of variance in the one-way classification. Annals of Mathematical Statistics 25:290–302.

Box, G. E. P., and C. R. Cox. 1964. An analysis of transformations. Journal of the Royal Statistical Society, Series B 26:211–243.

Box, G. E. P., and G. M. Jenkins. 1976. Time Series Analysis: Forecasting and Control, rev. ed. Holden-Day, San Francisco.

Box, G. E. P., and D. A. Pierce. 1970. Distributions of residual autocorrelations in autoregressive-integrated moving average time series models. Journal of the American Statistical Association 65:1509–1526.

Box, G. E. P., and G. C. Tiao. 1973. Bayesian Inference in Statistical Analysis. Wiley, New York.

Box, G. E. P., and G. C. Tiao. 1975. Intervention analysis with applications to economic and environmental parameters. Journal of the American Statistical Association 70: 70–79.

Box, G. E. P., W. G. Hunter, and J. S. Hunter. 1978. Statistics for Experimenters: An Introduction to Design, Data Analysis, and Model Building. Wiley, New York.

Boyce, M. S. 1978. Climatic variability and body size variation in the muskrats Ondatrazibethicus of North America. Oecologia 36:1–20.

Bradley, D. R., R. L. Russell, and C. P. Reeve. 1996. Statistical power in complex experimental designs. Behavior Research Methods, Instruments, and Computers 28:319–326.

Bradley, J. V. 1978. Robustness? British Journal of Mathematical and Statistical Psychology 31:144–152.

Bray, J. H., and S. E. Maxwell. 1985. Multivariate Analysis of Variance. Sage Publications, Inc., Newbury Park, California.

Breckler, S. J. 1990. Applications of covariance structure modeling in psychology: Cause for concern? Psychological Bulletin 107:260–273.

Brezonik, P. L., L. A. Baker, J. R. Eaton, T. M. Frost, P. Garrison, T. K. Kratz, J. J. Magnuson, J. E. Perry, W. J. Rose, B. K. Shepard, W. A. Swenson, C. J. Watras, and K. E. Webster. 1986. Experimental acidification of Little Rock Lake, Wisconsin. Water, Air, and Soil Pollution 31:115–121.

Brody, A. K. 1992. Oviposition choices by a pre-dispersal seed predator (*Hylemya* sp.): I. Correspondence with hummingbird pollinators, and the role of plant size, density and floral morphology. Oecologia 91:56–62.

Brown, J. H., D. W. Davidson, J. C. Munger, and R. S. Inouye. 1986. Experimental community ecology: The desert granivore system. Pages 41–61 in J. Diamond and T. J. Case, editors, Community Ecology. Harper & Row, New York.

Brownlee, K. A. 1965. Statistical Theory and Methodology in Science and Engineering, 2nd ed. John Wiley & Sons, New York.

Bruce, R. C., and N. G. Hairston Sr. 1990. Life history correlates of body size differences between two populations of the salamander *Desmognathus monticola*. Journal of Herpetology 24:124–134.

Buckland, S. T., and P. H. Garthwaite. 1991. Quantifying the precision of mark-recapture estimates using the bootstrap and related methods. Biometrics 47:255–268.

Buja, A., D. Cook, and D. F. Swayne. 1996. Interactive high-dimensional data visualization. Journal of Computational and Graphical Statistics 5:78–99.

Bullock, J. M., J. Silvertown, and B. Clear Hill. 1996. Plant demographic responses to environmental variation: Distinguishing between effects on age structure and effects on age-specific vital rates. Journal of Ecology 84:733–743.

Bunge, M. 1959. Causality. Harvard University Press, Cambridge, Massachusetts.

Buonaccorsi, J. P., and A. M. Liebhold. 1988. Statistical methods for estimating ratios and products in ecological studies. Environmental Entomology 17:572–580.

Burgman, M. A. 1987. An analysis of the distribution of plants on granite outcrops in southern Western Australia using Mantel tests. Vegetatio 71:79–86.

Burnham, K. P., and D. R. Anderson. 1998. Model Selection and Inference: A Practical Information-Theoretic Approach. Springer-Verlag, New York.

Burton, P. J., and D. Mueller-Dombois. 1984. Response of *Metrosideros polymorpha* seedlings to experimental canopy openings. Ecology 65:779–791.

Campbell, B. D., and J. P. Grime. 1992. An experimental test of plant strategy theory. Ecology 73:15–29.

Campbell, D. R., N. M. Waser, M. V. Price, E. A. Lynch, and R. J. Mitchell. 1991. Components of phenotypic selection: Pollen export and corolla width in *Ipomopsis aggregata*. Evolution 45:1458–1467.

Capaldi, E. A., and F. C. Dyer. 1999. The role of orientation flights on homing performance in honeybees. Journal of Experimental Biology 202:1655–1666.

Carey, J. R., P. Liedo, H.-G. Müller, J.-L. Wang, and J. W. Vaupel. 1998. Dual modes of aging in Mediterranean fruit fly females. Science 281:996–998.

Carpenter, S. R. 1989. Replication and treatment strength in whole-lake experiments. Ecology 70:453–463.

Carpenter, S. R. 1990. Large scale perturbations: Opportunities for innovation. Ecology 71:2038–2043.

Carpenter, S. R., T. M. Frost, D. Heisey, and T. Kratz. 1989. Randomized intervention analysis and the interpretation of whole ecosystem experiments. Ecology 70:1142–1152.

Carpenter, S. R., S. W. Chisholm, C. J. Krebs, D. W. Schindler, and R. F. Wright. 1995. Ecosystem experiments. Science 269:324–327.

Carpenter, S. R., J. J. Cole, T. E. Essington, J. R. Hodgson, J. N. Houser, J. F. Kitchell, and M. L. Pace. 1998. Evaluating alternative explanations in ecosystem experiments. Ecosystems 1:335–344.

Casas, J., and B. Hulliger. 1994. Statistical analysis of functional response experiments. Biocontrol Science and Technology 4:133–145.

Case, T. J. 1978. A general explanation for insular body size trends in terrestrial vertebrates. Ecology 59:1–18.

Caswell, H. 1989. Matrix Population Models. Sinauer, Sunderland, Massachusetts.

Cerato, R. M. 1990. Interpretable statistical tests for growth comparisons using parameters in the von Bertalanffy equation. Canadian Journal of Fisheries and Aquatic Sciences 47:1416–1426.

Chalmers, I., J. Hetherington, D. Elhourne, M. J. N. C. Keirse, and M. Enkin. 1989. Effective Care in Pregnancy and Childbirth. Oxford University Press, Oxford.

Cheng, C. S. 1983. Construction of optimal balanced incomplete block designs for correlated observations. Annals of Statistics 11:240–246.

Chernick, M. R. 1999. Bootstrap Methods: A Practioners Guide. John Wiley & Sons, New York.

Chow, S. C., and J. P. Liu. 1999. Design and Analysis of Bioavailability and Bioequivalence Studies, 2nd ed. Marcel Dekker, New York.

Chow, S. L. 1998. Précis of statistical significance: Rationale, validity, and utility (with commentary). Behavioral and Brain Sciences 21:169–239.

Christiansen, R. 1997. Log-linear Models and Logistic Regression. Springer-Verlag, New York.

Clark, J. S. 1989. Ecological disturbance as a renewal process: Theory and application to fire history. Oikos 56:17–30.

Clark, J. S. 1998. Why trees migrate so fast: Confronting theory with dispersal biology and the paleo record. American Naturalist 152:204–224.

Clark, J. S., B. Beckage, P. Camill, B. Cleveland, J. HilleRisLambers, J. Lichter, J. MacLachlan, J. Mohan, and P. Wyckoff. 1999a. Interpreting recruitment limitation in forests? American Journal of Botany 86:1–16.

Clark, J. S., M. Lewis, and L. Horvath. 2001. Invasion by extremes: variation of dispersal and reproduction retards population spread. American Naturalist (in press).

Clark, J. S., M. Silman, R. Kern, E. Macklin, and J. HilleRisLambers. 1999b. Seed dispersal near and far: Generalized patterns across temperate and tropical forests. Ecology 80:1475–1494.

Clarke, K. R. 1993. Non-parametric multivariate analyses of changes in community structure. Australian Journal of Ecology 18:117–143.

Clarke, K. R., and R. M. Warwick. 1994. Change in marine communities: an approach to statistical analysis and interpretation. Plymouth Marine Laboratory, Plymouth, UK.

Cleveland, W. S. 1979. Robust locally weighted regression and smoothing scatterplots. Journal of the American Statistical Association 74:829–836.

Cleveland, W. S. 1985. The Elements of Graphing Data. Wadsworth Advanced Books and Software, Monterey, California.

Cliff, A. D., and J. K. Ord. 1981. Spatial Processes: Models and Applications. Pion, London.

Cloninger, C. R., D. C. Rao, J. Rice, T. Reich, and N. E. Morton. 1983. A defense of path analysis in genetic epidemiology. American Journal of Human Genetics 35:733–756.

Cochran, W. G., and G. M. Cox. 1957. Experimental Designs, 2nd ed. Wiley, New York.

Cochran-Stafira, D. L., and C. N. von Ende. 1998. Integrating bacteria into food webs: Studies with *Sarracenia purpurea* inquilines. Ecology 79:880–898.

Cock, M. J. W. 1977. Searching behaviour of polyphagous predators. Ph.D. dissertation. Imperial College, London.

Cody, R. P., and J. K. Smith. 1997. Applied Statistics and the SAS Programming Language. Prentice Hall, Upper Saddle River, New Jersey.

Cohen, J. 1969. Statistical Power Analysis for the Behavioral Sciences. Academic Press, New York.

Cohen, J. 1988. Statistical Power Analysis for the Behavioral Sciences, 2nd ed. Lawrence Erlbaum Associates, Inc., Hillsdale, New Jersey.

Cohen, J., and P. Cohen. 1983. Applied Multiple Regression/Correlation Analysis for the Behavioral Sciences. Lawrence Erlbaum, Hillsdale, California.

Collett, D. 1994. Modelling survival data in medical research. Chapman and Hall, London.

Collier, R. O., Jr., F. B. Baker, G. K. Mandeville, and T. F. Hayes. 1967. Estimates of test size for several test procedures based on conventional variance ratios in the repeated measures design. Psychometrika 32:339–353.

Connell, J. H. 1983. On the prevalence and relative importance of interspecific competition: Evidence from field experiments. American Naturalist 122:661–696.

Connell, J. H. 1990. Apparent versus "real" competition in plants. Pages 9–26 in J. Grace and D. Tilman, editors, Perspectives on Plant Competition. Academic Press, San Diego.

Conner, J. K., S. Rush, and P. Jennetten. 1996. Measurements of natural selection on floral traits in wild radish (*Raphanus raphanistrum*). 1. Selection through lifetime female fitness. Evolution 50:1127–1136.

Connolly, J. 1986. On difficulties with replacement-series methodology in mixture experiments. Journal of Applied Ecology 23:125–137.

Cook, D., A. Buja, J. Cabrera, and C. Hurley. 1995. Grand tour and projection pursuit. Journal of Computational and Graphical Statistics 4:155–172.

Cooper, H. M. 1989. Integrating Research: A Guide for Literature Reviews, 2nd ed. Sage Publications, Newbury Park, California.

Cooper, H. M., and L. V. Hedges, editors. 1994. Handbook of Research Synthesis. Russell Sage Foundation, New York.

Cousins, R. D. 1995. Why isn't every physicist a Bayesian? American Journal of Physics 63:398–410.

Cox, D. R., and D. Oakes. 1984. Analysis of Survival Data. Chapman and Hall, London.

Crawley, M. J. 1993. GLIM for Ecologists. Blackwell Scientific, London.

Crespi, B. J. 1990. Measuring the effect of natural selection on phenotypic interaction systems. American Naturalist 135:32–47.

Crespi, B. J., and F. L. Bookstein. 1989. A path-analytic model for the measurement of selection on morphology. Evolution 43:18–28.

Cressie, N. 1985. Fitting variogram models by weighted least squares. Journal of the International Association for Mathematical Geology 17:563–586.

Cressie, N. 1991. Statistics for Spatial Data. Wiley, New York.

Cressie, N. 1992. Smoothing regional maps using empirical Bayes predictors. Geographical Analysis 24:75–95.

Cressie, N., and D. M. Hawkins. 1980. Robust estimation of the variogram. Journal of the International Association for Mathematical Geology 12:115–125.

Crome, F. H. J., M. R. Thomas, and L. A. Moore. 1996. A novel Bayesian approach to assessing impacts of rain forest logging. Ecological Applications 6:1104–1123.

Crowder, M. J., and D. J. Hand. 1990. Analysis of Repeated Measures. Chapman and Hall, London.

Crowley, P. H. 1992. Resampling methods for computation intensive data analysis in ecology and evolution. Annual Review of Ecology and Systematics 23:405–447.

Cryer, J. D. 1986. Time Series Analysis. Duxbury, Boston.

Cullis, B. R., and A. C. Gleeson. 1989. The efficiency of neighbour analysis for replicated variety trials in Australia. Journal of Agricultural Science, Cambridge 113:233–239.

Cullis, B. R., and A. C. Gleeson. 1991. Spatial analysis of field experiments—An extension to two dimensions. Biometrics 47:1449–1460.

Czárán, T., and S. Bartha. 1992. Spatiotemporal dynamic models of plant populations and communities. Trends in Ecology and Evolution 7:38–42.

Davis, J. A. 1985. The logic of causal order. Sage University Paper Series on Quantitative Applications in the Social Sciences 07–055. Sage, Thousand Oaks, California.

Davison, A. C., and D. V. Hinkley. 1997. Bootstrap Methods and their Application. Cambridge University Press, Cambridge.

Dawson, K. S., C. Gennings, and W. H. Carver. 1997. Two graphical techniques useful in detecting correlation structure in repeated measures data. American Statistician 51: 275–284.

Day, R. W., and G. P. Quinn. 1989. Comparisons of treatments after an analysis of variance in ecology. Ecological Monographs 59:433–463.

DeMaris, A. 1992. Logit modeling: Practical applications. Sage University Paper Series on Quantitative Applications in the Social Sciences 07-086. Sage, Thousand Oaks, California.

Dennis, B. 1996. Discussion: Should ecologists become Bayesians? Ecological Applications 6:1095–1103.

DeSteven, D. 1991a. Experiments on mechanisms of tree establishment in old-field succession: Seedling emergence. Ecology 72:1066–1075.

DeSteven, D. 1991b. Experiments on mechanisms of tree establishment in old-field succession: Seedling survival and growth. Ecology 72:1076–1089.

Diamond, J. 1986. Overview: Laboratory experiments, field experiments, and natural experiments. Pages 3–22 in J. Diamond and T. J. Case, editors, Community Ecology. Harper and Row, New York.

Digby, P. G. N., and R. A. Kempton. 1987. Multivariate Analysis of Ecological Communities. Chapman and Hall, London.

Diggle, P. J. 1990. Time Series: A Biostatistical Introduction. Clarendon Press, Oxford.

Dixon, P. M. 1993. The bootstrap and the jackknife: Describing the precision of ecological indices. Pages 290–318 in S. M. Scheiner and J. Gurevitch, editors, Design and Analysis of Ecological Experiments. Chapman and Hall, New York.

Dixon, P. M., and A. M. Ellison. 1996. Bayesian inference. Ecological Applications 6: 1034–1035.

Dixon, P. M., and K. A. Garrett. 1994. Statistical issues for field experimenters. Pages 439–450 in R. J. Kendall and T. E. Lacher Jr., editors, Wildlife Toxicology and Population Modeling: Integrated Studies of Agroecosystems. CRC Press, Boca Raton, Florida.

Dixon, P. M., and M. C. Newman. 1991. Analyzing toxicity data using statistical models of time-to-death: An introduction. Pages 207–242 in M. C. Newman and A. W. McIntosh, editors, Metal Ecotoxicology: Concepts and Applications. Lewis Publishers, Inc., Chelsea, Michigan.

Dixon, P. M., J. Weiner, T. Mitchell-Olds, and R. Woodley. 1987. Bootstrapping the Gini coefficient of inequality. Ecology 68:1548–1551.

Dobson, A. J. 1990. An Introduction to Generalized Linear Models. Chapman and Hall, London.

Douglas, M. E., and J. A. Endler. 1982. Quantitative matrix comparisons in ecological and evolutionary investigations. Journal of Theoretical Biology 99:777–795.

Dow, M. M., and J. M. Cheverud. 1985. Comparison of distance matrices in studies of population structure and genetic microdifferentiation: Quadratic assignment. American Journal of Physical Anthropology 68:367–373.

Draper, N. R., and H. Smith. 1981. Applied Regression Analysis, 2nd ed. Wiley, New York.

Dudewicz, E. J. 1972. Confidence intervals for power with special reference to medical trials. Journal of Statistics 14:211–216.

Dungan, M. L. 1986. Three-way interactions: Barnacles, limpets, and algae in a Sonoran desert rocky intertidal zone. American Naturalist 127:292–316.

Dunham, A. E. 1982. Demography and life history variation among populations of the iguanid lizard *Urosaurus ornatus*: Implications for the study of life history phenomena in lizards. Herpetologica 38:201–221.

Dunham, A. E., and S. J. Beaupre. 1998. Ecological experiments: Scale, phenomenology, mechanism, and the illusion of generality. Pages 27–49 in W. J. Resetarits Jr. and J. Bernardo, editors, Experimental Ecology—Issues and Prespectives. Oxford University Press, New York.

Dunham, A. E., B. W. Grant, and K. L. Overall. 1989. The interface between biophysical ecology and the population ecology of terrestrial ectotherms. Physiological Zoology 62:335–355.

Dutilleul, P. 1993. Spatial heterogeneity and the design of ecological field experiments. Ecology 74:1646–1658.

Dyer, L. A. 1995. Tasty generalists and nasty specialists? A comparative study of antipredator mechanisms in tropical lepidopteran larvae. Ecology 76:1483–1496.

Eberhardt, L. L. 1976. Quantitative ecology and impact assessment. Journal of Environmental Management 4:27–70.

Eberhardt, L. L., and J. M. Thomas. 1991. Designing environmental field studies. Ecological Monographs 61:53–73.

Edgington, E. S. 1995. Randomization Tests, 3rd ed. Marcel Dekker, Inc., New York.

Edwards, A. W. F. 1963. The measure of association in a 2×2 table. Journal of the Royal Statistical Society A126:109–114.

Edwards, A. W. F. 1972. Likelihood. Cambridge University Press, Cambridge.

Edwards, D. 1996. Comment: The first data analysis should be journalistic. Ecological Applications 6:1090–1094.

Edwards, W., H. Lindman, and L. J. Savage. 1983. Bayesian statistical inference for psychological research. Psychological Review 70:193–242.

Efron, B. 1982. The jackknife, the bootstrap, and other resampling plans. Society of Industrial and Applied Mathematics, CBMS-NSF Monograph 38.

Efron, B. 1987. Better bootstrap confidence intervals (with discussion). Journal of the American Statistical Association 82:171–200.

Efron, B. 1992. Six questions raised by the bootstrap. Pages 99–126 in R. LePage and L. Billard, editors, Exploring the Limits of Bootstrap. Wiley, New York.

Efron, B., and R. LePage. 1992. Introduction to bootstrap. Pages 3–10 in R. LePage and L. Billard, editors, Exploring the Limits of Bootstrap. Wiley, New York.

Efron, B., and R. Tibshirani. 1986. Bootstrap methods for standard errors, confidence intervals, and other measures of statistical accuracy. Statistical Science 1:54–77.

Efron, B., and R. Tibshirani. 1991. Statistical data analysis in the computer age. Science 253:390–395.

Efron, B., and R. J. Tibshirani. 1993. An Introduction to the Bootstrap. Chapman and Hall, London.

Ellison, A. M. 1992. Statistics for PCs. Bulletin of the Ecological Society of America 73: 74–87.

Ellison, A. M., and B. L. Bedford. 1991. Response of a wetland vascular plant community to disturbance: A simulation study. Ecosystem Research Center, publication ERC243, Cornell University, Ithaca, New York.

Ellison, A. M., J. S. Denslow, B. A. Loiselle, and M. D. Brens. 1993. Seed and seedling ecology of neotropical Melastomataceae. Ecology 74:1733–1749.

Estabrook, G. F., and B. Gates. 1984. Character analysis in the *Banisteriopsis campestris* complex (Malpighiaceae), using spatial auto-correlation. Taxon 33:13–25.

Evans, J. 1983. The ecology of *Ailanthus altissima*: An invading species in both early and late successional communities of the Eastern Deciduous Forest. B.S. thesis. Cornell University, Ithaca, New York.

Fan, Y., and F. L. Pettit. 1994. Parameter estimation of the functional response. Environmental Entomology 23:785–794.

Fan, Y., and F. L. Pettit. 1997. Functional responses, variance, and regression analysis: A reply to Williams and Juliano. Environmental Entomology 26:1–3.

Feinsinger, P., and H. M. I. Tiebout. 1991. Competition among plants sharing hummingbird pollinators: Laboratory experiments on a mechanism. Ecology 72:1946–1952.

Fienberg, S. E. 1979. The use of chi-squared statistics for categorical data problems. Journal of the Royal Statistical Society B41:54–64.

Fienberg, S. E., and S. H. Kim. 1999. Combining conditional log-linear structures. Journal of the American Statistical Association 94:229–239.

Firbank, L. G., and A. R. Watkinson. 1987. On the analysis of competition at the level of the individual plant. Oecologia 71:308–317.

Fisher, R. A. 1935. The Design of Experiments. Oliver and Boyd, Edinburgh.

Fisher, R. A. 1959. Statistical Methods and Scientific Inference. Hafner, New York.

Fisher, R. A. 1971. The Design of Experiments, 8th ed. Oliver and Boyd, London, UK.

Fortin, M.-J., P. Drapeau, and P. Legendre. 1989. Spatial autocorrelation and sampling design in plant ecology. Vegetatio 83:209–222.

Fowler, N. L. 1986. Density dependent population regulation in a Texas grassland. Ecology 67:545–554.

Fowler, N. 1990. The 10 most common statistical errors. Bulletin of the Ecological Society of America 71:161–164.

Fox, G. A. 1989. Consequences of flowering time in a desert annual: Adaptation and history. Ecology 70:1294–1306.

Fox, G. A. 1990a. Components of flowering time variation in a desert annual. Evolution 44:1404–1423.

Fox, G. A. 1990b. Perennation and the persistence of annual life histories. American Naturalist 135:829–840.

Fox, G. A. 1993. Failure-time analysis: Emergence, flowering, survivorship, and other waiting times. Pages 253–289 in S. M. Scheiner and J. Gurevitch, editors, Design and Analysis of Ecological Experiments. Chapman and Hall, New York.

Freckleton, R. P., and A. R. Watkinson. 1997. Measuring plant neighbour effects. Functional Ecology 11:532–536.

Freckleton, R. P., and A. R. Watkinson. 1999. The mis-measurement of plant competition. Functional Ecology 13:285–287.

Friedman, H. 1968. Magnitude of experimental effect and a table for its rapid estimation. Psychological Bulletin 70:245–251.

Frost, T. M., D. L. DeAngelis, T. F. H. Allen, S. M. Bartell, D. J. Hall, and S. H. Hurlbert. 1988. Scale in the design and interpretation of aquatic community research. Pages

229–258 in S. R. Carpenter, editor, Complex Interactions in Lake Communities. Springer-Verlag, New York.

Fry, J. D. 1992. The mixed-model analysis of variance applied to quantitative genetics: Biological meaning of the parameters. Evolution 46:540–550.

Gaines, S. D., and W. R. Rice. 1990. Analysis of biological data when there are ordered expectations. American Naturalist 135:310–317.

Gardner, M. J., and D. G. Altman. 1986. Confidence intervals rather than P values; estimation rather than hypothesis testing. British Medical Journal 292:746–750.

Garvey, J. E., E. A. Marschall, and R. A. Wright. 1998. From star charts to stoneflies: Detecting relationships in continuous bivariate data. Ecology 79:442–447.

Geisser, S., and S. W. Greenhouse. 1958. An extension of Box's results on the use of the F distribution in multivariate analysis. Annals of Mathematical Statistics 29:885–891.

Gelman, A., J. B. Carlin, H. S. Stern, and D. B. Rubin. 1995. Bayesian Data Analysis. Chapman and Hall, London.

Gerard, P. D., D. R. Smith, and W. Govinda. 1998. Limits of retrospective power. Journal of Wildlife Management 62:801–807.

Ghosh, M., and J. N. K. Rao. 1994. Small area estimation: An appraisal. Statistical Science 9:55–93.

Gibbs, J. P., and S. M. Melvin. 1997. Power to detect trends in waterbird abundance with call-response surveys. Journal of Wildlife Management 61:1262–1267.

Gibson, D. J., J. Connolly, D. D. Hartnett, and J. D. Weidenhamer. 1999. Designs for greenhouse studies of interactions between plants. Journal of Ecology 87:1–16.

Gittins, R. 1985. Canonical Analysis. Springer-Verlag, New York.

Glantz, S. A., and B. K. Slinker. 1990. Primer of Applied Regression and Analysis of Variance. McGraw-Hill, New York.

Glass, G. V., B. McGraw, and M. L. Smith. 1981. Meta-analysis in Social Research. Sage Publications, Beverly Hills, California.

Gnanadesikan, R. 1977. Methods for Statistical Data Analysis of Multivariate Observations. Wiley, New York.

Goldberg, D. E. 1994. On testing the importance of competition for community structure. Ecology 75:1503–1506.

Goldberg, D. E., and A. M. Barton. 1992. Patterns and consequences of interspecific competition in natural communities: A review of field experiments with plants. American Naturalist 139:771–801.

Goldberg, D. E., and G. Estabrook. 1998. A method for comparing diversity and species abundances among samples of different sizes and an experimental example with desert annuals. Journal of Ecology 86:983–988.

Goldberg, D. E., and K. Landa. 1991. Competitive effect and response: Hierarchies and correlated traits in the early stages of competition. Journal of Ecology 79:1013–1030.

Goldberg, D. E., and P. A. Werner. 1983. The equivalence of competitors in plant communities: A null hypothesis and a field experimental approach. American Journal of Botany 70:1098–1104.

Goldberg, D. E., R. Turkington, and L. Olsvig-Whittaker. 1995. Quantifying the community-level effects of competition. Folia Geobotanica and Phytotaxonomica 30:231–242.

Gonzalez, M. J., and T. M. Frost. 1994. Comparisons of laboratory bioassays and a whole-lake experiment: Rotifer responses to experimental acidification. Ecological Applications 4:69–80.

Gonzalez, M., T. M. Frost, and P. Montz. 1990. Effects of experimental acidification on rotifer population dynamics in Little Rock Lake, Wisconsin. Verhandlungen der

Internationale Vereinigung fur Theoretische und Angewandte Limnologie 24:449–456.

Goodall, D. W. 1978. Sample similarity and species correlation. Pages 99–149 in R. H. Whittaker, editor, Ordination of Plant Communities. Junk, The Hague, The Netherlands.

Goodman, S. N., and J. A. Berlin. 1994. The use of predicted confidence intervals when planning experiments and the misuse of power when interpreting results. Annals of Internal Medicine 121:201–206.

Gotway, C. A., and W. W. Stroup. 1997. A generalized linear model approach to spatial data analysis and prediction. Journal of Agricultural, Biological, and Environmental Statistics 2:157–178.

Gower, J. C., and P. Legendre. 1986. Metric and Euclidean properties of dissimilarity coefficients. Journal of Classification 3:5–48.

Grace, J. B., and B. H. Pugesek. 1997. Structural equation model of plant species richness and its application to a coastal wetland. American Naturalist 149:436–460.

Grace, J. B., and B. H. Pugesek. 1998. On the use of path analysis and related procedures for the investigation of ecological problems. American Naturalist 152:151–159.

Green, P. J., C. Jennison, and A. H. Seheult. 1985. Analysis of field experiments by least squares smoothing. Journal of the Royal Statistical Society, Series B 47:299–315.

Greenhouse, S. W., and S. Geisser. 1959. On methods in the analysis of profile data. Psychometrika 24:95–112.

Greenland, S. 1988. On sample-size and power calculations for studies using confidence intervals. American Journal of Epidemiology 128:231–237.

Grime, J. P. 1977. Evidence for the existence of three primary strategies in plants and its relevance to ecological and evolutionary theory. American Naturalist 111:1169–1194.

Grondona, M. O., and N. Cressie. 1991. Using spatial considerations in the analysis of experiments. Technometrics 33:381–392.

Grondona, M. O., and N. Cressie. 1993. Efficiency of block designs under stationary second-order autoregressive errors. Sankhya A 55:267–284.

Gross, K. L., and P. A. Werner. 1982. Colonizing abilities of "biennial" plant species in relation to ground cover: Implications for their distributions in a successional sere. Ecology 63:921–931.

Guo, Q., J. H. Brown, and B. J. Enquist. 1998. Using constraint lines to characterize plant performance. Oikos 83:237–245.

Gurevitch, J. 1986. Competition and the local distribution of the grass Stipa neomexicana. Ecology 67:46–57.

Gurevitch, J., and S. T. Chester Jr. 1986. Analysis of repeated experiments. Ecology 67:251–254.

Gurevitch, J., and S. L. Collins. 1994. Experimental manipulation of natural plant communities. Trends in Ecology and Evolution 9:94–98.

Gurevitch, J., and L. V. Hedges. 1993. Meta-analysis: Combining the results of independent experiments. Pages 378–398 in S. M. Scheiner and J. Gurevitch, editors, Design and Analysis of Ecological Experiments, 1st ed. Chapman & Hall, New York.

Gurevitch, J., and L. V. Hedges. 1999. Statistical issues in conducting ecological meta-analyses. Ecology 80:1142–1149.

Gurevitch, J., J. A. Morrison, and L. V. Hedges. 2000. The interaction between competition and predation: a meta-analysis of field experiments. American Naturalist 155:435–453.

Gurevitch, J., L. L. Morrow, A. Wallace, and J. S. Walsh. 1992. A meta-analysis of field experiments on competition. American Naturalist 140:539–572.

Haber, R. N., and L. Wilkinson. 1982. Perceptual components of computer displays. IEEE Computer Graphics and Applications 2:23–35.

Hairston, N. G., Sr. 1980. The experimental test of an analysis of field distributions: Competition in terrestrial salamanders. Ecology 61:817–826.

Hairston, N. G., Sr. 1989. Ecological Experiments: Purpose, Design, and Execution. Cambridge University Press, Cambridge.

Haldane, J. B. S. 1955. The estimation and significance of the logarithm of a ratio of frequencies. Annals of Human Genetics 20:309–311.

Hall, P. 1992. The Bootstrap and Edgeworth Expansion. Springer-Verlag, New York.

Hall, P., and S. R. Wilson. 1991. Two guidelines for bootstrap hypothesis testing. Biometrics 47:757–762.

Hamilton, L. C. 1992. Regression with Graphics. Duxbury, Belmont, California.

Hand, D., and M. Crowder. 1996. Practical Longitudinal Data Analysis. Chapman and Hall, New York.

Hand, D. J., and C. C. Taylor. 1987. Multivariate Analysis of Variance and Repeated Measures. Chapman and Hall, London.

Hannon, S. J., K. Martin, L. Thomas, and J. Scheick. 1993. Investigator disturbance and clutch predation in willow ptarmigan: Methods for evaluating impact. Journal of Field Ornithology 64:575–586.

Harlow, L. L., S. A. Mulaik, and J. A. Steiger, editors. 1997. What If There Were No Significance Tests? Lawrence Erlbaum Associates, London.

Harper, J. L. 1977. Population Biology of Plants. Academic Press, London.

Harris, R. J. 1985. A Primer of Multivariate Statistics. Academic Press, New York.

Harvey, L. E. 1996. Macroecological studies of species composition, habitat and biodiversity using GIS and canonical correspondence analysis. Proceedings, Third International Conference/Workshop on Integrating GIS and Environmental Modeling, Sante Fe, New Mexico, January 21–26, 1996. (http://www.ncgia.ucsb.edu/conf/SANTE *FECD*ROM/main.htlm).

Harvey, L. E., F. W. Davis, and N. Gale. 1988. The analysis of class dispersion patterns using matrix comparisons. Ecology 69:537–542.

Harville, D. A. 1985. Decomposition of prediction error. Journal of the American Statistical Association 80:132–138.

Hassell, M. P. 1978. The Dynamics of Arthropod Predator-Prey Systems. Princeton University Press, Princeton, New Jersey.

Hassell, M. P., J. H. Lawton, and J. R. Beddington. 1977. Sigmoid functional responses by invertebrate predators and parasitoids. Journal of Animal Ecology 46:249–262.

Hatcher, L., and E. J. Stepanski. 1994. A step-by-step approach to using the SAS system for univariate and multivariate statistics. SAS Institute, Inc., Cary, North Carolina.

Hatfield, J. S., W. R. Gould, B. A. Hoover, M. R. Fuller, and E. L. Lindquist. 1996. Detecting trends in raptor counts: Power and type I error rates of various statistical tests. Wildlife Society Bulletin 24:505–515.

Hayduk, L. A. 1987. Structural Equation Modeling with LISREL. Johns Hopkins University Press, Baltimore, Maryland.

Hayes, A. F. 1996. Permutation test is not distribution free. Psychological Methods 1:184–198.

Hayes, J. P. 1987. The positive approach to negative results in toxicology studies. Ecotoxicology and Environmental Safety 14:73–77.

Hays, W. L. 1981. Statistics, 3rd ed. Holt, Rinehart & Winston, New York.

Hedges, L. V. 1983. A random effects model for effect sizes. Psychological Bulletin 93:388–395.

Hedges, L. V., and I. Olkin. 1985. Statistical Methods for Meta-analysis. Academic Press, New York.

Hedges, L. V., and I. Olkin. 2000. Statistical methods for meta-analysis in the medical and social sciences, 2nd ed. Academic Press, New York.

Hedges, L. V., J. Gurevitch, and P. Curtis. 1999. The meta-analysis of response ratios in experimental ecology. Ecology 80:1150–1156.

Heffner, R. A., M. J. Butler IV, and C. K. Reilly. 1996. Pseudoreplication revisited. Ecology 77:2558–2562.

Heltshe, J. F. 1988. Jackknife estimate of the matching coefficient of similarity. Biometrics 44:447–460.

Heltshe, J. F., and N. E. Forrester. 1985. Statistical evaluation of the jackknife estimate of diversity when using quadrat samples. Ecology 66:107–111.

Herr, D. G. 1986. On the history of ANOVA in unbalanced, factorial design: The first 30 years. American Statistician 40:265–270.

Herrera, C. M. 1993. Selection on floral morphology and environmental determinants of fecundity in a hawk-moth-pollinated violet. Ecological Monographs 63:251–275.

Hicks, C. R. 1982. Fundamental Concepts in the Design of Experiments, 3rd ed. Holt, Rinehart & Winston, New York.

Hilborn, R., and M. Mangel. 1997. The Ecological Detective: Confronting Models with Data. Princeton University Press, Princeton, New Jersey.

Hocking, R. R. 1985. The Analysis of Linear Models. Brooks/Cole Publishing Co., Belmont, California.

Hoening, J. M., and D. M. Heisey. 2001. The abuse of power: the pervasive fallacy of power calculations for data analysis. American Statistician (in press).

Holling, C. S. 1966. The functional response of invertebrate predators to prey density. Memoirs of the Entomological Society of Canada 48:1–87.

Holomuzki, J. R. 1989. Salamander predation and vertical distributions of zooplankton. Freshwater Biology 21:461–472.

Holt, R. D. 1977. Predation, apparent competition, and the structure of prey communities. Theoretical Population Biology 12:197–229.

Horn, J. S., H. H. Shugart, and D. L. Urban. 1989. Simulators as models of forest dynamics. Pages 256–267 in J. Roughgarden, R. M. May, and S. A. Levin, editors, Perspectives in Ecology Theory. Princeton University Press, Princeton, New Jersey.

Horton, D. R., P. L. Chapman, and J. L. Capinera. 1991. Detecting local adaptation in phytophagus insects using repeated measures designs. Environmental Entomology 20:410–418.

Horvitz, C. C., and D. W. Schemske. 1994. Effects of dispersers, gaps, and predators on dormancy and seedling emergence in a tropical herb. Ecology 75:1949–1958.

Houck, M. A., and R. E. Strauss. 1985. The comparative study of functional responses: Experimental design and statistical interpretation. Canadian Entomologist 117:617–629.

Howard, T., and D. E. Goldberg. 2000. Competitive response hierarchies for different fitness components of herbaceous perennials. Ecology (in press).

Hubert, L. J. 1987. Assignment Methods in Combinatorial Data Analysis. Marcel Dekker, New York.

Huberty, C. J., and J. D. Morris. 1989. Multivariate analysis versus multiple univariate analyses. Psychological Bulletin 105:302–308.

Huey, R. B., and E. R. Pianka. 1974. Ecological character displacement in a lizard. American Zoologist 14:1127–1136.

Huey, R. B., E. R. Pianka, M. E. Egan, and L. W. Coons. 1974. Ecological shifts in sympatry: Kalahari fossorial lizards (*Typhlosaurus*). Ecology 55:304–316.

Hunt, R. L. 1976. A long-term evaluation of trout habitat development and its relation to improving management-oriented research. Transactions of the American Fisheries Society 105:361–364.

Hunter, J. E., F. L. Schmidt, and G. B. Jackson. 1982. Meta-analysis: Cumulating Research Findings Across Studies. Sage Publications, Beverly Hills, California.

Hunter, M. D., and P. W. Price. 1992. Playing chutes and ladders: Heterogeneity and the relative roles of bottom-up and top-down forces in natural communities. Ecology 73: 724–732.

Hurlbert, S. H. 1984. Pseudoreplication and the design of ecological field experiments. Ecological Monographs 54:187–211.

Huston, M. A. 1997. Hidden treatments in ecological experiments: Re-evaluating the ecosystem function of biodiversity. Oecologia 110:449–460.

Hutchings, M. J., K. D. Booth, and S. Waite. 1991. Comparison of survivorship by the logrank test: Criticisms and alternatives. Ecology 72:2290–2293.

Huynh, H., and L. S. Feldt. 1970. Conditions under which mean square ratios in repeated measurement designs have exact F-distributions. Journal of the American Statistical Association 65:1582–1589.

Huynh, H., and L. S. Feldt. 1976. Estimation of the Box correction for degrees of freedom from sample data in the randomized block and split-plot designs. Journal of Educational Statistics 1:69–82.

Huynh, H., and G. K. Mandeville. 1979. Validity conditions in repeated measures designs. Psychological Bulletin 86:964–973.

Hyde, J. S., and M. C. Linn. 1986. The Psychology of Gender: Advances Through Meta-analysis. Johns Hopkins University Press, Baltimore, Maryland.

Ishii-Kuntz, M. 1994. Ordinal log-linear models. Sage University Paper Series on Quantitative Applications in the Social Sciences, 07-097. Sage, Thousand Oaks, California.

Ivlev, V. S. 1961. The Experimental Ecology of the Feeding of Fishes. Yale University Press, New Haven, Connecticut.

Jaccard, P. 1901. Distribution de la flore alpine dans le Bassin des Dranes et dans quelques régions voisines. Bulletin Société Vaudoise des Sciences Naturelles 37:241–272.

Jackson, J. B. C. 1981. Interspecific competition and species' distributions: The ghosts of theories and data past. American Zoologist 21:889–901.

James, F. C., and C. E. McCulloch. 1990. Multivariate analysis in ecology and systematics: panacea or Pandora's box? Annual Review of Ecology and Systematics 21:129–166.

Jassby, A. D., and T. M. Powell. 1990. Detecting changes in ecological time series. Ecology 71:2044–2052.

Jenkins, G. M. 1979. Practical Experiences with Modelling and Forecasting Time Series. Gwylim Jenkins and Partners (Overseas), Ltd., St. Helier, Jersey, Channel Islands.

Jennrich, R. I., and M. D. Schluchter. 1986. Unbalanced repeated-measures models with structured covariance matrices. Biometrics 42:805–820.

Johnson, C. R., and K. H. Mann. 1988. Diversity, patterns of adaptation, and stability of Nova Scotian kelp beds. Ecological Monographs 58:129–154.

Johnson, D. H. 1999. The insignificance of statistical significance testing. Journal of Wildlife Management 63:763–772.

Johnson, M. L., D. G. Huggins, and F. DeNoyelles Jr. 1991. Ecosystem modeling with LISREL: A new approach for measuring direct and indirect effects. Ecological Applications 1:383–398.

Johnson, R. A., and D. W. Wichern. 1988. Applied Multivariate Statistical Analysis, 2nd ed. Prentice Hall, Englewood Cliffs, New Jersey.

Johnston, G. 1996. Repeated measures analysis with discrete data using the SAS system. Proceeding of the Twenty-first Annual SAS Users Group International Conference [available at www.sas.com].

Jones, D. 1984. Use, misuse, and role of multiple-comparison procedures in ecological and agricultural entomology. Environmental Entomology 13:635–649.

Jones, D., and N. Matloff. 1986. Statistical hypothesis testing in biology: A contradiction in terms. Journal of Economic Entomology 79:1156–1160.

Jones, R. G. 1999. TechDig, Version 2.0. Shareware [available at http://www.xnet.com/~ronjones].

Jones, R. H. 1980. Maximum likelihood fitting of ARMA models to time series with missing observations. Technometrics 22:389–395.

Jordano, P. 1995. Frugivore-mediated selection on fruit and seed size: Birds and St. Lucie's cherry, *Prunus mahaleb*. Ecology 76:2627–2639.

Jöreskog, K. G., and D. Sörbom. 1988. LISREL 7 A Guide to the Program and Applications. Scientific Software, Morresville, Illinois.

Journel, A. G., and C. J. Hiujbregts. 1978. Mining Geostatistics. Academic Press, London.

Juliano, S. A. 1988. Chrysomelid beetles on water lily leaves: Herbivore density, leaf survival, and herbivore maturation. Ecology 69:1294–1298.

Juliano, S. A. 1998. Species introduction and replacement among mosquitoes: Interspecific resource competition or apparent competition? Ecology 79:255–268.

Juliano, S. A., and J. H. Lawton. 1990. The relationship between competition and morphology: II. Experiments on co-occurring dytiscid beetles. Journal of Animal Ecology 59: 831–848.

Juliano, S. A., and F. M. Williams. 1985. On the evolution of handling time. Evolution 39:212–215.

Juliano, S. A., and F. M. Williams. 1987. A comparison of methods for estimating the functional response parameters of the random predator equation. Journal of Animal Ecology 56:641–653.

Kabissa, J. C. B., J. G. Yarro, H. Y. Kayumbo, and S. A. Juliano. 1996. Functional responses of two chrysopid predators feeding on *Helicoverpa armigera* (Lep.: Noctuidae) and *Aphis gossypii* (Hom.: Aphididae). Entomolophaga 41:141–151.

Kalbfleisch, J. D., and R. L. Prentice. 1980. The Statistical Analysis of Failure Time Data. Wiley, New York.

Kardia, S. 1998. The graphical interface. Nature software reviews on-line: http://www.nature.com/.

Kareiva, P. 1994. Higher order interactions as a foil to reductionist ecology. Ecology 75: 1527–1528.

Karlin, S., E. C. Cameron, and R. Chakraborty. 1983. Path analysis in genetic epidemiology: A critique. American Journal of Human Genetics 35:695–732.

Keller, E. F. 1983. A Feeling for the Organism: The Life and Work of Barbara McClintock. Freeman, New York.

Kempthorne, O. 1975. Fixed and mixed models in the analysis of variance. Biometrics 31: 473–486.

Kempthorne, O., and T. E. Doerfler. 1969. The behaviour of some significance tests under experimental randomization. Biometrika 56:231–248.

Kempton, R. A., and C. W. Howes. 1981. The use of neighboring plot values in the analysis of variety trials. Applied Statistics 30:59–70.

Kenny, D. A., and C. M. Judd. 1986. Consequences of violating the independence assumption in analysis of variance. Psychological Bulletin 99:422–431.

Keppel, G. 1991. Design and Analysis: A Researcher's Handbook. Prentice-Hall, Engle-wood Cliffs, New Jersey.

Keselman, H. J., K. C. Carriere, and L. M. Lix. 1993. Testing repeated measures hypotheses when covariance matrices are heterogeneous. Journal of Educational Statistics 18: 305–319.

Keselman, H. J., J. Algina, R. K. Kowalchuk, and R. D. Wolfinger. 1999. A comparison of recent approaches to the analysis of repeated measures. British Journal of Mathematical and Statistical Psychology 52:63–78.

Keyfitz, N. 1985. Applied Mathematical Demography, 2nd ed. Springer-Verlag, New York.

Khattree, R., and D. N. Naik. 1999. Applied Multivariate Statistics with SAS Software. SAS Institute, Inc., Cary, North Carolina.

Kiefer, J. 1975. Construction and optimality of generalized Youden designs. Pages 333–353 in J. N. Srivastava, editor, A Survey of Statistical Design and Linear Models. North-Holland, Amsterdam, The Netherlands.

Kiefer, J., and H. P. Wynn. 1981. Optimum balanced block and latin square designs for correlated observations. Annals of Statistics 9:737–757.

King, R. B. 1989. Body size variation among island and mainland snake populations. Herpetologica 45:84–88.

King, R. B., and B. King. 1991. Sexual differences in color and color change in wood frogs. Canadian Journal of Zoology 69:1963–1968.

Kingsolver, J. G., and D. G. Schemske. 1991. Analyzing selection with path analysis. Trends in Ecology and Evolution 6:276–280.

Knoke, D., and P. J. Burke. 1980. Log-linear models. Sage University Paper Series on Quantitative Applications in the Social Sciences 07–020. Sage, Thousand Oaks, California.

Koch, G. G., I. A. Amara, M. E. Stokes, and D. B. Gillings. 1980. Some views on parametric and non-parametric analyses for repeated measurements and selected bibliography. International Statistical Review 48:249–265.

Kohn, J. R., and N. M. Waser. 1986. The effect of *Delphinium nelsonii* pollen on seed set in *Ipomopsis aggregata*, a competitor for hummingbird pollination. American Journal of Botany 72:1144–1148.

Kot, M., M. A. Lewis, and P. van den Driessche. 1996. Dispersal data and the spread of invading organisms. Ecology 77:2027–2042.

Kraemer, H. C., and S. Thiemann. 1987. How Many Subjects? Sage Publications, London.

Krannitz, P. G., L. W. Aarssen, and J. M. Dow. 1991. The effect of genetically based differences on seedling survival in *Arabidopsis thaliana*. American Journal of Botany 78:446–449.

Krebs, C. J., S. Boutin, R. Boonstra, A. R. E. Sinclair, J. N. M. Smith, M. R. T. Dale, K. Martin, and R. Turkington. 1995. Impact of food and predation on the snowshoe hare cycle. Science 269:1112–1115.

Kucera, C. L., and S. C. Martin. 1957. Vegetation and soil relationships in the glade region of the southwestern Missouri Ozarks. Ecology 38:285–291.

Kuehl, R. O. 1994. Statistical Principles of Research Design and Analysis. Duxbury Press, Belmont, California.

Kuhn, T. S. 1962. The structure of scientific revolutions. University of Chicago Press, Chicago.

LaBarbara, M. 1989. Analyzing body size as a factor in ecology and evolution. Annual Review of Ecology and Systematics 20:97–117.

Laird, N. M., and J. H. Ware. 1982. Random effects models for longitudinal data. Biometrics 38:963–974.

Lakatos, I. 1974. Popper on demarcation and induction. Pages 241–273 in P. A. Schilpp, editor, The Philosophy of Karl Popper. Open Court Publishing Co., LaSalle, Illinois.

Laska, M. S., and J. T. Wootton. 1998. Theoretical concepts and empirical approaches to measuring interaction strength. Ecology 79:461–476.

Lauter, J. 1978. Sample size requirements for the T2 test of MANOVA (Tables for one-way classification). Biometrical Journal 20:389–406.

Lavine, M., and S. Lozier. 1999. A Markov random field spatio-temporal analysis of ocean temperature. Environmental and Ecological Statistics 6:249–273.

Law, R., and A. R. Watkinson. 1987. Response-surface analysis of two-species competition: An experiment on *Phleum arenarium* and *Vulpia fasciculata*. Journal of Ecology 75:871–886.

Lawless, J. F. 1982. Statistical Models and Methods for Lifetime Data. Wiley, New York.

Leduc, A., P. Drapeau, Y. Bergeron, and P. Legendre. 1992. Study of spatial components of forest cover using partial Mantel tests and path analysis. Journal of Vegetation Science 3:69–78.

Lee, C.-S., and J. O. Rawlings. 1982. Design of experiments in growth chambers. Uniformity trials in the North Carolina State University phytotron. Crop Science 22:551–558.

Lee, E. T. 1980. Statistical Methods for Survival Data Analysis. Lifetime Learning, Belmont, California.

Lee, P. M. 1989. Bayesian Statistics: An Introduction. Oxford University Press, Oxford.

Legendre, L., and P. Legendre. 1983. Numerical Ecology. Developments in Environmental Modelling, Vol. 3. Elsevier, Amsterdam, The Netherlands.

Legendre, P. 1993. Spatial autocorrelation: Trouble or new paradigm? Ecology 74:1659–1673.

Legendre, P., and M.-J. Fortin. 1989. Spatial pattern and ecological analysis. Vegetatio 80:107–138.

Legendre, P., and L. Legendre. 1998. Numerical ecology, 2nd English edition. Elsevier Science BV, Amsterdam.

Legendre, P., and M. Troussellier. 1988. Aquatic heterotrophic bacteria: Modeling in the presence of spatial autocorrelation. Limnology and Oceanography 33:1055–1067.

Legendre, P., and A. Vaudor. 1991. The R Package: Multidimensional Analysis, Spatial Analysis. Département de sciences biologiques, Université de Montreal, Québec, Canada.

Legendre, P., R. R. Sokal, N. L. Oden, A. Vaudor, and J. Kim. 1990. Analysis of variance with spatial autocorrelation in both the variable and the classification criterion. Journal of Classification 7:53–75.

Léger, C., D. N. Politis, and J. P. Romano. 1992. Bootstrap technology and applications. Technometrics 34:378–398.

Lehmann, E. L. 1975. Nonparametrics: Statistical Methods Based on Ranks. Holden-Day, Inc., San Francisco.

Lenski, R. E., and P. M. Service. 1982. The statistical analysis of population growth rates calculated from schedules for survivorship and fecundity. Ecology 63:655–662.

Lewis, M. A. 1997. Variability, patchiness and jump dispersal in the spread of an invading population. Pages 46–74 in D. Tilman and P. Kareiva, editors, Spatial Ecology. Princeton University Press, Princeton, New Jersey.

Li, C. C. 1975. Path Analysis. A Primer. The Boxwood Press, Pacific Grove, California.

Light, R. J., and D. B. Pillemer. 1984. Summing Up: The Science of Reviewing Research. Harvard University Press, Cambridge, Massachusetts.

Likens, G. E., F. H. Bormann, N. M. Johnson, D. W. Fisher, and R. S. Pierce. 1977. Effects of forest cutting and herbicide treatment on nutrient budgets in the Hubbard Brook watershed ecosystem. Ecological Monographs 40:23–47.

Lindsey, J. K. 1993. Models for repeated measurements. Oxford University Press, Oxford.

Lindsey, L. K. 1995. Introductory statistics: A modelling approach. Cambridge University Press, Cambridge.

Lipsey, M. W. 1990. Design Sensitivity: Statistical Power for Experimental Research. Sage Publications, Newbury Park, California.

Littell, R. C., G. A. Milliken, W. W. Stroup, and R. D. Wolfinger. 1996. SAS System for Mixed Models. SAS Institute, Inc., Cary, North Carolina.

Livdahl, T. P. 1979. Evolution of handling time: The functional response of a predator to the density of sympatric and allopatric strains of prey. Evolution 33:765–768.

Livdahl, T. P., and A. E. Stiven. 1983. Statistical difficulties in the analysis of functional response data. Canadian Entomologist 115:1365–1370.

Livshits, G., R. R. Sokal, and E. Kobyliansky. 1991. Genetic affinities of Jewish populations. American Journal of Human Genetics 49:131–146.

Ljung, G. M., and G. E. P. Box. 1978. On a measure of lack of fit in time series models. Biometrika 65:297–303.

Loehlin, J. C. 1987. Latent Variable Models. Lawrence Erlbaum, Hillsdale, New Jersey.

Lovvorn, J. R., and M. P. Gillingham. 1996. Food dispersion and foraging energetics—A mechanistic synthesis for field studies of avian benthivores. Ecology 77:435–451.

Ludwig, J. A., and J. F. Reynolds. 1988. Statistical Ecology. John Wiley & Sons, New York.

Lynch, M. 1977. Fitness and optimal body size in zooplankton populations. Ecology 58:763–774.

MacCullum, R. 1986. Specification searches in covariance structure modeling. Psychological Bulletin 100:107–120.

MacIsaac, H. J., T. C. Hutchinson, and W. Keller. 1987. Analysis of planktonic rotifer assemblages form Sudbury, Ontario, area lakes of varying chemical composition. Canadian Journal of Fisheries and Aquatic Sciences 44:1692–1701.

Maddox, G. D., and J. Antonovics. 1983. Experimental ecological genetics in *Plantago*: A structural equation approach to fitness components in *P. aristata* and *P. patagonica*. Ecology 64:1092–1099.

Madenjian, C. P., D. J. Jude, and F. J. Tesar. 1986. Intervention analysis of power plant impact on fish populations. Canadian Journal of Fisheries and Aquatic Sciences 43:819–829.

Magurran, A. E. 1988. Ecological Diversity and Its Measurement. Princeton University Press, Princeton, New Jersey.

Manly, B. F. J. 1986a. Multivariate Statistical Methods. Chapman and Hall, London.

Manly, B. F. J. 1986b. Randomization and regression methods for testing for associations with geographical, environmental, and biological distances between populations. Researches in Population Ecology 28:201–218.

Manly, B. F. J. 1990. Stage-Structured Populations. Chapman and Hall, New York.

Manly, B. F. J. 1997. Randomization, Bootstrap and Monte Carlo Methods in Biology, 2nd ed. Chapman and Hall, London.

Mantel, N. 1967. The detection of disease clustering and a generalized regression approach. Cancer Research 27:209–220.

Mardia, K. V., and R. J. Marshall. 1984. Maximum likelihood estimation of models for residual covariance in spatial regression. Biometrika 71:135–146.

Markham, J. H., and C. P. Chanway. 1996. Measuring plant neighbour effects. Functional Ecology 10:548–549.

Marshal, J. P., and S. Boutin. 1999. Power analysis of wolf-moose functional response. Journal of Wildlife Management 63:396–402.

Matheron, G. 1963. Principles of geostatistics. Economic Geology 58:1246–1266.

Matson, P. A., and S. R. Carpenter. 1990. Statistical analysis of ecological responses to large-scale perturbations. Ecology 71:2037.

Maxwell, S. E., and H. D. Delaney. 1990. Designing Experiments and Analyzing Data: A Model Comparison Perspective. Wadsworth, Belmont, California.

Maxwell, S. E., C. J. Camp, and R. D. Arvey. 1981. Measures of strength and association: A comparative examination. Journal of Applied Psychology 66:525–534.

Mayo, D. G. 1996. Error and the growth of experimental knowledge. University of Chicago Press, Chicago.

McCleary, R., and R. A. Hay. 1980. Applied Time Series Analysis for the Social Sciences. Sage Publications, Beverly Hills, California.

McCreary, N. J., S. R. Carpenter, and J. E. Chaney. 1983. Coexistence and interference in two submerged freshwater perennial plants. Oecologia 59:393–396.

McDonald, D. B., J. W. Fitzpatrick, and G. E. Woolfenden. 1996. Actuarial senescence and demographic heterogeneity in the Florida scrub jay. Ecology 77:2373–2381.

McGill, R., J. W. Tukey, and W. A. Larsen. 1978. Variations of box plots. American Statistician 32:12–16.

McGraw, J. B., and F. S. Chapin III. 1989. Competitive ability and adaptation to fertile and infertile soils in two *Eriophorum* species. Ecology 70:736–749.

McIntosh, R. P. 1985. The Background of Ecology. Cambridge University Press, Cambridge.

McLean, R. A., W. L. Sanders, and W. W. Stroup. 1991. A unified approach to mixed linear models. American Statistician 45:54–64.

McPeek, M. A., and B. L. Peckarsky. 1998. Life histories and the strengths of species interactions: combining mortality, growth, and fecundity effects. Ecology 79:867–879.

Mead, R. 1988. The Design of Experiments: Statistical Principles for Practical Application. Cambridge University Press, Cambridge.

Mead, R., and R. N. Curnow. 1983. Statistical Methods in Agricultural and Experimental Biology. Chapman and Hall, London.

Melton, R. H. 1982. Body size and island *Peromyscus*: A pattern and an hypothesis. Evolutionary Theory 6:113–126.

Menard, S. 1995. Applied Logistic Regression Analysis. Sage, Thousand Oaks, California.

Messina, F. J., and J. B. Hanks. 1998. Host plant alters the shape of the functional response of an aphid predator (Coleoptera: Coccinellidae). Environmental Entomology 27:1196–1202.

Meyer, J. S., C. G. Ingersoll, L. L. McDonald, and M. S. Boyce. 1986. Estimating uncertainty in population growth rates: Jackknife versus Bootstrap techniques. Ecology 67:1156–1166.

Mielke, P. W., K. J. Berry, P. J. Brockwell, and J. S. Williams. 1981. A class of nonparametric tests based on multiresponse permutation procedures. Biometrika 68:720–724.

Milan, L., and J. Whittaker. 1995. Application of the parametric bootstrap to models that incorporate a singular value decomposition. Applied Statistics 44:31–49.

Miller, R. G. 1974. The jackknife—A review. Biometrika 61:1–17.

Miller, T. E. 1996. On quantifying the intensity of competition across gradients. Ecology 77:978–981.

Milligan, G. W., D. S. Wong, and P. A. Thompson. 1987. Robustness properties of nonorthogonal analysis of variance. Psychological Bulletin 101:464–470.

Milliken, G. A., and D. E. Johnson. 1984. Analysis of Messy Data, Volume 1: Designed Experiments. Lifetime Learning Publications, Belmont, California.

Mitchell, R. J. 1992. Testing evolutionary and ecological hypotheses using path analysis and structural equation modeling. Functional Ecology 6:123–129.

Mitchell, R. J. 1993. Adaptive significance of *Ipomopsis aggregata* nectar production rate: Observational and experimental studies in the field. Evolution 47:25–35.

Mitchell, R. J. 1994. Effects of floral traits, pollinator visitation, and plant size on *Ipomopsis aggregata* fruit production. American Naturalist 143:870–889.

Mitchell, R. J., and N. M. Waser. 1992. Adaptive significance of *Ipomopsis aggregata* nectar production rate: Pollination success of single flowers. Ecology 73:633–638.

Mitchell-Olds, T. 1987. Analysis of local variation in plant size. Ecology 68:82–87.

Mitchell-Olds, T., and R. G. Shaw. 1987. Regression analysis of natural selection: Statistical inference and biological interpretation. Evolution 41:1149–1161.

Molenberghs, G., E. J. T. Goetghebeur, S. R. Lipsitz, and M. G. Kenward. 1999. Nonrandom missingness in categorical data: Strengths and limitations. American Statistician 53:110–118.

Mood, A. M., and F. A. Graybill. 1993. Introduction to the Theory of Statistics, 2nd ed. McGraw-Hill, New York.

Morrison, D. F. 1990. Multivariate Statistical Methods, 3rd ed. McGraw-Hill Publishing Co., New York.

Morrow, L. L. 1990. Field experiments on competition: A meta-analysis. Master's thesis. State University of New York at Stony Brook, Stony Brook, New York.

Mueller, L. D. 1979. A comparison of two methods for making statistical inferences on Nei's measure of genetic distance. Biometrics 35:757–763.

Muenchow, G. 1986. Ecological use of failure time analysis. Ecology 67:246–250.

Muir, W. M. 1986. Estimation of response to selection and utilization of control populations for additional information and accuracy. Biometrics 42:381–391.

Mueller, L. D., and L. Altenberg. 1985. Statistical inference on measures of niche overlap. Ecology 66:1204–1210.

Muller, K. E., and V. A. Benignus. 1992. Increasing scientific power with statistical power. Neurotoxicology and Teratology 14:211–219.

Muller, K. E., and V. B. Pasour. 1997. Bias in linear model power and sample size due to estimating variance. Communications in Statistics: Theory and Methods 26:839–851.

Murren, C. J., and A. M. Ellison. 1996. Effects of habitat, plant size, and floral display on male and female reproductive success of the neotropical orchid *Brassavola nodosa*. Biotropica 28:30–41.

Muthen, B., and D. Kaplan. 1985. A comparison of some methodologies for the factor analysis of non-normal Likert variables. British Journal of Mathematical and Statistical Psychology 38:171–193.

Nannini, M. A., and S. A. Juliano. 1998. Effects of the facultative predator *Anopheles barberi* on population performance of its prey *Aedes triseriatus* (Diptera: Culicidae). Annals of the Entomological Society of America 91:33–42.

Nelder, J. A., and R. W. M. Wedderburn. 1972. Generalised linear models. Journal of the Royal Statistical Society A135:370–384.

Nemac, A. F. L. 1991. Power Analysis Handbook for the Design and Analysis of Forestry Trials. Biometrics Information Handbook 2 BC Ministry of Forests. Victoria, Canada.

Nester, M. R. 1996. An applied statistician's creed. Applied Statistics Journal of the Royal Statistical Society 45:401–410.

Neter, J., and W. Wasserman. 1974. Applied Linear Statistical Models. Richard S Irwin, Homewood, California.

Nevo, E. 1973. Adaptive variation in the size of cricket frogs. Ecology 54:1271–1281.

Newman, J. A., J. Bergelson, and A. Grafen. 1997. Blocking factors and hypothesis tests in ecology: Is your statistics text wrong? Ecology 78:1312–1320.

Newman, M. C., and J. T. McCloskey. 1996. Time-to-event analysis of ecotoxicity data. Ecotoxicology 5:187–196.

Neyman, J., and E. S. Pearson. 1928. On the use and interpretation of certain test criteria for purposes of statistical inference. Biometrika 20A:175–240.

Noakes, D. 1986. Quantifying changes in British Columbia Dungeness crab landings using intervention analysis. Canadian Journal of Fisheries and Aquatic Sciences 43:634–639.

Norusis, M. J. 1990. SPSS Advanced Statistics User's Guide. SPSS Inc., Chicago, Illinois.

O'Brien, R. G. 1998. A tour of UnifyPow: A SAS module/macro for sample-size analysis. Pages 1346–1355 in Proceedings of the 23rd SAS Users Group International Conference. SAS Institute, Inc., Cary, North Carolina [http://www.bio.ri.ccf.org/power.html].

O'Brien, R. G., and M. K. Kaiser. 1985. MANOVA method for analyzing repeated measures designs: An extensive primer. Psychological Bulletin 97:316–333.

O'Brien, R. G., and K. E. Muller. 1993. Unified power analysis for t-tests through multivariate hypotheses. Pages 297–344 in L. K. Edwards, editor, Applied Analysis of Variance in Behavioral Science. Marcel Dekker, New York.

Oden, N. L., and R. R. Sokal. 1986. Directional autocorrelation: An extension of spatial correlograms to two dimensions. Systematic Zoology 35:608–617.

Oden, N. L., and R. R. Sokal. 1992. An investigation of three-matrix permutation tests. Journal of Classification 9:275–280.

Osenberg, C. W., O. Sarnelle, S. D. Cooper, and R. D. Holt. 1999. Resolving ecological questions through meta-analysis: Goals, metrics, and models. Ecology 80:1105–1117.

Osenberg, C. W., R. J. Schmitt, S. J. Holbrook, K. E. Abu-Saba, and A. R. Flegal. 1994. Detection of environmental impacts: Natural variability, effect size, and power analysis. Ecological Applications 4:16–30.

Pacala, S. W., and J. A. Silander. 1990. Field tests of neighborhood population dynamic models of two annual weed species. Ecological Monographs 60:113–134.

Packer, C., M. Tatar, and A. Collins. 1998. Reproductive cessation in female mammals. Nature 392:807–811.

Paine, R. T. 1976. Size-limited predation: An observational and experimental approach with the *Mytilus–Pisaster* interaction. Ecology 57:858–873.

Pallesen, L., P. M. Berthouex, and K. Booman. 1985. Environmental intervention analysis: Wisconsin's ban on phosphate detergents. Water Research 19:353–362.

Palmer, J. O. 1984. Environmental determinants of seasonal body size variation in the milkweed leaf beetle *Labidomera clivicollis* Coleoptera: Chrysomelidae. Annals of the Entomological Society of America 77:188–192.

Palmer, M. W. 1991. Estimating species richness: The second-order jackknife reconsidered. Ecology 72:1512–1513.

Palmer, M. W. 1993. Putting things in even better order: The advantages of canonical correspondence analysis. Ecology 74:2215–2230.

Parilla, G., A. Lavin, H. Bryden, and R. Millard. 1994. Rising temperatures in the subtropical North Atlantic Ocean over the past 35 years. Nature 369:48–51.

Pedhazur, E. J. 1982. Multiple Regression in Behavioral Research, 2nd ed. Holt, Rinehart & Winston, New York.

Peterman, R. M. 1990. Statistical power analysis can improve fisheries research and management. Canadian Journal of Fisheries and Aquatic Sciences 47:2–15.

Peterman, R. M., and M. J. Bradford. 1987. Statistical power of trends in fish abundance. Canadian Journal of Fisheries and Aquatic Sciences 44:1879–1889.

Peters, R. H. 1991. A Critique for Ecology. Cambridge University Press, Cambridge.

Petraitis, P. S. 1998. How can we compare the importance of ecological processes if we never ask "Compared to what?" Pages 183–201 in W. J. Resetarits Jr. and J. Bernardo, editors, Experimental Ecology—Issues and Prespectives. Oxford University Press, New York.

Petraitis, P. S. 1998. Rates of mussel mortality and predator activity in sheltered bays of the Gulf of Maine. Journal of Experimental Marine Biology and Ecology 231:47–62.

Petraitis, P. S., A. E. Dunham, and P. H. Niewiarowski. 1996. Inferring multiple causality: The limitations of path analysis. Functional Ecology 10:421–431.

Pickett, S. T. A., J. Kolasa, and C. G. Jones. 1994. Ecological Understanding. Academic Press, San Diego, California.

Platt, J. R. 1964. Strong inference. Science 146:347–353.

Platt, W. J., and I. M. Weis. 1985. An experimental study of competition among fugitive prairie plants. Ecology 66:708–720.

Pons, T. L., and J. van der Toorn. 1988. Establishment of *Plantago lanceolata* L. and *Plantago major* L. among grass. I. Significance of light for germination. Oecologia 75:394–399.

Poole, R. W. 1978. The statistical prediction of population fluctuations. Annual Review of Ecology and Systematics 9:427–448.

Popper, K. R. 1959. The Logic of Scientific Discovery. Hutchinson & Co., London.

Porter, K. G. G. J., and J. D. J. Orcutt. 1982. The effect of food concentration on swimming patterns, feeding behaviour, ingestion, assimilation and respiration by *Daphnia*. Limnology and Oceanography 27:935–949.

Porter, W. R., and W. F. Trager. 1977. Improved nonparametric statistical methods for estimation of Michaelis–Menten enzyme kinetics parameters by the direct linear plot. Biochemical Journal 161:293–302.

Portnoy, S., and M. F. Willson. 1993. Seed dispersal curves: Behavior of the tail of the distribution. Evolutionary Biology 7:25–44.

Posten, H. O. 1984. Robustness of the two-sample *t* test. Pages 92–99 in D. Rash and M. L. Tiku, editors, Robustness of Statistical Methods and Nonparametric Statistics. D. Reidel, Dordrecht, The Netherlands.

Potvin, C. 1993. ANOVA: Experiments in controlled environments. Pages 46–68 in S. M. Scheiner and J. Gurvitch, editors, Design and Analysis of Ecological Experiments, 1st ed. Chapman & Hall, New York.

Potvin, C., and S. Tardif. 1988. Sources of variability and experimental design in growth chambers. Functional Ecology 2:123–130.

Potvin, C., M. J. Lechowicz, G. Bell, and D. Schoen. 1990a. Spatial, temporal, and species-specific pattern of heterogeneity in growth chamber experiments. Canadian Journal of Botany 68:499–504.

Potvin, C., M. J. Lechowicz, and S. Tardif. 1990b. The statistical analysis of ecophysiological response curves obtained from experiments involving repeated measures. Ecology 71:1389–1400.

Prasad, N. G. N., and J. N. K. Rao. 1990. The estimation of mean squared errors of small-area estimators. Journal of the American Statistical Association 85:163–171.

Pugesek, B. H., and J. B. Grace. 1998. On the utility of path modelling for ecological and evolutionary studies. Functional Ecology 12:843–856.

Pugesek, B. H., and A. Tomer. 1995. Determination of selection gradients using multiple regression versus structural equation models (SEM). Biometry Journal 37:449–462.

Pugesek, B. H., and A. Tomer. 1996. The Bumpus house sparrow data: A reanalysis using structural equation models. Evolutionary Ecology 10:387–404.

Pusey, A., J. Williams, and J. Goodall. 1997. The influence of dominance rank on the reproductive success of female chimpanzees. Science 277:828–831.

Pyke, G. H., H. R. Pulliam, and E. L. Charnov. 1977. Optimal foraging: A selective review of theory and tests. Quarterly Review of Biology 52:137–156.

Quade, D. 1982. Nonparametric analysis of covariance by matching. Biometrics 38:597–611.

Quinn, J. F., and A. E. Dunham. 1983. On hypothesis testing in ecology and evolution. American Naturalist 122:602–617.

Raftery, A. E., and J. E. Zeh. 1993. Estimation of Bowhead whale, *Balaena mysticetus*, population size. Pages 163–240 in C. Gatsonis, J. Hodges, R. Kass, and N. Singpurwalla, editors, Case Studies in Bayesian Statistics. Springer-Verlag, New York.

Ralls, K., and P. H. Harvey. 1985. Geographic variation in size and sexual dimorphism of North American weasels. Biological Journal of the Linnean Society 25:119–167.

Rao, P. V. 1998. Statistical Research Methods in the Life Sciences. Duxbury Press, New York.

Raudenbush, S. W. 1991. Summarizing evidence: Crusaders for simplicity. Educational Researcher 20:33–37.

Reed, D. C., and M. S. Foster. 1984. The effects of canopy shading on algal recruitment and growth in a giant kelp forest. Ecology 65:937–948.

Rempel, R. S., A. R. Rodgers, and K. F. Abraham. 1995. Performance of a GPS animal location system under boreal forest canopy. Journal of Wildlife Management 59:543–551.

Rice, W. R. 1990. A consensus combined *P*-value test and the family-wide significance of component tests. Biometrics 46:303–308.

Rice, W. R., and S. D. Gaines. 1994a. Heads I win, tails you lose—Testing directional alternative hypotheses in ecological and evolutionary research. Trends in Ecology and Evolution 9:235–237.

Rice, W. R., and S. D. Gaines. 1994b. The ordered-heterogeneity family of tests. Biometrics 50:746–752.

Richardson, J. T. E. 1996. Measures of effect size. Behavior Research Methods, Instruments, and Computers 28:12–22.

Ricklefs, R. E. 1998. Evolutionary theories of aging: Confirmation of a fundamental prediction, with implications for the genetic basis and evolution of life span. American Naturalist 152:24–44.

Roff, D. A. 1980. Optimizing development time in a seasonal environment: The "ups and downs" of clinal variation. Oecologia 45:202–208.

Roff, D. A. 1992. The Evolution of Life Histories: Theory and Analysis. Chapman and Hall, London.

Rogers, D. J. 1972. Random search and insect population models. Journal of Animal Ecology 41:369–383.

Rohlf, F. J., and R. R. Sokal. 1995. Statistical Tables, 3rd ed. W. H. Freeman and Co., San Francisco.

Rosenberg, M. S., D. C. Adams, and J. Gurevitch. 1999. MetaWin 2.0. Statistical software for conducting meta-analysis: fixed effect models, mixed effect models, and resampling tests. Sinauer Associates, Sunderland, Massachusetts.

Rotenberry, J. T., and J. A. Wiens. 1985. Statistical power analysis and community-wide patterns. American Naturalist 125:164–168.

Royall, R. M. 1997. Statistical Evidence: A Likelihood Paradigm. International Thompson Publishing, New York.

Rudstam, L. G., R. C. Lathrop, and S. R. Carpenter. 1993. The rise and fall of a dominant planktivore: Direct and indirect effects on zooplankton. Ecology 74:303–319.

Russo, R. 1986. Comparison of predatory behavior in five species of *Toxorhynchites* (Diptera: Culicidae). Annals of the Entomological Society of America 79:715–722.

Sacks, H. S., J. Berrier, D. Reitman, V. A. Ancona-Berk, and T. C. Chalmer. 1987. Meta-analyses of randomized controlled trials. New England Journal of Medicine 316:450–455.

Salsburg, D. S. 1985. The religion of statistics as practiced in medical journals. American Statistician 39:220–223.

SAS Institute, Inc. 1988. SAS/ETS User's Guide, Version 6, 1st ed. SAS Institute, Inc., Cary, North Carolina.

SAS Institute, Inc. 1989a. SAS/STAT User's Guide, Version 6, 4th ed., Vol. 1. SAS Institute, Inc., Cary, North Carolina.

SAS Institute, Inc. 1989b. SAS/STAT User's Guide, Version 6, 4th ed., Vol. 2. SAS Institute, Inc., Cary, North Carolina.

SAS Institute, Inc. 1990. SAS Procedures Guide, Version 6, 3rd ed. SAS Institute, Inc., Cary, North Carolina.

SAS Institute, Inc. 1992. SAS Technical Report P-229, SAS/STAT Software: Changes and Enhancements, Release 6.07. SAS Institute, Inc., Cary, North Carolina.

SAS Institute, Inc. 1995. Jackboot Macro. SAS Institute, Inc., Cary, North Carolina.

SAS Institute, Inc. 1996. SAS/STAT Software: Changes and Enhancements, Release 6.12. SAS Institute, Inc., Cary, North Carolina.

Satterthwaite, F. E. 1946. An approximate distribution of estimates of variance components. Biometrics Bulletin 2:110–114.

Saville, D. J. 1990. Multiple comparison procedures: The practical solution. American Statistician 44:174–180.

Scharf, F. S., F. Juanes, and M. Sutherland. 1998. Inferring ecological relationships from the edges of scatter diagrams: Comparisons of regression techniques. Ecology 79:448–460.

Scheiner, S. M., and H. S. Callahan. 1999. Measuring natural selection on phenotypic plasticity. Evolution 53:1704–1713.

Scheiner, S. M., R. J. Mitchell, and H. Callahan. 2000. Using path analysis to measure natural selection. Journal of Evolutionary Biology 13:423–433.

Scheirer, C. J., W. S. Roy, and N. Hare. 1976. The analysis of ranked data derived from completely randomized factorial designs. Biometrics 32:429–434.

Schemske, D. W., and C. C. Horvitz. 1988. Plant–animal interactions and fruit production in a neotropical herb: A path analysis. Ecology 69:1128–1137.

Schenker, N. 1985. Qualms about bootstrap confidence intervals. Journal of the American Statistical Association 80:360–361.

Schindler, D. W. 1974. Eutrophication and recovery in experimental lakes: Implication for lake management. Science 184:897–898.

Schindler, D. W. 1987. Detecting ecosystem responses to anthropogenic stress. Canadian Journal of Fisheries and Aquatic Sciences 44(Suppl. 1):6–25.

Schindler, D. W. 1998. Replication versus realism: The need for ecosystem-scale experiments. Ecosystems 1:323–334.

Schindler, D. W., K. H. Mills, D. F. Malley, D. L. Findlay, J. A. Shearer, I. J. Davies, M. A. Turner, G. A. Linsey, and D. R. Cruikshank. 1985. Long-term ecosystem stress: The effects of years of experimental acidification on a small lake. Science 228: 1395–1401.

Schoener, T. W. 1983. Field experiments on interspecific competition. American Naturalist 122:240–285.

Schoener, T. W. 1986. Mechanistic approaches to community ecology: A new reductionism. American Zoologist 26:81–106.

Schwaner, T. D. 1985. Population structure of black tiger snakes, Notechis ater niger on offshore islands of south Australia. Pages 35–46 in G. Grigg, R. Shine, and H. Ehmann, editors, The Biology of Austraasian Frogs and Reptiles. Surrey Beatty and Sons Pty. Ltd., Sydney, Australia.

Schwaner, T. D., and S. D. Sarre. 1988. Body size of tiger snakes in Southern Australia, with particular reference to Notechis ater serventyi (Elapidae) on Chappell Island. Journal of Herpetology 22:24–33.

Scott, S. J., and R. A. Jones. 1990. Generation means analysis of right-censored response-time traits: Low temperature seed germination in tomato. Euphytica 48:239–244.

Searcy-Bernal, R. 1994. Statistical power and aquacultural research. Aquaculture 127:371–388.

Searle, S. R. 1971. Linear Models. John Wiley & Sons, New York.

Searle, S. R. 1987. Linear Models for Unbalanced Data. John Wiley & Sons, New York.

Sebens, K. P. 1982. The limits to indeterminate growth: An optimal size model applied to passive suspension feeders. Ecology 63:209–222.

Sedinger, J. S., and P. L. Flint. 1991. Growth rate is negatively correlated with hatch date in Black Brant. Ecology 72:496–502.

Sedlmeier, P., and G. Gigerenzer. 1989. Do studies of statistical power have an effect on the power of studies? Psychological Bulletin 105:309–316.

Service, P. M., C. A. Michieli, and K. McGill. 1998. Experimental evolution of senescence: An analysis using a "heterogeneity" mortality model. Evolution 52:1844–1850.

Shao, J., and D. Tu. 1995. The Jackknife and Bootstrap. Springer-Verlag, New York.

Shaw, R. G. 1987. Maximum-likelihood approaches applied to quantitative genetics of natural populations. Evolution 41:812–826.

Shaw, R. G., and T. Mitchell-Olds. 1993. ANOVA for unbalanced data: An overview. Ecology 74:1638–1645.

Sheffé, H. 1959. The Analysis of Variance. John Wiley & Sons, London.

Shipley, B. 1995. Structured interspecific determinants of specific leaf area in 34 species of herbaceous angiosperms. Functional Ecology 9:312–319.

Shipley, B. 1997. Exploratory path analysis with applications in ecology and evolution. American Naturalist 149:1113–1138.

Shirley, E. A. C. 1981. A distribution-free method for the analysis of covariance based on ranked data. Applied Statistics 30:158–162.

Shirley, E. A. C. 1987. Applications of ranking methods to multiple comparison procedures and factorial experiments. Applied Statistics 36:205–213.

Shrader-Frechette, K. S., and E. D. McCoy. 1992. Statistics, costs and rationality in ecological inference. Trends in Ecology and Evolution 7:96–99.

Siegel, S. 1956. Nonparametric Statistics for the Behavioral Sciences. McGraw-Hill Book Co., New York.

Sigurjonsdottir, H. 1984. Food competition among Scatophaga stercoraria larvae with emphasis on its effects on reproductive success. Ecological Entomology 9:81–90.

Silvertown, J. 1987. Introduction to Plant Population Ecology, 2nd ed. Longman, Harlow, UK.

Simes, R. J. 1986. An improved Bonferroni procedure for multiple tests of significance. Biometrika 73:751–754.

Simken, D., and R. Hastie. 1987. An information-processing analysis of graph perception. Journal of the American Statistical Association 82:454–465.

Simms, E. L., and D. S. Burdick. 1988. Profile analysis of variance as a tool for analyzing correlated responses in experimental ecology. Biometrical Journal 30:229–242.

Simpson, E. H. 1951. The interpretation of interaction in contingency tables. American Statistician 13:238–241.

Sinervo, B., and D. F. DeNardo. 1996. Costs of reproduction in the wild: Path analysis of natural selection and experimental tests of causation. Evolution 50:1299–1313.

Sitter, R. R. 1992. Comparing three bootstrap methods for survey data. Canadian Journal of Statistics 20:135–154.

Sjoberg, S. 1980. Zooplankton feeding and queueing theory. Ecological Modelling 10: 215–225.

Smith, E. P. 1985. Estimating the reliability of diet overlap measures. Environmental Biology of Fishes 13:125–138.

Smith, E. P., and G. van Belle. 1984. Nonparametric estimation of species richness. Biometrics 40:119–129.

Smith, E. P., R. B. Genter, and J. Cairns. 1986. Confidence intervals for the similarity between algal communities. Hydrobiologia 139:237–245.

Smith, F. A., J. H. Brown, and T. J. Valone. 1997. Path analysis: A critical evaluation using long-term experimental data. American Naturalist 149:29–42.

Smith, L. 1994. Temperature influences functional response of *Anisopteromalus calandrae* (Hymenoptera, Pteromalidae) parasitizing maize weevil larvae in shelled corn. Annals of the Entomological Society of America 87:849–855.

Smith, W. D., D. Kravitz, and J. F. Grassle. 1979. Confidence intervals for similarity measures using the two sample jackknife. Pages 253–262 in L. Orloci, C. R. Rao, and W. M. Stiteler, editors, Multivariate Methods in Ecological Work. International Cooperative Publishing House, Fairland, Maryland.

Smouse, P. E., J. C. Long, and R. R. Sokal. 1986. Multiple regression and correlation extensions of the Mantel test of matrix correspondence. Systematic Zoology 35:627–633.

Snedecor, G. W., and W. G. Cochran. 1989. Statistical Methods. Iowa State University Press, Ames, Iowa.

Sokal, R. R. 1979. Testing statistical significance of geographic variation patterns. Systematic Zoology 28:227–231.

Sokal, R. R., and F. J. Rohlf. 1995. Biometry. W. H. Freeman, New York.

Sokal, R. R., I. A. Lengyel, P. A. Derish, M. C. Wooten, and N. L. Oden. 1987. Spatial autocorrelation of ABO serotypes in medieval cemeteries as an indicator of ethnic and familial structure. Journal of Archeological Science 14:615–633.

Sokal, R. R., N. L. Oden, B. A. Thomson, and J. Kim. 1993. Testing for regional differences in means: Distinguishing inherent from spurious spatial autocorrelation by restricted randomization. Geographical Analysis 25:199–210.

Sokal, R. R., N. L. Oden, P. Legendre, M.-J. Fortin, J. Kim, B. A. Thomson, A. Vaudor, R. M. Harding, and G. Barbujani. 1990. Genetics and language in European populations. American Naturalist 135:157–175.

Solow, A. R., and K. Sherman. 1997. Testing for stability in a predator–prey system. Ecology 78:2624–2627.

Song, Y. H., and K. L. Heong. 1997. Changing searching responses with temperature of *Cyrtorhinus lividipennis* Reuter (Hemiptera: Miridae) on the eggs of the brown plant hopper, *Nilaparvata lugens* (Stal.) (Homoptera: Delphacidae). Researches in Population Ecology 39:201–206.

Stamps, J. A. 1995. Using growth-based models to study behavioral factors affecting sexual size dimorphism. Herpetological Monographs 9:75–87.

Stearns, S. C. 1992. The Evolution of Life Histories. Oxford University Press, Oxford.

Steel, R. G. D., J. H. Torrie, and D. A. Dickey. 1996. Principles and Procedures of Statistics: A Biometrical Approach. McGraw-Hill Book Company, London.

Steidl, R. J., C. R. Griffin, and L. J. Niles. 1991. Differential reproductive success of ospreys in New Jersey. Journal of Wildlife Management 55:270–279.

Steidl, R. J., J. P. Hayes, and E. Schauber. 1997. Statistical power in wildlife research. Journal of Wildlife Management 61:270–279.

Steiger, J. H., and R. T. Fouladi. 1997. Noncentrality interval estimation and the evaluation of statistical models. Pages 221–257 in L. L. Harlow, S. A. Muliak, and J. H. Steiger, editors, What If There Were No Significance Tests? Lawrence Erlbaum Associates, Hillsdale, New Jersey.

Stephens, D. W., and J. R. Krebs. 1986. Foraging Theory. Princeton University Press, Princeton, New Jersey.

Stevens, J. 1992. Applied Multivariate Statistics for the Social Sciences, 2nd ed. Lawrence Erlbaum, Hillsdale, New Jersey.

Stevens, J. 1986. Applied multivariate statistics for the social sciences. Lawrence Erlbaum, Mahway, New Jersey.

Stewart-Oaten, A., W. W. Murdoch, and K. R. Parker. 1986. Environmental impact assessment: "Pseudoreplication" in time. Ecology 67:929–940.

Stewart-Oaten, A., J. R. Bence, and C. W. Osenberg. 1992. Assessing effects of unreplicated perturbations: No simple solutions. Ecology 73:1396–1404.

Stigler, S. M. 1986. The History of Statistics. Harvard University Press, Cambridge, Massachusetts.

Still, A. W., and A. P. White. 1981. The approximate randomization test as an alternative to the *F* test in analysis of variance. British Journal of Mathematical and Statistical Psychology 34:243–252.

Stimson, J. 1985. The effect of shading by the table coral *Acropora lyacinthus* on understory corals. Ecology 66:40–53.

Stine, R. 1989. An introduction to bootstrap methods: Examples and ideas. Sociological Methods and Research 18:243–291.

Stoffer, D. S., and K. D. Wall. 1991. Bootstrapping state-space models: Gaussian maximum likelihood estimation and the Kalman filter. Journal of the American Statistical Association 86:1024–1033.

Stokes, M. E., C. S. Davis, and G. G. Koch. 1995. Categorical data analysis using the SAS system. SAS Institute, Inc., Cary, North Carolina.

Stow, C. A., S. R. Carpenter, K. E. Webster, and T. M. Frost. 1998. Long-term environmental monitoring: Some perspectives from lakes. Ecological Applications 8:269–276.

Tamura, R. N., L. A. Nelson, and G. C. Naderman. 1988. An investigation on the validity and usefulness of trend analysis for field plot data. Agronomy Journal 80:712–718.

Tanaka, J. S. 1987. "How big is big enough?": Sample size and goodness of fit in structural equation models with latent variables. Child Development 58:134–146.

Taylor, B. L., and T. Gerrodette. 1993. The uses of statistical power in conservation biology: The Vaquita and Northern Spotted Owl. Conservation Biology 7:489–500.

Taylor, D. J., and K. E. Muller. 1995. Computing confidence bounds for power and sample size of the general linear univariate model. American Statistician 49:43–47.

Taylor, R. J. 1984. Predation. Chapman and Hall, New York.

ter Braak, C. J. F. 1987. Ordination. Pages 91–173 in R. H. G. Jongman, C. J. F. ter Braak, and O. F. R. van Tongeran, editors, Community and Landscape Ecology. Pudoc, Wageningen, The Nertherlands.

ter Braak, C. J. F. 1988. Partial canonical correspondende analysis. Pages 551–558 in H. H. Bock, editor, Classification and Related Methods of Data Analysis. Elsevier, North Holland.

ter Braak, C. J. F. 1992. Permutation versus bootstrap significance tests in multiple regression and ANOVA. Pages 79–86 in K. H. Jöckel, editor, Bootstrapping and Related Techniques. Springer-Verlag, Berlin.

ter Braak, C. J. F. 1994. Canonical community ordination. Part I: Basic theory and linear methods. Ecoscience 1:127–140.

Thomas, L. 1997. Retrospective power analysis. Conservation Biology 11:276–280.

Thomas, L., and F. Juanes. 1996. The importance of statistical power analysis: An example from animal behaviour. Animal Behaviour 52:856–859.

Thomas, L., and C. J. Krebs. 1997. A review of statistical power analysis software. Bulletin of the Ecological Society of America 78:128–139.

Thompson, D. J. 1975. Towards a predator–prey model incorporating age structure: The effects of predator and prey size on the predation of *Daphnia magna* by *Ischnura elegans*. Journal of Animal Ecology 44:907–916.

Thompson, S. K. 1992. Sampling. John Wiley & Sons, New York.

Thomson, J. D., G. Weiblen, B. A. Thomson, S. Alfaro, and P. Legendre. 1996. Untangling multiple factors in spatial distributions: Lilies, gophers, and rocks. Ecology 77:1698–1715.

Tibshirani, R. 1992. Bootstrap hypothesis testing. Biometrics 48:969–970.

Tilman, D. 1987. The importance of the mechanisms of interspecific competition. American Naturalist 129:769–774.

Tilman, D. 1988. Plant Strategies and the Dynamics and Structure of Plant Communities. Princeton University Press, Princeton, New Jersey.

Tinkle, D. W. 1972. The role of environment in the evolution of life history differences within and between lizard species. University of Arkansas Museum Occasional Papers 4:77–100.

Toft, C. A., and P. J. Shea. 1983. Detecting community-wide patterns: Estimating power strengthens statistical inference. American Naturalist 122:618–625.

Trexler, J. C., and J. Travis. 1993. Nontraditional regression analysis. Ecology 74:1629–1637.

Trexler, J. C., C. E. McCulloch, and J. Travis. 1988. How can the functional response best be determined? Oecologia 76:206–214.

Tritton, L. M., and J. W. Hornbeck. 1982. Biomass equations for major tree species of the northeast. USDA Forest Service PL Northeastern Forest Experiment Station General Technical Report, NE-69.

Tufte, E. R. 1983. The Visual Display of Quantitative Information. Graphics Press, Cheshire, UK.

Tufte, E. R. 1990. Envisioning Information. Graphics Press, Cheshire, UK.

Tukey, J. W. 1977. Exploratory Data Analysis. Addison-Wesley, Reading, Massachusetts.

Turlings, T. C. J., J. H. Tumlinson, and W. J. Lewis. 1990. Exploitation of herbivore-induced plant odors by host-seeking parasitic wasps. Science 250:1251–1253.

Turner, P. R. 1989. Guide to Numerical Analysis. CRC Press, Boca Raton, Florida.

Turner, T. 1985. Stability of rocky intertidal surfgrass beds: Persistence, preemption and recovery. Ecology 66:83–92.

Twolan-Strutt, L., and P. A. Keddy. 1996. Above- and belowground competition intensity in two contrasting wetland plant communities. Ecology 77:259–270.

Underwood, A. J. 1994. On beyond BACI: Sampling designs that might reliably detect environmental disturbances. Ecological Applications 4:3–15.

Underwood, A. J. 1997. Experiments in Ecology: Their Logical Design and Interpretation Using Analysis of Variance. Cambridge University Press, Cambridge.

Upton, G. J. G., and B. Fingleton. 1985. Spatial Data Analysis by Example. Volume 1: Point Pattern and Quantitative Data. Wiley, New York.

van Es, H. M., C. L. van Es, and D. K. Cassel. 1989. Application of regionalized variable theory to large-plot field experiments. Soil Science Society of America Journal 53: 1178–1183.

van Groenendael, J., and P. Slim. 1988. The contrasting dynamics of two populations of *Plantago lanceolata* classified by age and size. Journal of Ecology 76:585–599.

van Latesteijn, H. C., and R. H. D. Lambeck. 1986. The analysis of monitoring data with the aid of time-series analysis. Environmental Monitoring and Assessment 7: 287–297.

van Lenteren, J. C., and K. Bakker. 1976. Functional responses in invertebrates. Netherlands Journal of Zoology 26:567–572.

Van Voorhies, W. A. 1992. Production of sperm reduces nematode lifespan. Nature 360: 456–458.

Vaupel, J. W., J. R. Carey, K. Christensen, T. E. Johnson, A. I. Yashin, N. V. Holm, I. A. Iachine, V. Kannisto, A. A. Khazaeli, P. Liedo, V. D. Longo, Y. Zeng, K. G. Manton, and J. W. Curtsinger. 1998. Biodemographic trajectories of longevity. Science 280:855–860.

Ver Hoef, J. M. 1996. Parametric empirical Bayes methods for ecological applications. Ecological Applications 6:1047–1055.

Ver Hoef, J. M. 1993. Universal kriging for ecological data. Pages 447–453 in M. F. Goodchild, B. O. Parks, and L. T. Steyaert, editors, Environmental Modeling with GIS. Oxford University Press, New York.

Ver Hoef, J. M., N. Cressie, R. N. Fisher, and T. J. Case. (2001). Uncertainty and spatial linear models for ecological data. Pages in C. T. Hunsaker, M. F. Goodchild, M. Friedl, and T. J. Case, editors, Uncertainty in Spatial Data for Ecological Analyses. Springer-Verlag, New York.

Vermunt, J. K. 1997. Log-Linear Models for Event Histories. Sage, Thousand Oaks, California.

Watras, C. J., and T. M. Frost. 1989. Little Rock Lake (Wisconsin): Perspectives on an experimental ecosystem approach to seepage lake acidification. Archives of Environmental Contamination and Toxicology 18:157–165.

Watt, A. S. 1947. Pattern and process in the plant community. Journal of Ecology 35: 1–22.

Webster, R. 1985. Quantitative spatial analysis of soil in the field. Advances in Soil Science 3:1–70.

Wei, W. W. S. 1990. Time Series Analysis: Univariate and Multivariate Methods. Addison-Wesley, New York.

Weiner, J., and O. T. Solbrig. 1984. The meaning and measurement of size hierarchies in plant populations. Oecologia 61:334–336.

Weis, A. E., and A. Kapelinski. 1994. Variable selection on *Eurosta*'s gall size. II. A path analysis of the ecological factors behind selection. Evolution 48:734–745.

Weis, A. E., C. L. Wolfe, and W. L. Gorman. 1989. Genotypic variation and integration in histological features of the goldenrod ball gall. American Journal of Botany 76: 1541–1550.

Weller, D. E. 1987. A reevaluation of the -3/2 power rule of self-thinning. Ecological Monographs 57:23–43.

Welsh, A. H., A. T. Peterson, and S. A. Altmann. 1988. The fallacy of averages. American Naturalist 132:277–288.

Werner, E. E., and B. R. Anholt. 1993. Ecological consequences of the tradeoff between growth and mortality rates mediated by foraging activity. American Naturalist 142: 242–272.

Wilkinson, G. N., S. R. Eckert, T. W. Hancock, and O. Mayo. 1983. Nearest neighbor (NN) analysis of field experiments. Journal of the Royal Statistical Society, Series B 45:152–212.

Wilkinson, L. 1988. SYSTAT: The System for Statistics. SYSTAT, Inc., Evanston, Illinois.

Wilkinson, L. 1990. SYGRAPH: The System for Graphics. SYSTAT, Inc., Evanston, Illinois.

Williams, F. M. 1980. On understanding predator–prey interactions. Pages 349–376 in D. C. Ellwood, J. N. Hedger, M. J. Latham, L. M. Lynch, and J. H. Slater, editors, Contemporary Microbial Ecology. Academic Press, London.

Williams, F. M., and S. A. Juliano. 1985. Further difficulties in the analysis of functional-response experiments, and a resolution. Canadian Entomologist 117:631–640.

Williams, F. M., and S. A. Juliano. 1996. Functional responses revisited. Environmental Entomology 25:549–550.

Wilson, S. D., and D. Tilman. 1995. Competitive responses of eight old-field plant species in four environments. Ecology 76:1159–1180.

Winer, B. J., D. R. Brown, and K. M. Michaels. 1991. Statistical Principles in Experimental Design. McGraw-Hill, New York.

Wolf, L. L., and F. R. Hainsworth. 1990. Non-random foraging by hummingbirds: Patterns of movement between Ipomopsis aggregata (Pursch) V. Grant inflorescences. Functional Ecology 4:149–158.

Wolfinger, R. D. 1996. Heterogenous variance-covariance structure for repeated measures. Journal of Agricultural, Biological, and Environmental Statistics 1:205–230.

Wolfinger, R. D., and M. Chang. 1995. Comparing the SAS GLM and MIXED procedures for repeated measures. Proceeding of the Twentieth Annual SAS Users Group International Conference [available at www.sas.com].

Wolfinger, R. D., R. D. Tobias, and J. Sall. 1994. Computing Gaussian Likelihoods and their derivatives for the general linear mixed models. SIAM Journal of Scientific Computing 15:1294–1310.

Wolfson, L. J., J. B. Kadane, and M. J. Small. 1996. Bayesian environmental policy decisions: two case studies. Ecological Applications 6:1056–1066.

Wootton, J. T. 1990. Direct and indirect effects of bird predation and excretion on the spatial and temporal patterns of intertidal species. Ph.D. dissertation. University of Washington, Seattle, Washington.

Wootton, J. T. 1994. Putting the pieces together: Testing the independence of interactions among organisms. Ecology 75:1544–1551.

Wright, S. 1920. The relative importance of heredity and environment in determining the piebald pattern of guinea pigs. Proceedings of the National Academy of Science 6: 320–332.

Wright, S. 1921. Correlation and causation. Journal of Agricultural Research 20:557–585.

Wright, S. 1934. The method of path coefficients. Annals of Mathematics and Statistics 5: 161–215.

Wright, S. 1960. Path coefficients and path regression: Alternative or complementary concepts? Biometrics 16:189–202.

Wright, S. 1983. On "path analysis in genetic epidemiology: A critique." American Journal of Human Genetics 35:733–756.

Wyckoff, P., and J. S. Clark. 2000. Predicting tree mortality from diameter growth: A comparison of approaches. Canadian Journal of Forest Research 30:156–167.

Yates, F. 1938. The comparative advantages of systematic and randomized arrangements in the design of agricultural and biological experiments. Biometrika 30:444–466.

Yoccoz, N. G. 1991. Use, overuse, and misuse of significance tests in evolutionary biology and ecology. Bulletin of the Ecological Society of America 72:106–111.

Young, A. 1994. Bootstrap: More than a stab in the dark? (with discussion). Statistical Science 9:382–415.

Young, L. J., and J. H. Young. 1991. Alternative view of statistical hypothesis testing. Environmental Entomology 20:1241–1245.

Yule, G. U. 1912. On the methods of measuring association between two attributes (with discussion). Journal of the Royal Statistical Society 75:579–642.

Zar, J. H. 1996. Biostatistical Analysis, 3rd ed. Prentice Hall, Upper Saddle River, New Jersey.

Zimmerman, D. L., and N. Cressie. 1992. Mean squared prediction error in the spatial linear model with estimated covariance parameters. Annals of the Institute of Statistical Mathematics 44:227–243.

Zimmerman, D. L., and D. A. Harville. 1991. A random field approach to the analysis of field-plot experiments and other spatial experiments. Biometrics 47:223–239.

Zimmerman, G. M., H. Goetz, and P. W. Mielke. 1985. Use of an improved statistical method for group comparisons to study effects of prairie fire. Ecology 66:606–611.

Zumbo, B. D., and A. M. Hubley. 1998. A note on misconceptions concerning prospective and retrospective power. Journal of the Royal Statistical Society Series D 47:385–388.

Index